Simulation Technologies for Optoelectronics

Volume 1: Light Waves

Tutorial Texts Series Related Title

- *Computational Fourier Optics: A MATLAB Tutorial*, D. G. Voelz, Vol. TT89

(For a complete list of Tutorial Texts, see http://spie.org/publications/books/tutorial-texts.)

Other related SPIE Press books:

- *Field Guide to Light-Matter Interaction*, G. Nemova, Vol. FG51
- *Field Guide to Optoelectronics and Photonics*, J. Hernández-Cordero and M. Hautefeuille, Vol. FG50
- *Fourier Transforms Using* Mathematica®, J. W. Goodman, Vol. PM322
- *Modeling and Simulating Optical Waveguides*, R. A. Motes, Vol. PM338
- *Simulating Fourier Optics Using* Mathematica®, J. W. Goodman, Vol. PM384
- *Wave Optics: Lectures in Optics, Volume 3*, G. Asimellis, Vol. PM296

Simulation Technologies for Optoelectronics
Volume 1: Light Waves

Gen-ichi Hatakoshi

Tutorial Texts in Optical Engineering
Volume TT129

Jessica DeGroote Nelson, Series Editor

SPIE PRESS
Bellingham, Washington USA

Library of Congress Cataloging-in-Publication Data

Names: Hatakoshi, Gen-ichi author.
Title: Simulation technologies for optoelectronics / Gen-ichi Hatakoshi.
Description: Bellingham, Washington : SPIE Press [2026] | Includes index
Identifiers: LCCN 2025031234 | ISBN 9781510693401 paperback | ISBN 9781510693418 pdf | ISBN 9781510693425 epub
Subjects: LCSH: Optoelectronics–Mathematical models | Optoelectronics–Computer simulation
Classification: LCC TA1750 .H37 2026
LC record available at https://lccn.loc.gov/2025031234

Published by

SPIE
P.O. Box 10
Bellingham, Washington 98227-0010 USA
Phone: +1 360.676.3290
Fax: +1 360.647.1445
Email: books@spie.org
Web: www.spie.org

Printed in the United States of America.
First printing 2026.
For updates to this book, visit http://spie.org and type "TT129" in the search field.

SPIE.

Introduction to the Series

The *Tutorial Text* series provides readers with an introductory reference text to a particular field or technology. The books in the series are different from other technical monographs and textbooks in the manner in which the material is presented. True to their name, they are tutorial in nature, and graphical and illustrative material is used whenever possible to better explain basic and more-advanced topics. Heavy use of tabular reference data and numerous examples further explain the presented concept. A grasp of the material can be deepened and clarified by taking corresponding SPIE short courses.

The initial concept for the series came from Jim Harrington (1942–2018) in 1989. Jim served as Series Editor from its inception to 2018. The *Tutorial Texts* have grown in popularity and scope of material covered since 1989. They are popular because they provide a ready reference for those wishing to learn about emerging technologies or the latest information within a new field. The topics in the series have grown from geometrical optics, optical detectors, and image processing to include the emerging fields of nanotechnology, biomedical optics, engineered materials, data processing, and laser technologies. Authors contributing to the series are instructed to provide introductory material so that those new to the field may use the book as a starting point to get a basic grasp of the material.

The publishing time for *Tutorial Texts* is kept to a minimum so that the books can be as timely and up-to-date as possible. When a proposal for a text is received, it is evaluated to determine the relevance of the proposed topic. This initial reviewing process helps authors identify additional material or changes in approach early in the writing process, which results in a stronger book. Once a manuscript is completed, it is peer reviewed by multiple experts in the field to ensure that it accurately communicates the key components of the science and technologies in a tutorial style.

It is my goal to continue to maintain the style and quality of books in the series and to further expand the topic areas to include new emerging fields as they become of interest to our readers.

Jessica DeGroote Nelson
Edmund Optics

Contents

Preface

This book is the English version of the Japanese book *Simulation Technologies for Optoelectronics* published by the Optronics Co., Ltd. in 2022. It is published by SPIE Press in 2026 with the permission of Optronics Co., Ltd.

Simulation technologies for analyzing or designing the characteristics of optoelectronic devices are used in various fields. In recent years, fairly advanced simulation analysis has become easy to perform on a PC, and various simulation software packages have become available as free software. When using these software packages, the physical models and numerical calculation methods are not necessarily visible to the user. However, in reality, there is a big difference between understanding and not understanding this. For example, when an unexpected calculation result is obtained, in order to determine that it is a new discovery, it is necessary to be able to qualitatively differentiate it from the original physical model. This is because strange results can sometimes occur due to the influence of calculation errors. This book introduces an overview of the physical models that are the basis of various simulations in optoelectronics, as well as simple simulation examples. It also introduces calculation examples that can be done at the spreadsheet level so that you can actually see the operation of the simulation.

Actual simulations are often written in a compilable programming language such as C++, but some simulations can also be done using Excel's VBA (Visual Basic for Applications). Since results are immediately obtained by inputting data into a table, this is one way to understand the basic principles. This book particularly uses VBA animations in several chapters; these simulations are useful for understanding the time-dependent changes in physical phenomena, such as the propagation of light. Some of the Excel files used in the simulations in the text are included in Appendix I-A. Refer to "About the Online Supplemental Appendices" at the end of the book to download the appendix and to make use of the files. Appendix I-B also provides examples of animations related to the figures in each chapter.

This book is based on 37 articles serialized in the monthly magazine "OPTRONICS" in Japan from April 2018 to April 2021. The first half mainly discusses simulation technologies related to light, and the second half mainly discusses simulation technologies related to electrons. *Simulation Technologies for Optoelectronics, Volume I: Light Waves* starts with the propagation of light in free space and covers simulations of reflection and refraction, the behavior of Gaussian beams, optical multilayer films, planar optical waveguides, optical fibers, and more. Most of these discussions start with Maxwell's equations.

The contents of this book are based on the author's research during his university days, the author's research and development at Toshiba Corporation, where he was employed, and topics discussed at academic conferences and research meetings. The author would like to express his deep gratitude to the many people who have helped during this time and to the editorial staff at Optronics Co., Ltd., who helped with the serialization that formed the basis of this book. The author would also like to express his deepest gratitude to the editorial staff at SPIE Press, who helped with the publication of this book.

Gen-ichi Hatakoshi
May 2026

Symbols and Units for Constants and Variables

Mathematical Constants and Physical Constants

i	imaginary unit	
π	pi	
e	base of the natural logarithm, also known as Napier's constant	
c	speed of light	$[\mathrm{m \cdot s^{-1}}]$
e	elementary charge	[C]
h	Planck constant	[J]
$\hbar\ (= h/(2\pi))$	reduced Planck constant, Dirac constant	[J]
ε_0	vacuum permittivity	$[\mathrm{F\ m^{-1}}]$
μ_0	vacuum (magnetic) permeability	$[\mathrm{H\ m^{-1}}]$

Operators

∇	$= \left(\dfrac{\partial}{\partial x}, \dfrac{\partial}{\partial y}, \dfrac{\partial}{\partial z}\right)$	$[\mathrm{m^{-1}}]$
∇^2	$= \dfrac{\partial^2}{\partial x^2} + \dfrac{\partial^2}{\partial y^2} + \dfrac{\partial^2}{\partial z^2}$	$[\mathrm{m^{-2}}]$
∇A	$= \left(\dfrac{\partial A}{\partial x}, \dfrac{\partial A}{\partial y}, \dfrac{\partial A}{\partial z}\right)$	
$\nabla \cdot A$	$= \dfrac{\partial A}{\partial x} + \dfrac{\partial A}{\partial y} + \dfrac{\partial A}{\partial z}$	
$\nabla \times A$	$= \left(\dfrac{\partial A_z}{\partial y} - \dfrac{\partial A_y}{\partial z}, \dfrac{\partial A_x}{\partial z} - \dfrac{\partial A_z}{\partial x}, \dfrac{\partial A_y}{\partial x} - \dfrac{\partial A_x}{\partial y}\right)$	
$\hat{k}_x = i\dfrac{\partial}{\partial x}$	angular wavenumber operator	$[\mathrm{rad\ m^{-1}}]$

Variables

t	time	[s]
T	temperature	[K]
$\boldsymbol{E}$	electric field	[V m^{-1}]
$\boldsymbol{H}$	magnetic field	[A m^{-1}]
$\boldsymbol{D}$	electric flux density	[C m^{-2}]
$\boldsymbol{B}$	magnetic flux density	[T], [Wb m^{-2}]
$\boldsymbol{P}$	dielectric polarization	[C m^{-2}]
$\boldsymbol{S}$	time-averaged Poynting vector	[W m^{-2}]
ρ	charge density	[C m^{-3}]
j, $\boldsymbol{J}$	current density	[A m^{-2}]
σ	electrical conductivity	[Ω^{-1} m^{-1}]
ε	permittivity	[F m^{-1}]
ε_r	relative permittivity	
μ	(magnetic) permeability	[H m^{-1}]
μ_r	relative (magnetic) permeability	
χ	electric susceptibility	
$\boldsymbol{k}$	angular wavenumber vector	[rad m^{-1}]
ν	frequency	[s^{-1}]
ω	angular frequency	[rad s^{-1}]
λ	wavelength	[m]
n_r	refractive index	
n_c	complex refractive index	
κ	extinction coefficient	
n_{eq}	equivalent refractive index	
n_g	group refractive index	
Δ	relative refractive index	
v_p	phase velocity	[m s^{-1}]
v_g	group velocity	[m s^{-1}]
β	effective propagation constant	[rad m^{-1}]
V	normalized frequency	
b	normalized propagation constant	
h_{eff}	effective film thickness	[m]
θ_B	Brewster's angle	[rad]
θ_c	critical angle for total reflection	[rad]

z_R	Rayleigh length	[m]
$d_\sigma(z)$	beam width	[m]
Θ_σ	beam divergence angle	[rad]
f	focal length	[m]
R	reflectivity	
T	transmittance	

Special Functions

$J_\nu(\xi)$	Bessel function
$Y_\nu(\xi)$	Neumann function
$H_\nu^{(k)}(\xi)$	Hankel function
$I_\nu(\xi)$	modified Bessel function of the first kind
$K_\nu(\xi)$	modified Bessel function of the second kind
$j_\nu(\xi)$	spherical Bessel function
$y_\nu(\xi)$	spherical Neumann function
$h_\nu^{(k)}(\xi)$	spherical Hankel function
$P_\nu^\mu(\xi)$	associated Legendre polynomials
$H_m(\xi)$	Hermite polynomials
$L_n^m(\xi)$	associated Laguerre polynomials

Chapter 1
Propagation of Light: Plane Waves

1.1 Introduction

As mentioned in the preface, simulation technology plays an important role in various fields related to optoelectronics.[1] This book covers the optical and electronic simulation technologies that form the basis of optoelectronics and provides an overview of the underlying physical models and simple simulation examples.

In this chapter, we will first discuss plane waves as the basis of light propagation.

1.2 Electromagnetic Waves and Maxwell's Equations

Since light[†] is an electromagnetic wave, let's start with Maxwell's equations,[2–4] which are the basis of electromagnetic waves. The electric field $\boldsymbol{E}$ and magnetic field $\boldsymbol{H}$ of electromagnetic waves are given as solutions of the following Maxwell's equations:

$$\nabla \times \boldsymbol{E} = -\frac{\partial \boldsymbol{B}}{\partial t}, \tag{1.1}$$

$$\nabla \times \boldsymbol{H} = \boldsymbol{j} + \frac{\partial \boldsymbol{D}}{\partial t}, \tag{1.2}$$

$$\nabla \cdot \boldsymbol{D} = \rho, \tag{1.3}$$

$$\nabla \cdot \boldsymbol{B} = 0, \tag{1.4}$$

[†]Note: International standards such as ISO (International Organization for Standardization) and IEC (International Electrotechnical Commission) stipulate that the term "light" should only be used for visible radiation. However, in this book, we will use the term "light" in a broad sense to refer to electromagnetic waves, including infrared radiation and ultraviolet radiation.

where $\boldsymbol{D}$ is the electric flux density, $\boldsymbol{B}$ is the magnetic flux density, $\boldsymbol{j}$ is current density, and ρ is charge density. The relationship between $\boldsymbol{D}$, $\boldsymbol{E}$, $\boldsymbol{B}$, and $\boldsymbol{H}$ is expressed using the permittivity ε and magnetic permeability μ, as follows:

$$\boldsymbol{D} = \varepsilon \boldsymbol{E}, \tag{1.5}$$

$$\boldsymbol{B} = \mu \boldsymbol{H}. \tag{1.6}$$

First, considering the case where $\boldsymbol{j} = 0$ and $\rho = 0$, the following wave equation can be obtained from Eqs. (1.1) to (1.4):

$$\nabla^2 \boldsymbol{E} - \varepsilon\mu \frac{\partial^2 \boldsymbol{E}}{\partial t^2} = 0, \tag{1.7}$$

$$\nabla^2 \boldsymbol{H} - \varepsilon\mu \frac{\partial^2 \boldsymbol{H}}{\partial t^2} = 0. \tag{1.8}$$

A sine wave is a solution of Eq. (1.7) or Eq. (1.8). If the angular frequency of light (electromagnetic waves) is ω, then

$$E = \cos(\boldsymbol{k} \cdot \boldsymbol{r} - \omega t), \quad \sin(\boldsymbol{k} \cdot \boldsymbol{r} - \omega t), \quad \exp(\pm i[\boldsymbol{k} \cdot \boldsymbol{r} - \omega t]) \tag{1.9}$$

are all solutions of Eq. (1.7). The parameter $\boldsymbol{k}$ in the above equation is the angular wavenumber vector and is given from Eq. (1.7) as follows:

$$k \equiv |\boldsymbol{k}| = \omega\sqrt{\varepsilon\mu}. \tag{1.10}$$

From Eq. (1.9), the speed u of electromagnetic waves is given by

$$u = \omega/k = 1/\sqrt{\varepsilon\mu}. \tag{1.11}$$

In vacuum,

$$u = c \equiv 1/\sqrt{\varepsilon_0\mu_0} = 299792458\,\mathrm{m/s}. \tag{1.12}$$

This is the speed of light. The unit of length, m (meter), is defined as the distance that light travels in 1/299792458 seconds.

1.3 Animation

Now that we have a sine wave in Eq. (1.9), let's draw a diagram using a simulation. Since it is a sine wave, it can be easily drawn using an Excel spreadsheet, and it is not so much a simulation. Here, Excel will be used to calculate how the waves move over time.

In preparation for creating an animation in Excel, create the following four buttons using VBA (Visual Basic for Applications):

Figure 1.1 Excel screen for animation showing clickable buttons in cells F1, G1, H1, I1.

(a) Decrease the value of a specific cell by 1,
(b) Increase the value of a specific cell by 1,
(c) Continuously change the value of a specific cell from 0 to $N - 1$, and
(d) Repeat (c) M times.

Figure 1.1 shows the Excel screen where the above button was created. The input values are M (cell A2) and N (cell B2). Clicking the buttons in cells F1, G1, H1, and I1 executes (a), (b), (c), and (d), respectively. The specific cell here refers to cell C2, and the same value is copied to C3. Also, the button in cell E1 is a reset button that returns the numerical value to its initial value.

The VBA for creating these buttons is not so difficult; therefore, the explanation is omitted here. Please refer to the Excel file in Appendix I-A (a separate file). If you run the macro named "Button" once, the button shown on the screen above will be created. Do this once and you won't need to do it again. When creating other Excel animation files, just copy this sheet.

Although the above macro is extremely simple, it can be used for various animations. For example, if you draw a graph of the function of Eq. (1.9) with time t as a value proportional to the value in cell C3 and click the [Scan1] or [ScanM] button in Fig. 1.1, the graph you created will move. Furthermore, if the rotation angle of a solid figure is given by C3, the figure can be rotated (see Fig. 1.2). Both examples are included in the Excel files. This animation button will be used many times later in this book.

In Figs. 1.2(b) and (c), a coordinate transformation is performed by projecting a solid in three-dimensional space onto a two-dimensional plane. There are various methods for such coordinate transformation.[5] This method will be used in this book hereafter, including in the plotting of electromagnetic waves. A detailed explanation of the coordinate transformation is omitted here.

1.4 Plane Waves

1.4.1 Traveling waves and standing waves

A homogeneous medium with no boundaries in the xyz space is considered. The sine wave in Eq. (1.9) is a plane wave[2] that travels in such an infinitely

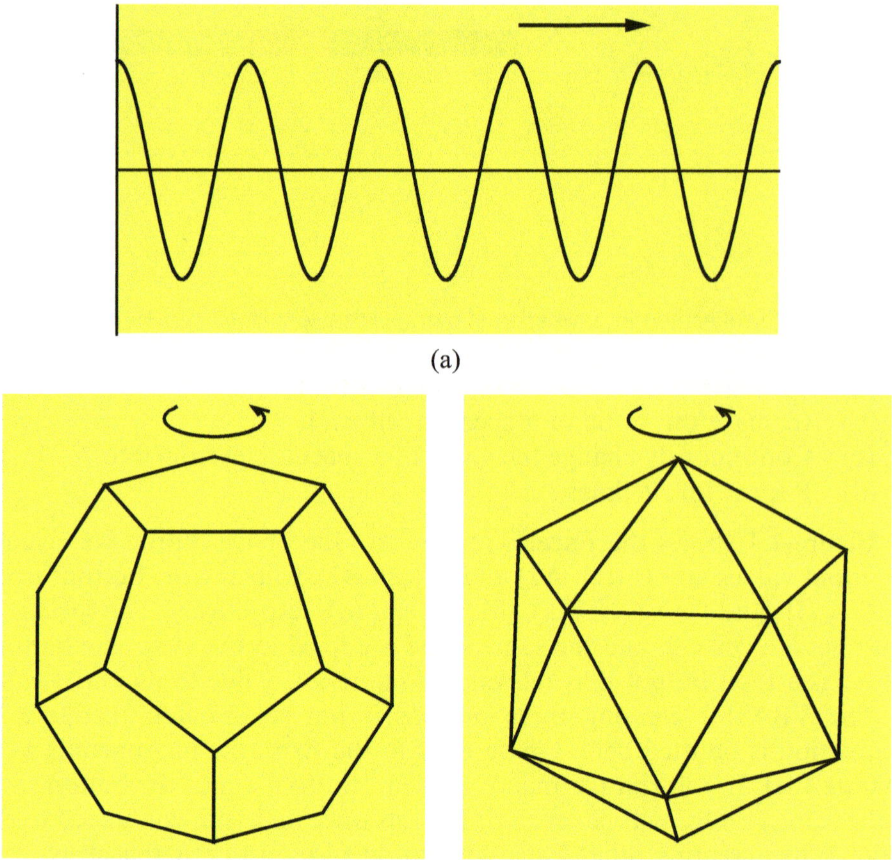

(a)

(b) (c)

Figure 1.2 Example of animation using Excel. (a) Propagation of a sine wave, (b) rotation of a dodecahedron, and (c) rotation of an icosahedron.

expanding uniform space. Although the functions cos, sin, or exp can be used, it is convenient to use the complex function $\exp(\pm i[\boldsymbol{k}\cdot\boldsymbol{r}-\omega t])$ in expressions, which will be used hereafter. Either double sign can be taken. There are methods that set the time dependence to $+i\omega t$ and $-i\omega t$. In this book, $+i\omega t$ dependence is adopted. Then the electric field $\boldsymbol{E}$ is represented by the following equation:†

$$\boldsymbol{E}=\boldsymbol{E}_0\exp(-i\boldsymbol{k}\cdot\boldsymbol{r}+i\omega t). \tag{1.13}$$

†Note: In this book, the electric and magnetic fields are expressed as complex vectors, $\boldsymbol{A}\exp(-i\boldsymbol{k}\cdot\boldsymbol{r}+i\omega t)$; the notation is called "phasor" representation. Actually, the electric or magnetic field in real space is given by the real or imaginary part of this phasor. Note that, as shown in Eq. (1.9), if Eq. (1.13) is a solution to Maxwell's equations, then its real or imaginary part is also a solution to Maxwell's equations.

This wave is a traveling wave that propagates in the direction of the angular wave vector $\boldsymbol{k}$. For this reason, $\boldsymbol{k}$ is also called a propagation vector. The way the sign is taken will affect the sign of the extinction coefficient of the complex refractive index. Let the permittivity ε be expressed by the product of the relative permittivity ε_r and the permittivity ε_0 in vacuum as follows:

$$\varepsilon = \varepsilon_r \varepsilon_0 = {n_r}^2 \varepsilon_0, \tag{1.14}$$

where n_r is the refractive index. In most materials, permeability is $\mu \approx \mu_0$ in the visible wavelength range. Therefore, from Eqs. (1.10) and (1.12),

$$\varepsilon = \varepsilon_r \varepsilon_0 = {n_r}^2 \varepsilon_0. \tag{1.15}$$

In the case wherein absorption or gain exist, n_r becomes a complex value. Let the complex refractive index in that case be n_c, as follows:

$$n_c = n_r \pm i\kappa. \tag{1.16}$$

The complex refractive index will be described in Chapters 5 and 6.

Let the propagation direction of light be z, and by substituting Eq. (1.16) into Eq. (1.13), $\boldsymbol{E}$ is expressed by the following equation:

$$\boldsymbol{E} = \boldsymbol{E}_0 \exp(-ikz + i\omega t) = \boldsymbol{E}_0 \exp(-ik_0 n_r z + i\omega t) \exp(\pm \kappa z). \tag{1.17}$$

If there is absorption, the wave must be attenuated when propagating in the $+z$ direction. Therefore, in the case of $\kappa > 0$, the negative sign of $\exp(\pm\kappa z)$ in Eq. (1.17) must be taken. It follows that, in the case of absorption, n_c is represented by

$$n_c = n_r - i\kappa \quad (\kappa > 0), \tag{1.18}$$

and in the case of gain, n_c is represented by

$$n_c = n_r + ig \quad (g > 0). \tag{1.19}$$

It should be noted that, if $-i\omega t$ is chosen as the time dependence, the signs of Eqs. (1.18) and (1.19) will be reversed.

The vectorial solutions satisfying Maxwell's equations for the electric field $\boldsymbol{E}$ and the magnetic field $\boldsymbol{H}$ of the traveling wave propagating in the $+z$ direction are represented by the following equations:

$$\boldsymbol{E} = E_0 \begin{pmatrix} 0 \\ 1 \\ 0 \end{pmatrix} \exp(-ikz + i\omega t), \tag{1.20}$$

$$\boldsymbol{H} = -\frac{E_0 k}{\omega\mu}\begin{pmatrix}1\\0\\0\end{pmatrix}\exp(-ikz + i\omega t), \tag{1.21}$$

where the coordinate axes were chosen so that the electric field is in the y direction.

The electromagnetic waves expressed by Eqs. (1.20) and (1.21) are illustrated in Fig. 1.3(a). The propagation of these electromagnetic waves can be viewed as an animation in the Excel file.

Here, set the coefficient of Eq. (1.21) as follows:

$$H_0 = \frac{E_0 k}{\omega\mu}. \tag{1.22}$$

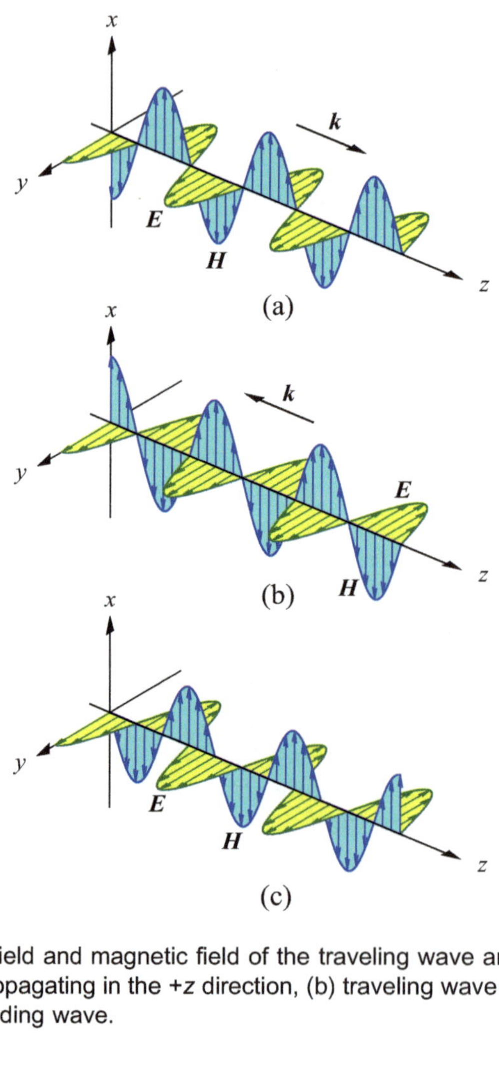

Figure 1.3 Electric field and magnetic field of the traveling wave and the standing wave. (a) Traveling wave propagating in the +z direction, (b) traveling wave propagating in the $-z$ direction, and (c) standing wave.

Then, from Eqs. (1.11) and (1.12), the ratio of E_0 to H_0 is given by

$$\frac{E_0}{H_0} = \sqrt{\frac{\mu}{\varepsilon}}. \tag{1.23}$$

E_0/H_0 expressed by Eq. (1.23) is called optical impedance, and its unit is Ω.[3]

Now, how can we express the equation for electromagnetic waves traveling in the $-z$ direction? It seems that the solution is simply to reverse the sign of ωt in Eqs. (1.20) and (1.21), that is, to reverse only the direction of propagation in Fig. 1.3(a). However, this is incorrect. Equations where ωt is replaced by $-\omega t$ in Eqs. (1.20) and (1.21) do not satisfy Maxwell's equations. Similarly, equations where z is replaced by $-z$ in Eqs. (1.20) and (1.21), do not satisfy Eqs. (1.1) and (1.2).†

Equations for the electromagnetic wave propagating in the $-z$ direction are obtained by replacing k with $-k$ and are represented as follows:

$$\boldsymbol{E} = E_0 \begin{pmatrix} 0 \\ 1 \\ 0 \end{pmatrix} \exp(ikz + i\omega t), \tag{1.24}$$

$$\boldsymbol{H} = \frac{E_0 k}{\omega \mu} \begin{pmatrix} 1 \\ 0 \\ 0 \end{pmatrix} \exp(ikz + i\omega t). \tag{1.25}$$

The forward traveling waves expressed by Eqs. (1.20) and (1.21) and the backward traveling waves expressed by Eqs. (1.24) and (1.25) are both solutions of Maxwell's equations. Therefore, their linear combination is also a solution. The sum or difference of the wave traveling in the $+z$ direction and the wave traveling in the $-z$ direction becomes a standing wave. In the case of a sum process, E_y and H_x are represented by the following equations:

$$E_y = 2E_0 \cos(kz) \exp(i\omega t), \tag{1.26}$$

$$H_x = 2iH_0 \sin(kz) \exp(i\omega t). \tag{1.27}$$

As seen from the form of the equation, the function of the spatial coordinate z is independent of the time t; therefore, this wave does not propagate in the $\pm z$ direction but is a standing wave that oscillates at that position. Another notable point is that the spatial parts of E_y and H_x are sin(kz) and cos(kz), respectively; therefore, $\boldsymbol{E}$ and $\boldsymbol{H}$ are out of spatial phase by $\pi/2$. Also, since $\boldsymbol{H}$

†Note: Magnetic flux density and current density are reversed by the time reversal operation. Maxwell's equations themselves preserve both time-reversal symmetry and space-reversal symmetry.

is multiplied by the imaginary unit i [$= \exp(i\pi/2)$], the time dependence is also shifted by $\pi/2$. The real parts of Eqs. (1.26) and (1.27) are represented, respectively, by

$$\mathrm{Re}(E_y) = 2E_0 \cos(kz)\cos(\omega t), \tag{1.28}$$

$$\mathrm{Re}(H_x) = -2H_0 \sin(kz)\sin(\omega t). \tag{1.29}$$

Equations (1.28) and (1.29) expressly show that the temporal phase is shifted by $\pi/2$.

It is easier to understand how the standing waves change by watching an animation. Refer to the Excel file.

1.4.2 Standing waves at medium boundaries

Standing waves appear in a cavity structure or in the case where a reflected wave exists at a medium boundary. Now, if we look at Fig. 1.3(c) again, the question arises: what will happen to this standing wave at the medium boundary? If $z = 0$ is the boundary, then in the region of $z < 0$ there should be a traveling wave propagating in the $-z$ direction. However, as can be seen from Eqs. (1.28) and (1.29), at $z = 0$, $\boldsymbol{E}$ oscillates with the cos function, but $\boldsymbol{H}$ is always 0. At the boundary, both tangential components of $\boldsymbol{E}$ and $\boldsymbol{H}$ should be continuous, so in this case they will not connect to the traveling wave.

As the explanation of the reflection will be described in another chapter, details will not be discussed here, but considering the boundary conditions that the electric field and magnetic field components parallel to the boundary surface are continuous at the medium interface,[2] the reflectance is always less than unity, so in general forward traveling waves and backward traveling waves of the same amplitude cannot be added. Therefore, the wave becomes a mixture of a traveling wave and a standing wave.

Figure 1.4(a) shows the situation when light is emitted from a semiconductor ($n_1 = 3.5$) to air ($n_2 = 1$). The phase shift between $\boldsymbol{E}$ and $\boldsymbol{H}$ is not constant at $\pi/2$ as shown in Fig. 1.3(c) but changes with time. Refer to the Excel file in the appendix as an example of animation. By executing the Excel animation, it can be seen that the wave behaves very interestingly. Since it is possible to input the refractive index of the two media, it is a good idea to change them and execute the program. If the value of the refractive index of one side is increased to, for example, ~20 (although this is unlikely), the reflectance becomes approximately unity, so the traveling wave component is almost eliminated and the standing wave component, wherein the phase difference between $\boldsymbol{E}$ and $\boldsymbol{H}$ is a constant of $\pi/2$, becomes dominant. For example, if $n_1 = 20$ and $n_2 = 1$, then the situation will be close to that where $\boldsymbol{H}$ has a node and $\boldsymbol{E}$ has an antinode at the boundary. On the other hand, if

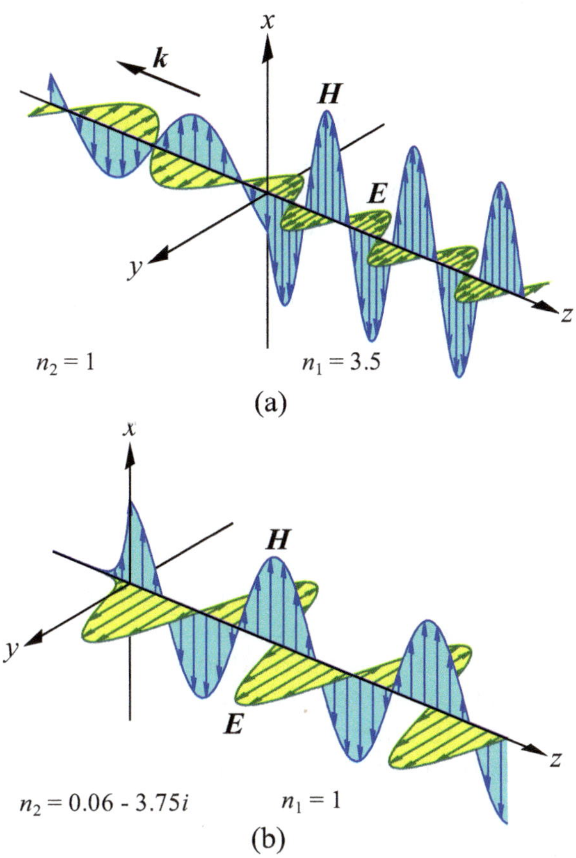

Figure 1.4 Standing waves at a medium boundary. (a) Light emission from semiconductor to air and (b) light reflection by metal (Ag).

$n_1 = 1$, $n_2 = 20$, then the situation will be close to that where ***E*** has a node and ***H*** has an antinode at the boundary.†

As seen from Eq. (1.23), the value of E_0/H_0 (optical impedance) is proportional to the reciprocal of the refractive index. Therefore, E_0/H_0 becomes large in a medium with a high refractive index. Furthermore, as can be seen from Eqs. (1.11) and (1.15), in a medium with a high refractive index, the speed of electromagnetic waves decreases in proportion to $1/n_r$, and the wavelength also becomes $1/n_r$. This situation is also shown in Fig. 1.4(a). Note that the $1/n_r$ magnification has been slightly changed to make it easier to see.

Figure 1.4(b) shows a standing wave formed by reflection on a metal surface. The regions of $z > 0$ and of $z < 0$ are air and the metal (Ag),

†Note: The terms fixed end and open end (free end) are used, but this causes some confusion when it comes to electromagnetic waves. The state in which the standing wave becomes a node at the boundary is called a fixed end, but in electromagnetic waves, the node of ***E*** is the antinode of ***H***. Furthermore, as shown in Fig. 1.4, complete nodes are not generally formed. If the boundary is completely fixed, light will not propagate outside the boundary.

respectively, and the refractive index of Ag was set as $n_2 = 0.06 - 3.75i$ (@ 600 nm). As seen in this example, the reflectance on a metal surface is close to unity when the real part of the refractive index of the metal is small and the imaginary part (extinction coefficient) is large. Therefore, the standing wave in air is close to the state shown in Fig. 1.3(c). Strictly speaking, since traveling waves are partly mixed, the phase shift between $\boldsymbol{E}$ and $\boldsymbol{H}$ is slightly different from $\pi/2$. The light attenuates rapidly in metals with large extinction coefficients.

One more supplemental item to be noted in Fig. 1.4 is that E_y and H_x are continuous at the boundary, and the z differential of E_y is continuous, but the z-differential of H_x is discontinuous. Reflection and refraction at boundaries will be dealt with again in Chapter 4.

Side Topic: Illustration of a plane wave

Figure 1.3 is a diagram that is often seen when explaining plane waves, but it can be somewhat misleading. First, the coordinate system is a Cartesian coordinate with x, y, and z orthogonal axes, but what is shown is E_y and H_x; the unit of electric field is V/m (or N/C), and the unit of magnetic field is A/m. In other words, objects with different unit systems are drawn in the spatial coordinate system.

To avoid this confusion, use the H_x and E_y axes instead of the x and y axes, as shown in Fig. 1.S1. Such a diagram is often seen. In this case, the diagram shows the z dependence of E_y and H_x, so the vectorial representation of the electric and magnetic fields as in Fig. 1.3 is not necessary. However, in the case of this diagram, there may be some objections that it is not clear that $\boldsymbol{E}$ has only a y component and $\boldsymbol{H}$ only has an x component. In other words, it is difficult to know what happens to the other components.

From the perspective of the wave oscillations, it would be better to have vectorial representations of the electric and magnetic fields, so we will return to Fig. 1.3(a), but this diagram seems to be difficult to understand in another respect. This is because although Fig. 1.3 appears

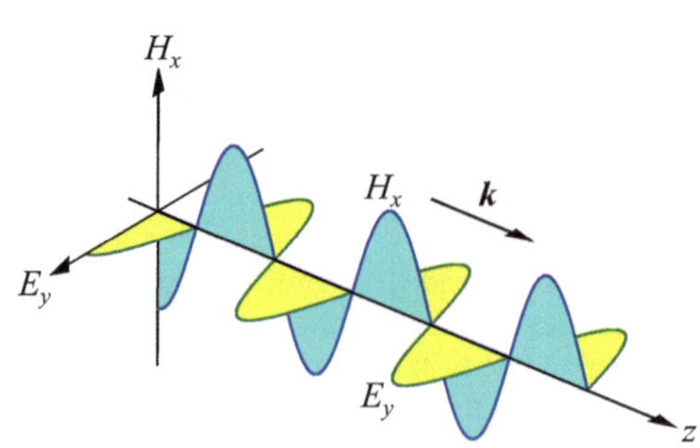

Figure 1.S1 Illustration of a plane wave (part 1).

to represent the electric and magnetic fields on the z-axis, we do not know what happens at points other than $(x, y, z) = (0, 0, z)$. It is true that the image of a "plane wave" cannot be seen from Fig. 1.3.

Based on the above reason, Fig. 1.S2 was drawn. There may be other ways to draw it, but in short, the vectors of the electric and magnetic fields are drawn over the entire space. Doesn't this somehow represent a "plane wave"? However, if you draw a diagram like this, it will probably become even more difficult to understand. The subject "plane waves" was taken in this chapter with the intention of being the easiest subject for simulation, but there is a lot of trouble when trying to draw diagrams.

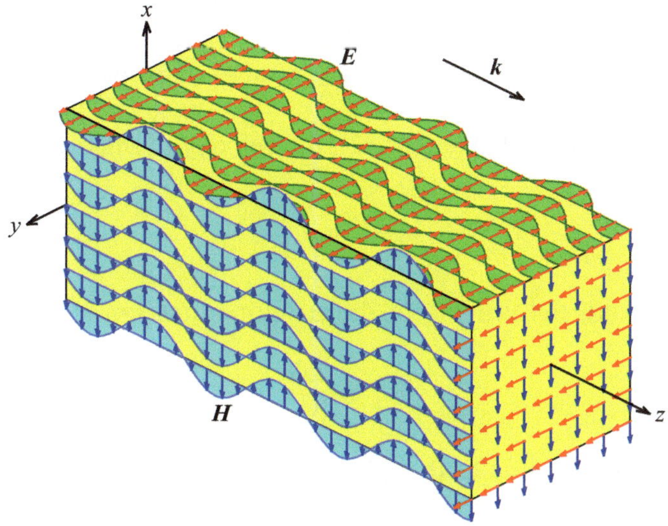

Figure 1.S2 Illustration of a plane wave (part 2).

1.5 Wave Packet and Group Velocity

Let's take a look at the group velocity of a wave packet,[2] which is a superposition of multiple sine waves, through a simulation.

First, consider two plane waves as a simple case:

$$E_1(z,t) = E_0 \exp(-ik_1 z + i\omega_1 t), \tag{1.30}$$

$$E_2(z,t) = E_0 \exp(-ik_2 z + i\omega_2 t). \tag{1.31}$$

Let k and ω be the average angular wave number and average angular frequency, respectively, and let Δk and $\Delta\omega$ be the respective differences, represented by the following equations:

$$k_1 = k - \Delta k/2, \;\; k_2 = k + \Delta k/2, \tag{1.32}$$

$$\omega_1 = \omega - \Delta\omega/2, \;\; \omega_2 = \omega + \Delta\omega/2. \tag{1.33}$$

The superposition of two plane waves is given by

$$E(z,t) = E_1(z,t) + E_2(z,t) = 2E_0 \exp(-ikz + i\omega t)\cos\left(\frac{\Delta k}{2}z - \frac{\Delta\omega}{2}t\right). \tag{1.34}$$

Equation (1.34) represents a state in which a plane wave represented by $E_0\exp(-ikz + i\omega t)$ is modulated by the envelope function of $\cos([\Delta k/2]z - [\Delta\omega/2]t)$. The real part of Eq. (1.34) is given by

$$\mathrm{Re}[E(z,t)] = 2E_0\cos(kz - \omega t)\cos\left(\frac{\Delta k}{2}z - \frac{\Delta\omega}{2}t\right). \tag{1.35}$$

Figure 1.5 shows a schematic diagram of the wave packet represented by Eq. (1.35).

As shown by Eqs. (1.11) and (1.12), the velocity of the original plane wave is given by

$$u_\mathrm{p} = \frac{\omega}{k} = \frac{c}{n_\mathrm{r}}. \tag{1.36}$$

On the other hand, the velocity of the envelope function is represented by

$$u_\mathrm{g} = \frac{\Delta\omega}{\Delta k}. \tag{1.37}$$

If Δk is sufficiently small, the above equation can be replaced by the differential form:

$$u_\mathrm{g} = \frac{\partial\omega}{\partial k}. \tag{1.38}$$

Equations (1.36) and (1.38) represent the phase velocity and the group velocity, respectively.

From Eq. (1.36),

$$\omega = ck/n_\mathrm{r}. \tag{1.39}$$

By substituting Eq. (1.39) into Eq. (1.38), the following formula is obtained:

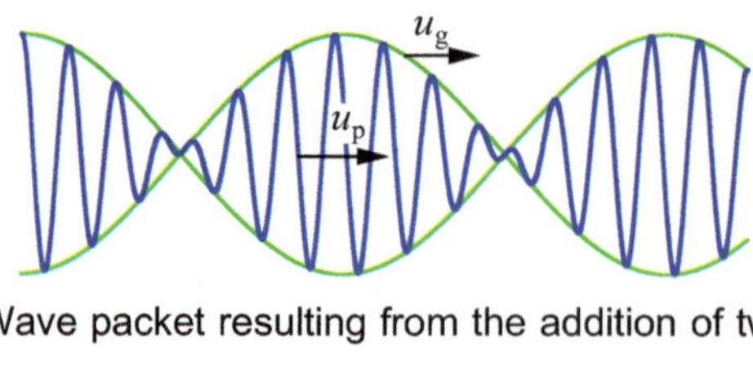

Figure 1.5 Wave packet resulting from the addition of two sine waves.

$$u_g = u_p\left(1 - \frac{k}{n_r}\frac{\partial n_r}{\partial k}\right). \tag{1.40}$$

Let the wavelength in vacuum be

$$\lambda = \frac{2\pi c}{\omega}. \tag{1.41}$$

Then the quantity $1/u_g$ is transformed as follows:

$$\begin{aligned}\frac{1}{u_g} &= \frac{\partial k}{\partial \omega} = \frac{\partial}{\partial \omega}\left(\frac{n_r \omega}{c}\right) = \frac{n_r}{c} + \frac{\omega}{c}\frac{\partial n_r}{\partial \omega} = \frac{1}{u_p} + \frac{\omega}{c}\frac{\partial \lambda}{\partial \omega}\frac{\partial n_r}{\partial \lambda} \\ &= \frac{1}{u_p} - \frac{\omega}{c}\frac{2\pi c}{\omega^2}\frac{\partial n_r}{\partial \lambda} = \frac{1}{u_p} - \frac{\lambda}{c}\frac{\partial n_r}{\partial \lambda} = \frac{1}{c}\left(n_r - \lambda\frac{\partial n_r}{\partial \lambda}\right).\end{aligned} \tag{1.42}$$

From the above relation, the following formulas are obtained:

$$u_g = \frac{c}{n_g}, \tag{1.43}$$

$$n_g \equiv n_r - \lambda\frac{\partial n_r}{\partial \lambda}. \tag{1.44}$$

The quantity n_g defined by Eq. (1.44) is called the group refractive index. In normal materials, $\partial n_r/\partial \lambda_0 < 0$ in the visible wavelength range; therefore, n_g takes a value larger than the refractive index n_r.

The relationship between the phase velocity and the group velocity can be understood by changing the value of $\partial \omega/\partial k$ in the animation of Fig. 1.5 (see Excel file).

The above example is the addition of two plane waves. In general, the wave packet can be expressed as an integral as shown below:

$$E(z,t) = \int a(k)\exp[-ikz + i\omega(k)t]dk. \tag{1.45a}$$

An optical pulse, for example, can be represented by this equation. The integral in Eq. (1.45a) can also be approximated by a sum as follows:

$$E(z,t) = \sum_{m=1}^{M} a_m(k)\exp[-ikz + i\omega(k)t]. \tag{1.45b}$$

If $\omega(k) = \omega_0$ is constant in Eq. (1.45), and $a(k)$ is assumed to be a Gaussian distribution, $E(z, t)$ becomes a Gaussian optical pulse. This is clear from the Fourier transform relationship. For comparison, with the case of adding two plane waves, Fig. 1.6 shows an example of a Gaussian wave packet obtained

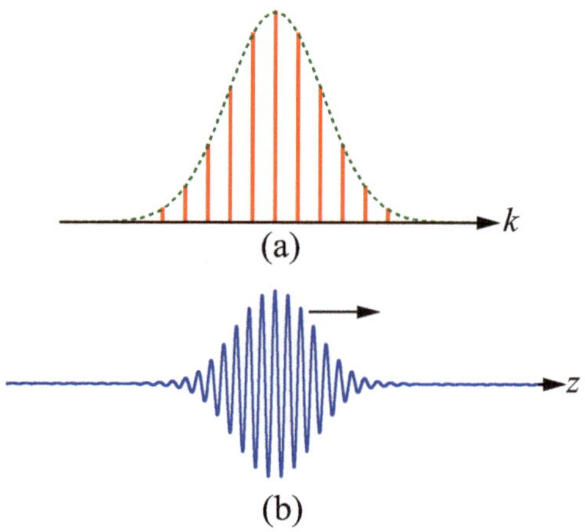

Figure 1.6 Gaussian wave packet: (a) assumed $a_m(k)$ distribution with Gaussian envelope and (b) resulting electric field distribution given by Eq. (1.45b).

by adding a finite number of plane waves. The coefficient $a_m(k)$ of each plane wave is added together, assuming a Gaussian distribution. It can be seen that when the number M of plane waves to be added in the Excel file is increased, the pulse shape becomes closer to a Gaussian distribution.

1.6 Conclusion

Regarding the propagation of plane waves derived from Maxwell's equations, simulation examples of traveling waves, standing waves, and wave packets were introduced. Animation using Excel is a convenient method for understanding changes in various phenomena over time, so the method is used for various simulation examples in the following chapters.

References

1. K. Iga and G. Hatakoshi, *Treasure Microbox of Optoelectronics I–IV*, Advanced Communication Media (2018, 2019) [in Japanese].
2. M. Born and E. Wolf, *Principles of Optics, 7th Ed.*, Cambridge University Press (1999).
3. K. Kuroda, *Physical Optics—Propagation of light waves in a medium*, Asakura Publishing (2011) [in Japanese].
4. K. Iga and G. Hatakoshi, Chapter 5, *Treasure Microbox of Optoelectronics I*, Advanced Communication Media (2018) [in Japanese].
5. M. Mori, *Curves and Surfaces: Drawing and Tracing by Computer*, Kyoiku Shuppan (1974) [in Japanese].

Appendix 1A Excel Files

Appendix I-A (online) includes the following Excel files, the tabs of which correspond to the figures in the text.

01-1_Animation.xlsm	
Instruction	
Sin wave	
Dodecahedron	Fig. 1.2(b)
Icosahedron	Fig. 1.2(c)
01-2_Plane wave.xlsm	
Instruction	
Forward traveling wave	Fig. 1.3(a)
Backward traveling wave	Fig. 1.3(b)
Standing wave (1)	Fig. 1.3(c)
Standing wave (2)	Fig. 1.4(a)
Standing wave (3)	Fig. 1.4(b)
Plane wave	Fig. 1.S2
01-3_Wave packet.xlsm	
Instruction	
Wave packet	Fig. 1.5
Gaussian wave packet	Fig. 1.6

The value of N entered in cell B2 is the number of divisions in one period, so in most of the examples above, increasing this number will cause the animation to move slowly, and decreasing this number will cause the animation to move quickly. Also, the cells marked in yellow in each file are input parameters, so if you change the value, you can run the animation using that parameter.

1A.1 About VBA code

As the animation VBA depends on the Excel version and OS, modify the commands below accordingly.

```
Application.Screen Updating = True
Application.Wait [Now()] + 10/86400000
DoEvents: DoEvents:
```

Normally the command "DoEvents" should be sufficient, but if your PC is too fast, the screen may not update. In that case, insert multiple "DoEvents," but if that doesn't solve the problem, force a pause using "Application.Wait." The 10/86400000 above represents a 10-ms pause.

Chapter 2
Propagation of Light: Cylindrical Waves

2.1 Introduction

The plane waves discussed in the previous chapter are solutions of Maxwell's equations and the wave equation derived from them. In this chapter, cylindrical waves obtained as solutions of the same Maxwell's equations in a cylindrical coordinate system will be introduced.

2.2 The Wave Equation and Optical Waveguide

2.2.1 The wave equation and coordinate system

As described in the previous chapter [see Eqs. (1.7) and (1.8)], the following wave equation is derived from Maxwell's equations:

$$\nabla^2 U - \varepsilon\mu \frac{\partial^2 U}{\partial t^2} = 0. \tag{2.1}$$

When there is no electric charge or current ($\rho = 0, \boldsymbol{j} = 0$), the components E_x, E_y, E_z, H_x, H_y, and H_z of the electric field $\boldsymbol{E}$ and the magnetic field $\boldsymbol{H}$ are all solutions of Eq. (2.1). Now consider the case where the time dependence is represented as $\exp(i\omega t)$:

$$U = U_1(x, y, z)\exp(i\omega t). \tag{2.2}$$

Then, Eq. (2.1) becomes the following Helmholtz equation:

$$\nabla^2 U_1 + k^2 U_1 = 0, \tag{2.3}$$

where

$$k^2 \equiv \omega^2 \varepsilon\mu = k_0^2 n_r^2. \tag{2.4}$$

The plane wave discussed in the previous chapter corresponds to the solution of Eq. (2.3) with the separation of variables by an orthogonal coordinate system.

What happens in a different coordinate system? In this chapter, a cylindrical coordinate system (r, ϕ, z) is considered. The relationship with the orthogonal coordinate system (Cartesian coordinate system) (x, y, z) is given by the following equations (see Fig. 2.1):

$$x = r\cos\phi, \tag{2.5a}$$

$$y = r\sin\phi. \tag{2.5b}$$

If U_1 in Eq. (2.3) is separated into variables using cylindrical coordinates, a cylindrical wave solution can be obtained. The simplest example of applying cylindrical coordinates is a uniform free space similar to the case of plane waves. Other examples include optical waveguides[1–5] that have a layered structure in the z direction, as shown in Fig. 2.1, where the structure has no boundaries in the xy plane but has boundaries in the z direction. In this case, the boundary condition that the electric field and magnetic field components parallel to the boundary surface are continuous is applied.

2.2.2 Guided modes in optical waveguide

Before considering the cylindrical coordinate system, consider a solution for a structure that is uniform in the xy plane and has a layered structure in the z direction, such as the two-dimensional optical waveguide shown in Fig. 2.1. In this case, k^2 in Eq. (2.4) is a function of z only as shown by the following equation:

$$k^2 = k_0^2 n_r^2(z). \tag{2.6}$$

The optical waveguide will be dealt with separately; therefore, its specific structure will not be discussed here. The layered structure can be a multi-layer step-index structure or a graded-index structure.

For such a structure, U_1 in Eq. (2.3) is expressed as the product of the function $\Psi(x, y)$ and the function $F(z)$, as follows:

$$U_1(x, y, z) = \Psi(x, y)F(z). \tag{2.7}$$

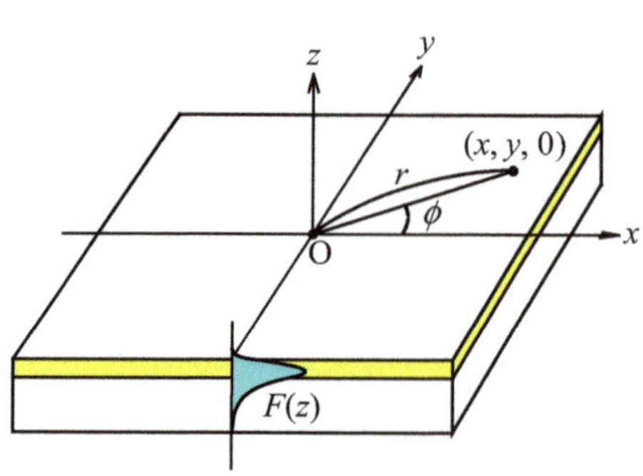

Figure 2.1 Optical waveguide and coordinate system.

In the case of an optical waveguide, $F(z)$ in the above equation is the waveguide mode function. It is not discussed how to obtain the guided mode function here, but it has a distribution as shown in Fig. 2.1, for example.

If the effective propagation constant of the guided mode propagating in the xy plane is β, then $\Psi(x, y)$ in Eq. (2.7) is the solution of the following equation:

$$\frac{\partial^2\Psi}{\partial x^2}+\frac{\partial^2\Psi}{\partial y^2}=-\beta^2\Psi. \tag{2.8}$$

From Eqs. (2.3), (2.7), and (2.8), the equation that the guided mode function $F(z)$ should satisfy is given by

$$\frac{d^2F}{dz^2}+(k^2-\beta^2)F=0. \tag{2.9}$$

In the following, the case where the electric field vector is in the xy plane is considered. Then, using the solution Ψ of Eq. (2.8) and the solution F of Eq. (2.9), the electric field $\boldsymbol{E}$ and the magnetic field $\boldsymbol{H}$ can be expressed as follows:

$$\boldsymbol{E}=a_1\begin{pmatrix}\partial\Psi/\partial y\\ -\partial\Psi/\partial x\\ 0\end{pmatrix}F\exp(i\omega t), \tag{2.10}$$

$$\boldsymbol{H}=\frac{ia_1}{\omega\mu}\begin{pmatrix}(\partial\Psi/\partial x)(d/dz)\\ (\partial\Psi/\partial y)(d/dz)\\ \beta^2\Psi\end{pmatrix}F\exp(i\omega t). \tag{2.11}$$

From the conditions of Eqs. (2.8) and (2.9), it can be confirmed that Eqs. (2.10) and (2.11) satisfy all of Maxwell's equations [Eqs. (1.1) to (1.4) in the previous chapter].

Up to this point, the cylindrical coordinates have not been introduced. An example of the solution of Eq. (2.8) is

$$\Psi=\exp[-i\beta(x\cos\theta+y\sin\theta)]. \tag{2.12}$$

This represents a guided plane wave that propagates in the θ direction in the xy plane, and Eqs. (2.10) and (2.11) represent a TE (transverse-electric) mode solution. Therefore, $F(z)$ satisfying Eq. (2.9) should be the solution for the TE mode.

If there is no layer structure in the z direction (i.e., $F=1$), then $\beta=k$, so in that case it becomes a plane wave traveling in free space in the direction of $(k_x, k_y, k_z)=(k\cos\theta, k\sin\theta, 0)$.

2.3 Cylindrical Waves

2.3.1 Cylindrical functions

Now that the preparations are complete, let's consider cylindrical coordinates. In cylindrical coordinates (r, ϕ, z), the second derivative is expressed as follows:

$$\frac{\partial^2}{\partial x^2} + \frac{\partial^2}{\partial y^2} = \frac{\partial^2}{\partial r^2} + \frac{1}{r}\frac{\partial}{\partial r} + \frac{1}{r^2}\frac{\partial^2}{\partial \phi^2}. \tag{2.13}$$

The function Ψ can be represented in the following form by the separation of variables:

$$\Psi = R(r)\Phi(\phi). \tag{2.14}$$

From the periodic boundary condition, $\Phi(\phi)$ is represented by

$$\Phi(\phi) = \cos(\nu\phi),\ \sin(\nu\phi),\ \exp(\pm i\nu\phi), \nu\text{: integer.} \tag{2.15}$$

This results in the following equation for $\Phi(\phi)$:

$$\frac{d^2}{d\phi^2}\Phi(\phi) = -\nu^2\Phi(\phi). \tag{2.16}$$

From Eqs. (2.13), (2.14), and (2.16), the equation that R should satisfy is

$$\frac{d^2R}{dr^2} + \frac{1}{r}\frac{dR}{dr} + (\beta^2 - \nu^2)R = 0. \tag{2.17}$$

This is the Bessel differential equation[6,7] of order ν, and its solutions are cylindrical functions such as $J_\nu(\beta r)$ and $Y_\nu(\beta r)$. Here, $J_\nu(\rho)$ and $Y_\nu(\rho)$ are the Bessel function (a cylindrical function of the first kind) and the Neumann function (a cylindrical function of the second kind), respectively.

The functions $J_\nu(\rho)$ or $Y_\nu(\rho)$ can be used as solutions for R. However, it is better to express the equation so as to represent a propagating wave like a plane wave of $\exp(\pm i\beta r)$. The function that satisfies this requirement is the Hankel function, as represented by the following equation:[6–8]

$$R(r) = H_\nu^{(k)}(\beta r) = \begin{cases} J_\nu(\beta r) + iY_\nu(\beta r) & (k=1) \\ J_\nu(\beta r) - iY_\nu(\beta r) & (k=2) \end{cases}. \tag{2.18}$$

The Hankel function is one of the cylindrical functions. In Eq. (2.18), $k=1$ and $k=2$ represent the Hankel functions of the first and second kind, respectively. In the region where $\beta r > 1$, these functions are approximated by

$$H_\nu^{(1)}(\beta r) \approx \sqrt{\frac{2}{\pi\beta r}} \exp\left(i\left[\beta r - \frac{\nu\pi}{2} - \frac{\pi}{4}\right]\right), \tag{2.19}$$

$$H_\nu^{(2)}(\beta r) \approx \sqrt{\frac{2}{\pi\beta r}} \exp\left(-i\left[\beta r - \frac{\nu\pi}{2} - \frac{\pi}{4}\right]\right). \tag{2.20}$$

If we define $\exp(-i\beta r)$ to represent a wave propagating in the $+r$ direction, which is similar to the case for the plane wave, then the Hankel functions of the first kind and the second kind correspond to converging cylindrical waves and diverging cylindrical waves, respectively.

It was mentioned above that $J_\nu(\rho)$ or $Y_\nu(\rho)$ may be used as the solution to R, but when the time dependence is set as $\exp(i\omega t)$, as in Eq. (2.2), the Hankel function should be used because without the Hankel function, it will not become a propagating wave. There is a time-dependent animation in the Excel file in the appendix, so you can see how the propagation progresses. Figure 2.2 is not an animation, so it may be difficult to understand, but it is the same diagram as in the Excel file. If we look at the real part of the function, that is, $J_0(\beta r)$ or $Y_0(\beta r)$, multiplied by $\exp(i\omega t)$ as shown in Figs. 2.2(a) and (b), it does not propagate and just oscillates in that position. In this case, the solution is an axially symmetric, oscillating standing wave. On the other hand, the real part of $H_0^{(1)}(\beta r)\exp(i\omega t)$ in Fig. 2.2(c) propagates toward the origin, and the real part of $H_0^{(2)}(\beta r)\exp(i\omega t)$ in Fig. 2.2(d) propagates from the origin outward.

Note that when the time dependence is set as $\exp(-i\omega t)$, $H_\nu^{(1)}(\beta r)$ becomes a divergent cylindrical wave and $H_\nu^{(2)}(\beta r)$ becomes a convergent cylindrical wave, which is the opposite of the case of $\exp(i\omega t)$.

2.3.2 Cylindrical waves in an optical waveguide

Based on the above discussion, we will show how the electric and magnetic fields of cylindrical waves are expressed. First, if we consider $\nu = 0$ (that is, $\Phi = 1$) as a simple case, then $R(r) = H_0^{(k)}(\beta r)$. In this case, from Eqs. (2.10) and (2.11), $\boldsymbol{E}$ and $\boldsymbol{H}$ can be expressed as follows:[8–10]

$$\boldsymbol{E} = aH_1^{(k)}(\beta r)e^{i\omega t}\begin{pmatrix} -\sin\phi \\ \cos\phi \\ 0 \end{pmatrix} F(z), \tag{2.21}$$

$$\boldsymbol{H} = \frac{iae^{i\omega t}}{\omega\mu_0}\begin{pmatrix} -H_1^{(k)}(\beta r)\cos\phi\dfrac{d}{dz} \\ -H_1^{(k)}(\beta r)\sin\phi\dfrac{d}{dz} \\ \beta H_0^{(k)}(\beta r) \end{pmatrix} F(z), \tag{2.22}$$

where $a = \beta a_1$ and $\mu = \mu_0$.

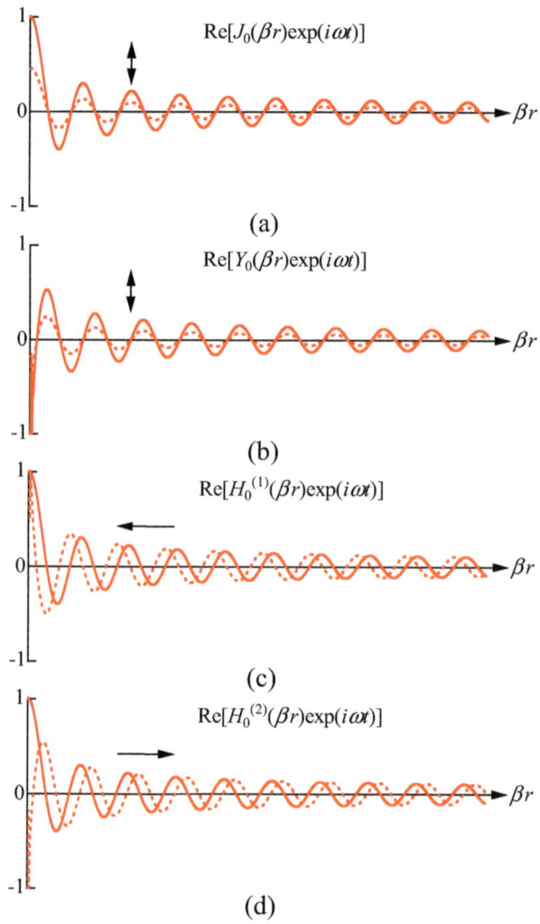

Figure 2.2 Time dependence of a cylindrical function × exp($i\omega t$).

Let's derive the Poynting vector of the cylindrical wave from Eqs. (2.21) and (2.22). The time-averaged Poynting vector $\boldsymbol{S}$ is given by

$$\boldsymbol{S} = \frac{1}{4}(\boldsymbol{E} \times \boldsymbol{H}^* + \boldsymbol{E}^* \times \boldsymbol{H}). \tag{2.23}$$

Substituting Eqs. (2.21) and (2.22), we obtain

$$\boldsymbol{S} = \frac{c_k a^2 F^2(z)}{\pi \omega \mu_0 r} \begin{pmatrix} \cos\phi \\ \sin\phi \\ 0 \end{pmatrix}, \tag{2.24}$$

$$c_k = \begin{cases} -1 & (k=1) \\ 1 & (k=2) \end{cases}. \tag{2.25}$$

In deriving Eq. (2.24), the following relationships were used:

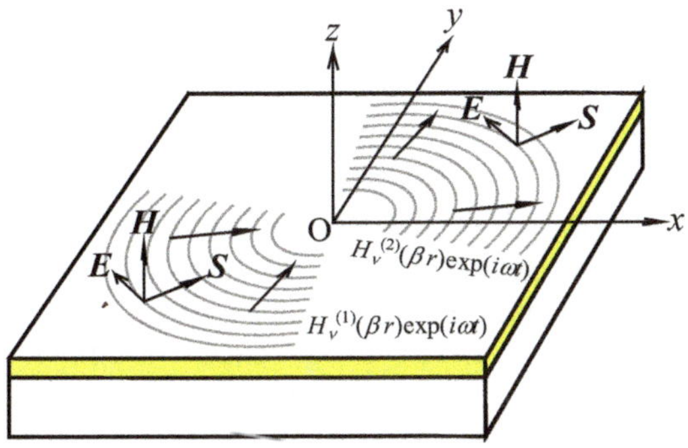

Figure 2.3 Electric field, magnetic field, and Poynting vector of a cylindrical wave in an optical waveguide.

$$H_\nu^{(1)*} = H_\nu^{(2)},\ H_\nu^{(2)*} = H_\nu^{(1)} \tag{2.26}$$

$$J_{\nu+1}(\xi)Y_\nu(\xi) - J_\nu(\xi)Y_{\nu+1}(\xi) = \frac{2}{\pi\xi}. \tag{2.27}$$

Equation (2.27) is called the Lommel equation.[7]

As can be seen from Eqs. (2.24) and (2.25), when $k = 1$, the Poynting vector is a vector toward the origin O, and when $k = 2$, it is a vector toward the outside from the origin. This also shows that $k = 1$ and $k = 2$ correspond to converging cylindrical waves and diverging cylindrical waves, respectively.

From Eqs. (2.21) and (2.24), it is obvious that the electric field vector $\boldsymbol{E}$ is orthogonal to the time-averaged Poynting, whose direction coincides with the propagation direction of the guided mode. In other words, $\boldsymbol{E}$ has only a transverse component, which is the definition of the TE (transverse-electric) mode.

Figure 2.3 shows the relationship between the electric field, magnetic field, and the time-averaged Poynting vector of converging cylindrical waves and diverging cylindrical waves in an optical waveguide. It is clear from Eq. (2.24) that the Poynting vector is a vector in the radial direction in the xy plane. As mentioned above, it is noteworthy that the solutions for the cylindrical waves expressed by Eqs. (2.21) and (2.22) satisfy not only the wave equation but also Maxwell's equations as a vector solution.

Although the case for the TE mode is shown here, the cylindrical wave of the TM mode can also be found in the same way. In the TM mode, $\boldsymbol{E}$ and $\boldsymbol{H}$ in Eqs. (2.21) and (2.22) are almost interchanged, but Eq. (2.9), which calculates the mode function, is slightly different. Furthermore, for the cylindrical wave in free space described below, a form in which the expressions of $\boldsymbol{E}$ and $\boldsymbol{H}$ are swapped (however, the coefficients change) is also a solution.

Side Topic: Transverse and longitudinal waves

Light is a transverse wave, and sound waves are longitudinal waves. The Japanese translations of "transverse" and "longitudinal" are the same terminology as "horizontal" and "vertical," respectively. This often causes confusion, for example for the terms "transverse mode" and "longitudinal mode" used in semiconductor lasers and optical waveguides.

"Transverse" means a cross section perpendicular to the propagation direction, and a transverse wave has its vibration direction within this cross section. Electromagnetic waves, gravitational waves, and seismic S waves are transverse waves. On the other hand, "longitudinal" indicates the direction of propagation. Sound waves and seismic P waves are longitudinal waves.

The "TE mode" in a waveguide is a transverse-electric mode, and the electric field vector $\boldsymbol{E}$ exists only in the plane perpendicular to the propagation direction. In this case, the magnetic field vector $\boldsymbol{H}$ actually has a component in the propagation direction. This seems to contradict the fact that $\boldsymbol{E} \times \boldsymbol{H}$ is the direction of the Poynting vector. In reality, the direction of $\boldsymbol{E} \times \boldsymbol{H}$ changes depending on time, but the time-average Poynting vector defined by Eq. (2.23) has a constant direction, which is the propagation direction.

Similarly, the "TM mode" is a transverse-magnetic mode in which the magnetic field vector $\boldsymbol{H}$ exists only in a plane perpendicular to the propagation direction. In this case, $\boldsymbol{E}$ has a component in the propagation direction.

2.3.3 Cylindrical wave in free space

If the z dependence in free space is $F(z) = \exp(-ik_z z)$ and the general solution of the cylindrical function $Z_\nu(\beta r)$ is used as the solution for R, the equations equivalent to (2.21) and (2.22) are as follows:

$$\boldsymbol{E} = aZ_1(\beta r)\begin{pmatrix} -\sin\phi \\ \cos\phi \\ 0 \end{pmatrix}\exp(-ik_z z + i\omega t), \tag{2.28}$$

$$\boldsymbol{H} = \frac{a}{\omega\mu_0}\begin{pmatrix} -k_z Z_1(\beta r)\cos\phi \\ -k_z Z_1(\beta r)\sin\phi \\ i\beta Z_0(\beta r) \end{pmatrix}\exp(-ik_z z + i\omega t), \tag{2.29}$$

where

$$k_z = (k^2 - \beta^2)^{1/2}. \tag{2.30}$$

The parameter k in the above equation is given by Eq. (2.4), and in this case there is no z dependence as in Eq. (2.6), and it is a constant. Note that k, representing the angular wave number in this case, resulted in the same symbol as (k), which represents the first and second kind of Hankel function, but it is different from that.

When $Z_\nu(\beta r)$ is a Bessel function, i.e., $Z_\nu(\beta r) = J_\nu(\beta r)$ $(\nu = 0,1)$, Eqs. (2.28) and (2.29) represent the so-called Bessel beam. The equations for a Bessel beam have the same form as the equations for the distribution in the core region in TE mode of an optical fiber, which will be described in Chapter 17 on optical fibers. Note that there is also a solution of the Bessel beam in the TM mode with $\boldsymbol{E}$ and $\boldsymbol{H}$ interchanged.

If the Hankel function is given as $Z_\nu(\beta r)$, the equations represent a strange cylindrical wave that converges toward or diverges from the z-axis while traveling in the z direction.

As a specific case, an example where $\beta = 0$, i.e., $F(z) = 1$, is given below:

$$\boldsymbol{E} = aH_1^{(k)}(kr)e^{i\omega t}\begin{pmatrix} -\sin\phi \\ \cos\phi \\ 0 \end{pmatrix}, \tag{2.31}$$

$$\boldsymbol{H} = \frac{iae^{i\omega t}}{\omega\mu_0}\begin{pmatrix} 0 \\ 0 \\ kH_0^{(k)}(kr) \end{pmatrix}. \tag{2.32}$$

From Eq. (2.28), the r component and ϕ component of $\boldsymbol{E}$ are given by the following equations:

$$E_r = 0, \tag{2.33}$$

$$E_\phi = aH_1^{(k)}(kr)e^{i\omega t}. \tag{2.34}$$

Figure 2.4 shows the amplitude distribution of the H_z component of the cylindrical wave given by Eqs. (2.31) and (2.32). Figure 2.5 shows the r dependence of E_ϕ and H_z. In this figure, the real part of E_ϕ and the real part of H_z are compared, but the relative relationship is exactly the same for the imaginary part, only the time phase is shifted by $\pi/2$.

As can be seen from Fig. 2.5, E_ϕ and H_z are almost in phase (or out of phase by π). As shown in Eqs. (2.19) and (2.20), at a sufficiently far distance, $H_1^{(k)}(kr)$ in E_ϕ and $H_0^{(k)}(kr)$ in H_z are out of phase by $\pi/2$. However, H_z is multiplied by i [$= \exp(i\pi/2)$], so they are in phase. Note that the situation is different near the origin, and the phase shifts slightly. Refer to the animation example in the Excel file of the appendix.

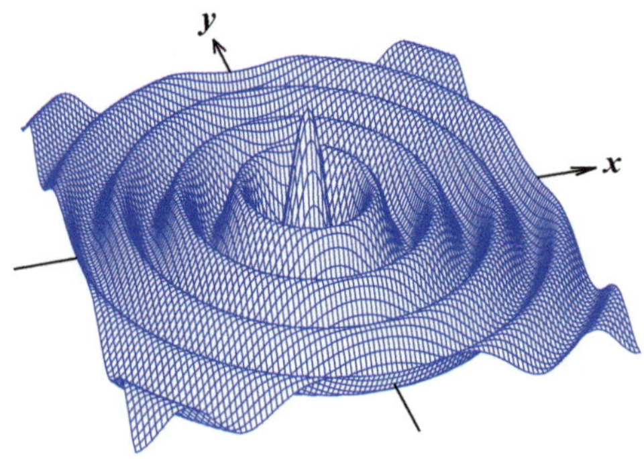

Figure 2.4 Calculated amplitude distribution of the cylindrical wave.

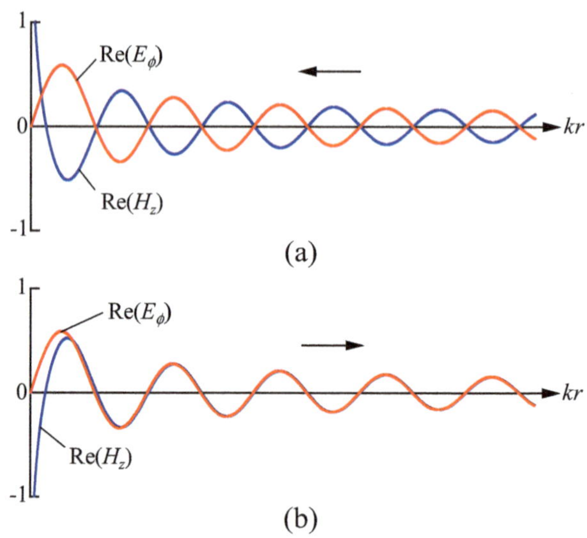

Figure 2.5 Relationship between E_ϕ and H_z in cylindrical waves: (a) converging cylindrical waves and (b) diverging cylindrical waves.

2.3.4 Electric flux line

Next, let's draw the electric flux lines of the cylindrical wave. As shown in Eqs. (2.30) and (2.31), in the cylindrical wave, the electric field $\boldsymbol{E}$ is always in the ϕ direction when $\nu = 0$. The direction of $\boldsymbol{E}$ is tangential to the electric flux lines, so in this case the electric flux lines become concentric circles. Therefore, although it is not worth the trouble of plotting it, it is necessary to find the radius of each concentric circle. Although an infinite number of the electric flux lines can be drawn, the electric flux lines along the zero point, maximum point, and minimum point of the electric field will be plotted. In other words, the radius can be found by finding the point where Eq. (2.31) becomes 0 and the point where the r differential of Eq. (2.31) becomes 0. Specifically, let $kr = \xi$, and derive the zero point of the following functions:

$$f_1(\xi) = \mathrm{Re}[H_1^{(k)}(\xi)e^{i\omega t}] = J_1(\xi)\cos(\omega t) \mp Y_1(\xi)\sin(\omega t), \tag{2.35}$$

$$\begin{aligned} f_2(\xi) &= \frac{d}{d\xi}\mathrm{Re}[H_1^{(k)}(\xi)e^{i\omega t}] \\ &= \frac{J_0(\xi)-J_2(\xi)}{2}\cos(\omega t)\mp\frac{Y_0(\xi)-Y_2(\xi)}{2}\sin(\omega t) \\ &= \frac{J_0(\xi)-J_2(\xi)}{2}\cos(\omega t)\mp\frac{Y_0(\xi)-Y_2(\xi)}{2}\sin(\omega t). \end{aligned} \tag{2.36}$$

A commonly used method to find the zero point of any function is Newton's method (the Newton–Raphson method). Newton's method requires a convergence evaluation, so normal programming is required to prepare it, but in Excel there is no need to create a macro, and in the case of a one-variable function, the zero point can be found just by using a spreadsheet. Refer to Appendix 2A.

The electric flux lines of the cylindrical wave and the corresponding amplitude distribution on the r-axis are shown in Fig. 2.6. In the Excel file in the appendix, an animation is provided, so you can see what happens with time. At a sufficiently far distance, they are equally spaced concentric circles, but as in Fig. 2.5, the appearance is different near the origin.

2.3.5 High-order cylindrical waves

Next, consider the case for $\nu = 1$. The function $\Phi(\phi)$ in this case is given, for example, by

$$\Phi(\phi) = \cos\phi. \tag{2.37}$$

In this case, $R(r) = H_1^{(k)}(\beta r)$; therefore, $\boldsymbol{E}$ and $\boldsymbol{H}$ are represented as follows. Note that since the equation is long, it is set as $H_\nu^{(k)} \equiv H_\nu^{(k)}(\beta r)$.

$$\boldsymbol{E} = ae^{i\omega t}\begin{pmatrix} \dfrac{\sin(2\phi)}{2}\left[\beta H_0^{(k)} - \dfrac{2}{r}H_1^{(k)}\right] \\ -\beta\cos^2\phi H_0^{(k)} + \dfrac{\cos(2\phi)}{r}H_1^{(k)} \\ 0 \end{pmatrix} F(z), \tag{2.38}$$

$$\boldsymbol{H} = \frac{iae^{i\omega t}}{\omega\mu_0}\begin{pmatrix} \left[\beta\cos^2\phi H_0^{(k)} - \dfrac{\cos(2\phi)}{r}H_1^{(k)}\right]\dfrac{d}{dz} \\ \dfrac{\sin(2\phi)}{2}\left[\beta H_0^{(k)} - \dfrac{2}{r}H_1^{(k)}\right]\dfrac{d}{dz} \\ \beta H_1^{(k)}\cos\phi \end{pmatrix} F(z). \tag{2.39}$$

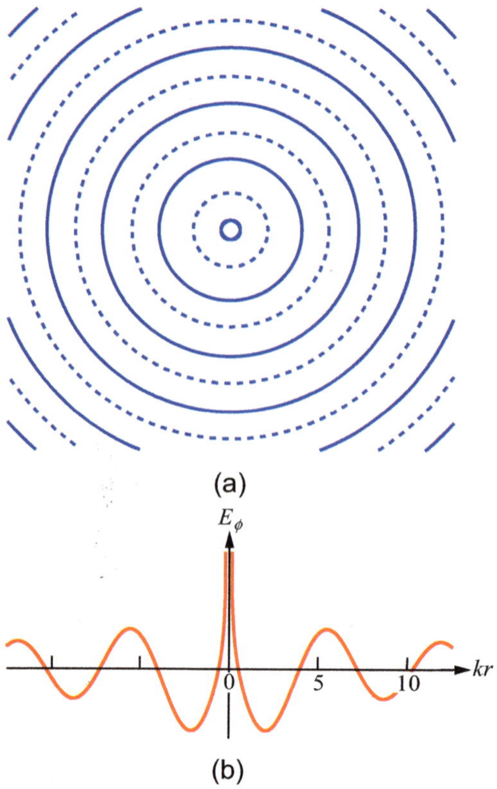

Figure 2.6 (a) Electric flux lines for cylindrical wave and (b) corresponding amplitude distribution. Solid lines and dashed lines in (a) represent zero points and maximum/minimum points of E_ϕ, respectively.

Equations (2.38) and (2.39) are cylindrical waves in a two-dimensional optical waveguide. If $F(z) = 1$ as in Eqs. (2.31) and (2.32), they become high-order cylindrical waves in free space. Figure 2.7 shows the amplitude distribution of the H_z component of this cylindrical wave. Figure 2.7(a) represents the case where $\Phi(\phi) = \cos\phi$, and Fig. 2.7(b) represents the case where $\Phi(\phi) = \exp(i\phi)$ [see Eq. (2.15)]. The latter corresponds to the "optical vortex."

For such higher modes, the electric field $\boldsymbol{E}$ and the time-averaged Poynting vector are no longer orthogonal to each other; therefore, the solutions are not "perfect" TE modes. However, it is important to note that $F(z)$ is the solution for the same TE mode as for the zeroth mode.

As shown in Eq. (2.21) or Fig. 2.3, in the case of a cylindrical wave for $\nu = 0$, the electric field $\boldsymbol{E}$ is always in the ϕ direction. In this case, the electric flux lines were concentric circles because the direction of $\boldsymbol{E}$ is tangential to the electric flux lines. What about the electric flux lines for $\boldsymbol{E}$ expressed by Eq. (2.38)?

As mentioned above, since the tangential direction of the electric flux lines is in the direction $\boldsymbol{E}$, the point (r, ϕ) on the electric flux lines satisfies the following equation:

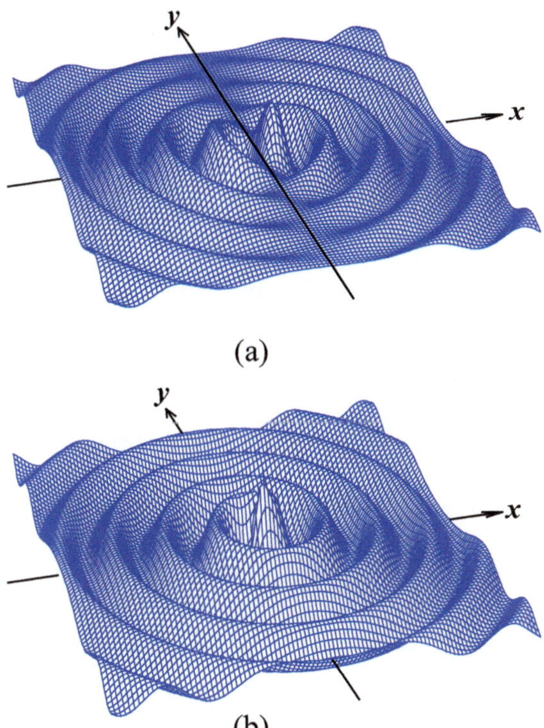

Figure 2.7 Amplitude distribution of the cylindrical wave with $\nu = 1$.

$$r\frac{d\phi}{dr} = \frac{\mathrm{Re}(E_\phi)}{\mathrm{Re}(E_r)}. \tag{2.40}$$

Let's find the radial direction (r) component and ϕ direction component of the electric field $\boldsymbol{E}$ from Eq. (2.38).

$$E_r = E_x \cos\phi + E_y \sin\phi = -ae^{i\omega t}\frac{\sin\phi}{r}H_1^{(k)}(\beta r)F(z), \tag{2.41}$$

$$E_\phi = -E_x \sin\phi + E_y \cos\phi = ae^{i\omega t}\cos\phi\left\{-\beta H_0^{(k)} + \frac{1}{r}H_1^{(k)}\right\}F(z). \tag{2.42}$$

The real part of each is represented by

$$\mathrm{Re}(E_r) = -aF(z)\frac{\sin\phi}{r}\{\cos(\omega t)J_1(\beta r) \mp \sin(\omega t)Y_1(\beta r)\}, \tag{2.43}$$

$$\begin{aligned}\mathrm{Re}(E_\phi) = aF(z)\cos\phi[&-\beta\{\cos(\omega t)J_0(\beta r) \mp \sin(\omega t)Y_0(\beta r)\}\\ &+ (1/r)\{\cos(\omega t)J_1(\beta r) \mp \sin(\omega t)Y_1(\beta r)\}].\end{aligned} \tag{2.44}$$

From Eqs. (2.43) and (2.44),

$$\frac{\mathrm{Re}(E_\phi)}{\mathrm{Re}(E_r)} = -\cot\phi\left[\frac{-\beta r\{\cos(\omega t)J_0(\beta r)\mp\sin(\omega t)Y_0(\beta r)\}}{\cos(\omega t)J_1(\beta r)\mp\sin(\omega t)Y_1(\beta r)}+1\right]. \tag{2.45}$$

Substituting this into Eq. (2.40) yields the following:

$$\tan\phi d\phi = \left[\frac{\frac{d}{dr}\{\cos(\omega t)J_1(\beta r)\mp\sin(\omega t)Y_1(\beta r)\}}{\cos(\omega t)J_1(\beta r)\mp\sin(\omega t)Y_1(\beta r)}\right]dr. \tag{2.46}$$

By integrating this, the following equation is obtained:

$$-\ln(\cos\phi)+C_1 = \ln[\cos(\omega t)J_1(\beta r)\mp\sin(\omega t)Y_1(\beta r)]. \tag{2.47}$$

Therefore,

$$C_2\sec\phi = \cos(\omega t)J_1(\beta r)\mp\sin(\omega t)Y_1(\beta r). \tag{2.48}$$

In Eqs. (2.47) and (2.48), C_1 and C_2 are integral constants. From Eq. (2.48), ϕ can be found as a function of r for a certain t. Therefore, the electric flux lines can be drawn by plotting this function. This may seem easy to draw, but it is not so easy because we need to find the upper and lower limits of r for one electric flux line. An example of the electric flux lines of a cylindrical wave with $\nu=1$ is shown in the Excel file in the appendix. The same is shown in Fig. 2.8(a). The Newton method mentioned earlier is used to find the upper and lower limits of r.

The file in Appendix I-A shows how the electric flux lines change with time. What is interesting is the movement near the origin. In the case of a diverging cylindrical wave, the electric flux lines appear to "tear off" from the center and spread outward. In fact, this first-order cylindrical wave is very similar to the electric dipole radiation, which will be discussed in the next chapter. Of course, there is a difference between waves that spread two-dimensionally and waves that spread three-dimensionally, and the formulas are different.

Equation (2.48) can be rewritten as follows:

$$\mathrm{Re}[H_1^{(k)}(\beta r)\cos\phi\exp(i\omega t)] = C_2. \tag{2.49}$$

As seen from Eq. (2.39), which represents the $\boldsymbol{H}$ vector, this is the equation representing $\mathrm{Im}(H_z)$ [$=\mathrm{Re}(H_z/i)$] =constant. In other words, the electric flux lines in Fig. 2.8(a) correspond to the contour lines of $\mathrm{Im}(H_z)$. When the number of contour lines (corresponding to the number of C_2 values) is increased, the

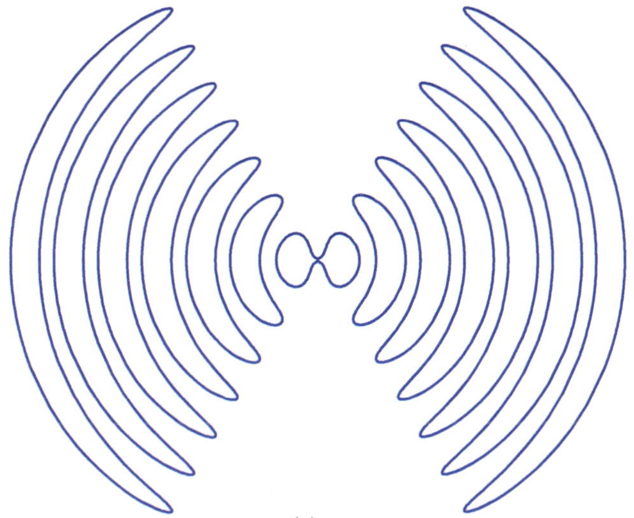

(a)

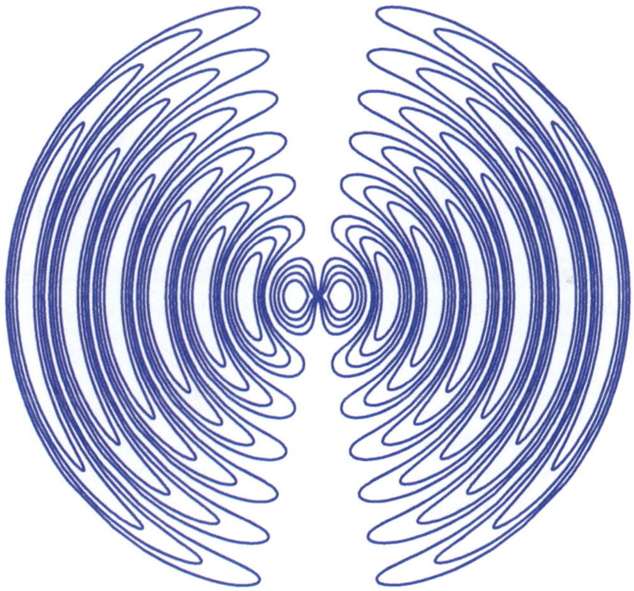

(b)

Figure 2.8 Electric flux lines of the cylindrical wave with $\nu = 1$.

electric flux lines become as shown in Fig. 2.8(b). In the Excel file, it is possible to input the value of C_2 (value normalized to the local maximum and minimum values; therefore, $0 < C_2 < 1$).

Next, consider the example of $\nu = 2$. The function $\Phi(\phi)$ for $\nu = 2$ is given, for example, by

$$\Phi(\phi) = \cos(2\phi). \tag{2.50}$$

In this case, $R(r) = H_2^{(k)}(\beta r)$; therefore, $\boldsymbol{E}$ and $\boldsymbol{H}$ can be found in the same way using Eqs. (2.10) and (2.11). As listing all the equations would be too long, only H_z and the equations corresponding to Eqs. (2.41), (2.42), and (2.48) for $\nu = 1$ will be shown below.

$$H_z = \frac{iae^{i\omega t}}{\omega\mu_0}\beta F(z)H_2^{(k)}\cos(2\phi)F(z), \tag{2.51}$$

$$E_r = -\frac{2a}{r}e^{i\omega t}\sin(2\phi)H_2^{(k)}F(z), \tag{2.52}$$

$$E_\phi = ae^{i\omega t}\cos(2\phi)\left\{\frac{2}{r}H_2^{(k)} - \beta H_1^{(k)}\right\}F(z), \tag{2.53}$$

$$C_2\sec(2\phi) = \cos(\omega t)J_2(\beta r) \mp \sin(\omega t)Y_2(\beta r). \tag{2.54}$$

Figure 2.9 shows the amplitude distribution of H_z [Eq. (2.51)] for this cylindrical wave, and Fig. 2.10 shows the electric flux lines given by Eq. (2.54). Similar to Eq. (2.49), Eq. (2.54) can be rewritten as follows:

$$\mathrm{Re}[H_2^{(k)}(\beta r)\cos(2\phi)\exp(i\omega t)] = C_2. \tag{2.55}$$

Therefore, in this case as well, the electric flux lines correspond to the contour lines of $\mathrm{Im}(H_z)$.

From Figs. 2.6 to 2.10, you can roughly predict what shape higher-order cylindrical waves will take.

2.4 Conclusion

Cylindrical waves obtained as solutions to wave equations in a cylindrical coordinate system were introduced. These cylindrical waves satisfy not only the wave equation but also Maxwell's equations as vectorial solutions.

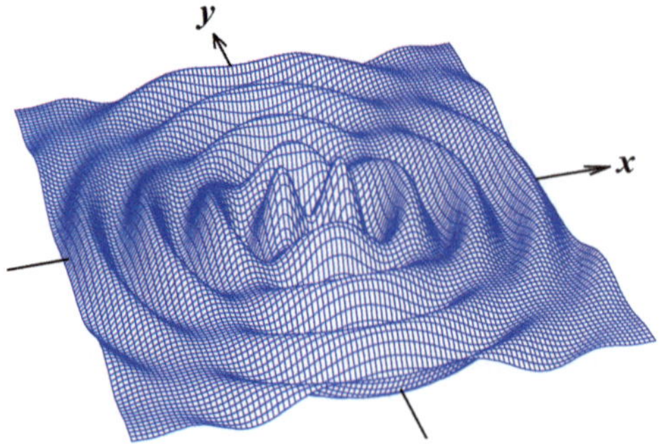

Figure 2.9 Amplitude distribution of the cylindrical wave with $\nu = 2$.

Figure 2.10 Electric flux lines of the cylindrical wave with $\nu = 2$.

References

1. G. Hatakoshi, "Optical waveguide," Chapter III-4, *Optical Engineering Handbook*, T. Ose, H. Saito, S. Tanaka, J. Tsujiuchi, and T. Namioka eds., Asakura Publishing, Tokyo (1986) [in Japanese].
2. K. Iga and G. Hatakoshi, Chapter 7, *Treasure Microbox of Optoelectronics I*, Advanced Communication Media, Tokyo (2018) [in Japanese].
3. H. Kogelnik, "Theory of dielectric waveguides," *Integrated Optics: Topics in Applied Physics, Vol. 7*, T. Tamir, et al., eds., Springer-Verlag, Berlin (1975).
4. D. Marcuse, *Theory of Dielectric Optical Waveguides*, Academic Press, New York (1974).
5. H.-G. Unger, *Planar Optical Waveguides and Fibres*, Oxford University Press (1977).
6. F. Bowman, *Introduction to Bessel Functions*, Dover Publications, New York (1958).
7. T. Ishizu, M. Sato, and N. Kaneko, eds., *Applied Mathematics Exercise 2*, Baifukan, Tokyo (1969) [in Japanese].
8. G. Hatakoshi and S. Tanaka, "Coupled-mode theory for a plane and a cylindrical wave in optical waveguide Bragg grating lenses," *J. Opt. Soc. Am.* **71**(1), 40–48 (1981).

9. G. Hatakoshi and S. Tanaka, "Coupling of two cylindrical guided waves in optical waveguide Bragg grating lenses," *J. Opt. Soc. Am.* **71**(1), 121–123 (1981).
10. G. Hatakoshi, "Coupling of guided mode by diffraction grating in optical waveguide," *Microoptics Handbook*, Chapter II-3, Asakura Publishing, Tokyo (1995) [in Japanese].

Appendix 2A Newton's Method in Excel

As shown in Fig. 2.AP1, Newton's method finds the solution to x that satisfies $f(x) = 0$ by the repeated process where the variable x is updated to a new x_{i+1} by using the following equation:

$$x_{i+1} = x_i - f(x_i)/f'(x_i). \tag{2.A1}$$

This is an extremely powerful method, and in most cases, it converges within a few to 10 cycles. Originally, it would be necessary to prepare a program to determine convergence, but in the case of Excel, the method can be performed by a spreadsheet calculation like the one shown in Table 2.AP1, assuming that convergence will occur after a few times. It usually converges after about 10 lines. If it does not converge, it is better to consider that the input of $f(x)$ or

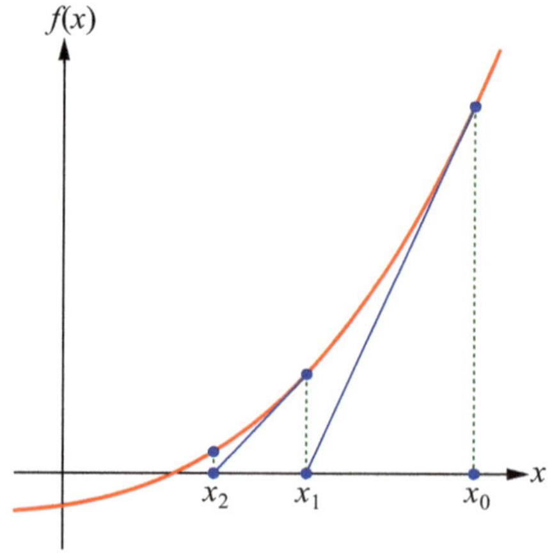

Figure 2.AP1 Newton's method in a plot.

Table 2.AP1 Newton's method in Excel.

	A	B	C	D
1	x_0	$=f(x_0)$	$=f'(x_0)$	$=x_0-f(x_0)/f'(x_0)$
2	$x_1 = \mathrm{D1}$	$=f(x_1)$	$=f'(x_1)$	$=x_1-f(x_1)/f'(x_1)$
3	$x_2 = \mathrm{D2}$	$=f(x_2)$	$=f'(x_2)$	$=x_2-f(x_2)/f'(x_2)$
4	$x_3 = \mathrm{D3}$	$=f(x_3)$	$=f'(x_3)$	$=x_3-f(x_3)/f'(x_3)$

$f'(x)$ is wrong. Table 2.AP1 is written with some redundancy. For example, in the case that column D does not exist, write the right-hand side of Eq. (2.A1) in column A. There is nothing that cannot be done with just one column, but it is better to have column B in order to confirm convergence.

Appendix 2B Excel File

The following file, whose tabs correspond to the figures in the text, is available in Appendix I-A (online). Refer to the operation instructions for running the animation.

02_Cylindrical wave.xlsm

Instruction	
Cylindrical function 1	Fig. 2.2
Cylindrical function 2	Fig. 2.5
Cylindrical wave ($\nu = 0$)	Fig. 2.6
Cylindrical wave ($\nu = 1$)	Fig. 2.8
Cylindrical wave ($\nu = 2$)	Fig. 2.10

Chapter 3
Propagation of Light: Spherical Waves

3.1 Introduction

As solutions to Maxwell's equations, plane waves obtained in a Cartesian coordinate system and cylindrical waves obtained in a cylindrical coordinate system were introduced in Chapters 1 and 2, respectively. This chapter describes spherical waves obtained in a spherical coordinate system.

3.2 Spherical Waves

3.2.1 Spherical coordinate system

Using cylindrical coordinates as in the previous chapter, solutions for cylindrical waves were obtained. So next, let's take a look at what happens in spherical coordinates. As shown in Fig. 3.1, the relationship between the spherical coordinate system (r, θ, ϕ) and the Cartesian coordinate system is as follows:

$$x = r \sin\theta \cos\phi, \tag{3.1a}$$

$$y = r \sin\theta \sin\phi, \tag{3.1b}$$

$$z = r \cos\theta. \tag{3.1c}$$

The wave equation and the Helmholtz equation described in the previous chapter are shown again as follows:

$$\nabla^2 U - \varepsilon\mu \frac{\partial^2 U}{\partial t^2} = 0, \tag{3.2}$$

$$\nabla^2 U_1 + k^2 U_1 = 0, \tag{3.3}$$

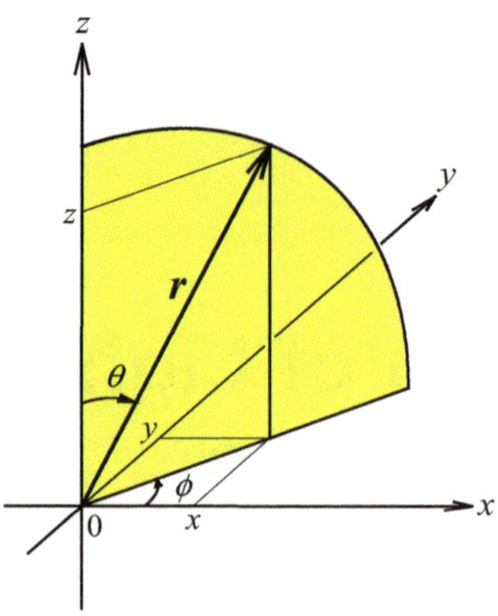

Figure 3.1 Spherical coordinate system.

where

$$U = U_1 \exp(i\omega t). \tag{3.4}$$

Equation (3.3) expressed in a spherical coordinate system is as follows:

$$\frac{\partial^2 U_1}{\partial r^2} + \frac{2}{r}\frac{\partial U_1}{\partial r} + \frac{1}{r^2}\left\{\frac{1}{\sin\theta}\frac{\partial}{\partial\theta}\left(\sin\theta\frac{\partial U_1}{\partial\theta}\right) + \frac{1}{\sin^2\theta}\frac{\partial^2 U_1}{\partial^2\phi}\right\} + k^2 U_1 = 0. \tag{3.5}$$

Here, U_1 is separated into variables using spherical coordinates as follows:

$$U_1 = R(r)\Theta(\theta)\Phi(\phi). \tag{3.6}$$

As in the case of the cylindrical coordinate system, the following equation, which is similar to Eq. (2.16), holds for $\Phi(\phi)$ based on the periodic boundary condition:

$$\frac{d^2}{d\phi^2}\Phi(\phi) = -\mu^2\Phi(\phi) \quad (\mu\text{:integer}). \tag{3.7}$$

Regarding $R(r)$ and $\Theta(\theta)$, the following formulas hold with the separation constant set to $\nu(\nu+1)$:

$$\frac{d^2 R}{dr^2} + \frac{2}{r}\frac{dR}{dr} + \left\{k^2 - \frac{\nu(\nu+1)}{r^2}\right\}R = 0, \tag{3.8}$$

$$\frac{1}{\sin\theta}\frac{d}{d\theta}\left(\sin\theta\frac{d\Theta}{d\theta}\right) + \left\{\nu(\nu+1) - \frac{\mu^2}{\sin^2\theta}\right\}\Theta = 0. \tag{3.9}$$

By setting $\cos\theta = \zeta$ in Eq. (3.9), the following equation is obtained:

$$\frac{d}{d\zeta}\left\{(1-\zeta^2)\frac{d\Theta(\zeta)}{d\zeta}\right\} + \left\{\nu(\nu+1) - \frac{\mu^2}{1-\zeta^2}\right\}\Theta(\zeta) = 0. \tag{3.10}$$

This is the Legendre differential equation of order ν, and its basic solutions are the associated Legendre polynomials of the first and second kinds.[1,2]

3.2.2 Spherical waves

Considering the simplest case of $\mu = \nu = 0$ in Eqs. (3.7) to (3.10), Φ and Θ are constants (= 1), and the equation that R must satisfy is as follows:

$$\frac{d^2R}{dr^2} + \frac{2}{r}\frac{dR}{dr} + k^2 R = 0. \tag{3.11}$$

A particular solution for Eq. (3.11) is represented by

$$R = \frac{1}{r}\exp(\pm ikr). \tag{3.12}$$

This is a well-known spherical wave. Letting the time dependence be $\exp(+i\omega t)$, a spherical wave Ψ_1 converging towards the origin and a spherical wave Ψ_2 diverging from the origin are represented by the following equations:

$$\Psi_1 = \frac{1}{r}\exp(ikr + i\omega t), \tag{3.13}$$

$$\Psi_2 = \frac{1}{r}\exp(-ikr + i\omega t). \tag{3.14}$$

The wavefront (equiphase surface) of the spherical wave in Eqs. (3.12) to (3.14) is a concentric spherical surface with equal periods in the radial direction, so there is no need to illustrate it, but just to be sure, a cross section of the equiphase surface is shown in Fig. 3.2. The Excel file in Appendix I-A enables the performance of converging or diverging animations.

Equations (3.13) and (3.14) are scalar waves as solutions to the wave equation of Eq. (3.3). Is there such a spherically symmetric solution to Maxwell's equations? In order to examine this problem, consider a vector potential $\boldsymbol{A}$ whose z component is expressed by Eq. (3.13) or Eq. (3.14), and whose x and y components are 0:

$$\boldsymbol{A} = \begin{pmatrix} 0 \\ 0 \\ 1 \end{pmatrix} \frac{1}{r}\exp(\pm ikr + i\omega t). \tag{3.15}$$

The scalar potential ψ can be derived so as to satisfy the Lorentz condition (see Appendix 3A). By using the relation

$$\frac{\partial \psi}{\partial t} = -\frac{1}{\varepsilon\mu}\nabla \cdot \boldsymbol{A} = \frac{1}{\varepsilon\mu}\left(\frac{1}{r^2} \mp \frac{ik}{r}\right)\cos\theta \exp(\pm ikr + i\omega t), \tag{3.16}$$

Figure 3.2 Cross section of a spherical wave as a scalar wave.

ψ is represented as follows:

$$\psi = \frac{-i}{\varepsilon\mu\omega}\left(\frac{1}{r^2} \mp \frac{ik}{r}\right)\cos\theta\exp(\pm ikr + i\omega t). \tag{3.17}$$

The electric field $\boldsymbol{E}$ and the magnetic field $\boldsymbol{H}$ can be expressed as functions of $\boldsymbol{A}$ and ψ, as follows:

$$\boldsymbol{E} = -\nabla\psi - \frac{\partial \boldsymbol{A}}{\partial t}, \tag{3.18}$$

$$\boldsymbol{H} = \frac{1}{\mu}\nabla \times \boldsymbol{A}. \tag{3.19}$$

Since $\boldsymbol{A}$ in Eq. (3.15) and ψ in Eq. (3.17) satisfy the Lorenz condition, $\boldsymbol{E}$ and $\boldsymbol{H}$ given by Eqs. (3.18) and (3.19) are solutions of Maxwell's equations. However, as shown in Eq. (3.17), ψ has a θ dependence, so unfortunately this solution is not spherically symmetric. In any case, if the scalar solution in the spherical coordinate system is found, the vector solution that satisfies Maxwell's equations can be found using the above method.

The solutions given by Eqs. (3.15), (3.17), (3.18), and (3.19) are actually the same as the equation for electric dipole radiation described in Section 3.4. However, since electric dipole radiation is an electromagnetic wave that diverges from the origin, only one of the double signs is selected. The specific formula will be described in Section 3.4.

Side Topic: Lorentz and Lorenz

Lorentz and Lorenz are both names of historically famous scholars in the field of physics. The two names are similar but can be distinguished by the presence or absence of the "t."

Hendrik Antoon Lorentz is a Dutch physicist who, together with Zeeman, won the Nobel Prize in Physics for the discovery of the Zeeman effect and its theoretical interpretation. Many terms in the field of physics include Lorentz's name. The Lorentz contraction (Fitzgerald–Lorentz contraction), Lorentz transformation, Lorentz factor, etc., are widely known in connection with Einstein's special relativity theory. In addition, there are Lorentz forces, the Lorentz distribution (Cauchy–Lorentz distribution), Drude–Lorentz model, Cody–Lorentz model, etc.

Ludvig Valentin Lorenz is a Danish physicist and mathematician, known for his work on the Lorenz–Mie theory of Mie scattering. The spherical waves based on spherical coordinates in this chapter also appear in Mie scattering theory. Also, the Lorenz gauge and Lorenz conditions that appear in this chapter were named after Ludvig Lorenz. The Wiedemann–Franz–Lorenz law, which expresses the correlation between electrical conductivity σ and thermal conductivity K, is expressed in the form $K/(T\sigma) = L$, where L on the right-hand side is called the Lorenz number.

The Lorentz–Lorenz equation is a relational equation named after Hendrik Lorentz and Ludvig Lorenz. This is an equation that connects refractive index, particle density, and polarizability.

3.2.3 General solutions in spherical coordinates

In Eq. (3.8), let's define

$$x = kr. \tag{3.20}$$

Then, the following equation is obtained:

$$\frac{d^2R}{dx^2} + \frac{2}{x}\frac{dR}{dx} + \left\{1 - \frac{\nu(\nu+1)}{x^2}\right\}R = 0. \tag{3.21}$$

This is a spherical Bessel differential equation, and its solutions are the spherical Bessel function, spherical Neumann function, and spherical Hankel function.[2,3] Here the following variable is introduced:

$$y = r^{1/2}R. \tag{3.22}$$

Equation (3.21) is transformed into the following equation:

$$\frac{\partial^2 y}{\partial x^2} + \frac{1}{x}\frac{\partial y}{\partial x} + \left\{1 - \frac{(\nu + 1/2)^2}{x^2}\right\} y = 0. \tag{3.23}$$

This is the Bessel differential equation of order $(\nu + 1/2)$. Therefore, the spherical Bessel function, spherical Neumann function, and spherical Hankel function, which are the solutions to Eq. (3.21), are formally defined as half-odd order cylindrical functions as follows:

$$j_\nu(x) = \sqrt{\frac{\pi}{2x}} J_{\nu+1/2}(x), \tag{3.24}$$

$$y_\nu(x) = \sqrt{\frac{\pi}{2x}} Y_{\nu+1/2}(x), \tag{3.25}$$

$$h_\nu^{(k)}(x) = \sqrt{\frac{\pi}{2x}} H_{\nu+1/2}^{(k)}(x). \tag{3.26}$$

The associated Legendre polynomials, which are the solutions to Eq. (3.10), are covered in many textbooks, so a detailed explanation is omitted here. Among $P_\nu^\mu(\zeta)$ and $Q_\nu^\mu(\zeta)$, the latter diverges at $\zeta = \pm 1$, corresponding to $\theta = 0$ and π, so the finite solution is $P_\nu^\mu(\zeta)$. From the above, the general solution in spherical coordinates can be expressed as follows:[2,3]

$$U_1 = \sum_{\nu=0}^{\infty} \sum_{\mu=-\nu}^{\nu} a_{\mu\nu} \left\{ \begin{array}{c} h_\nu^{(1)}(kr) \\ h_\nu^{(2)}(kr) \end{array} \right\} P_\nu^\mu(\cos\theta) \left\{ \begin{array}{c} \cos(\mu\phi) \\ \sin(\mu\phi) \end{array} \right\}. \tag{3.27}$$

Here, { } represents a linear combination of two independent functions written within the brackets. $\{h_\nu^{(1)}(kr),\ h_\nu^{(2)}(kr)\}$ may be a combination of other independent functions $\{j_\nu(kr),\ y_\nu(kr)\}$. Similarly, $\{\cos(\mu\phi),\ \sin(\mu\phi)\}$ may be $\{\exp(i\mu\phi),\ \exp(-i\mu\phi)\}$.

A simple case $(0 \le \mu,\ \nu \le 2)$ is given below as an example of the solution to the spatial part expressed by Eq. (3.27). Note that the subscripts represent the values of μ and ν.

$$U_{00} = \frac{1}{kr}\exp(\pm ikr), \tag{3.28}$$

$$U_{01} = \frac{\cos\theta}{kr}\left(\mp\frac{i}{kr} - 1\right)\exp(\pm ikr), \tag{3.29}$$

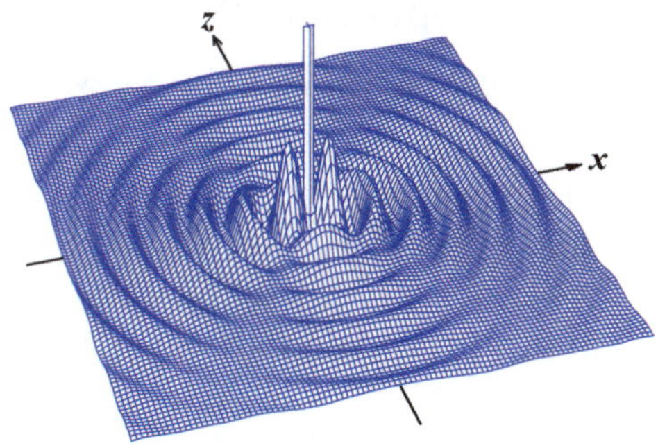

Figure 3.3 Example of the solution in spherical coordinates [Eq. (3.31)].

$$U_{11} = \frac{\cos\phi \sin\theta}{kr}\left(\mp\frac{i}{kr} - 1\right)\exp(\pm ikr), \tag{3.30}$$

$$U_{02} = \frac{3\cos(2\theta) + 1}{4kr}\left\{\mp\left(\frac{3}{k^2r^2} - 1\right)i - \frac{3}{kr}\right\}\exp(\pm ikr). \tag{3.31}$$

Equation (3.28) is the same so-called spherical wave as Eq. (3.12). Also, Eq. (3.29) has the same form as Eq. (3.17). Therefore, as stated in Section 3.2.2, the equation is similar to the equation for electric dipole radiation described in Section 3.4. The distribution in the xz plane expressed by Eq. (3.31) is shown in Fig. 3.3.

Both of these solutions are scalar solutions to the Helmholtz equation. To find $\boldsymbol{E}$ and $\boldsymbol{H}$ that satisfy Maxwell's equations, for example, as mentioned in the section for spherical waves, give the vector potential and scalar potential that satisfy the Lorenz condition based on the solution of the Helmholtz equation, and use the Eqs. (3.18) and (3.19) to find $\boldsymbol{E}$ and $\boldsymbol{H}$. Alternatively, $\boldsymbol{E}$ and $\boldsymbol{H}$ can be found by using the Debye potential[4,5] as the solution to the Helmholtz equation. Note that if a function of only r, such as Eq. (3.28), is used as the Debye potential, all components of $\boldsymbol{E}$ and $\boldsymbol{H}$ become 0, and no meaningful solution can be obtained. The solutions given by Eqs. (3.15) to (3.19) also have θ dependence. After all, in a spherical coordinate system, it is impossible to obtain a solution to Maxwell's equations expressed as a function of r alone.†

†Note: For cylindrical waves, there was a solution that satisfied Maxwell's equations using only a function of r [see Eqs. (2.32) to (2.34)]. On the other hand, a vector field on a three-dimensional sphere always has a singular point. This is related to the hairy ball theorem (hair has a singular point called a whorl), which is sometimes introduced as a special case of the Poincaré–Hopf theorem.

3.3 Mutual Expansion of Spherical Waves and Plane Waves

3.3.1 Plane wave expansion of spherical waves

Any function can be expanded by a completely orthogonal function system. The Fourier transform is an example.

The spatial part of the spherical wave expressed by Eqs. (3.13) and (3.14) can be expanded into a plane wave. This is known as the Weyl representation of a spherical wave[6,7] and can be expressed in an extremely simple form as shown by the following equation:

$$\frac{\exp(\pm ikr)}{r} = \pm\frac{i}{2\pi}\int_{-\infty}^{\infty}\int_{-\infty}^{\infty}\frac{\exp(-ik_x x - ik_y y \pm ik_z|z|)}{k_z}dk_x dk_y, \tag{3.32}$$

$$k_z = \begin{cases} \sqrt{k^2 - k_x^2 - k_x^2} & (k^2 > k_x^2 + k_x^2) \\ \pm i\sqrt{k_x^2 + k_x^2 - k^2} & (k^2 < k_x^2 + k_x^2) \end{cases}. \tag{3.33}$$

In Eqs. (3.32) and (3.33), the signs + and – correspond to a converging spherical wave and a diverging spherical wave, respectively, and the sign on the right-hand side and the sign for k_z are also selected accordingly. As explained many times, this is the case when the time dependence is $\exp(+i\omega t)$, and in the case of $\exp(-i\omega t)$, converging and diverging are opposite. In the equation, it does not matter which one is converging or diverging. Also, the right-hand side of Eq. (3.32) is expanded as $\exp(-ik_x x - ik_y y)$, but the same is true even if this is expanded as $\exp(+ik_x x + ik_y y)$. In other words, if the variables are set as $k_x \to -k_x$ and $k_y \to -k_y$, the latter expression is obtained (note that it is a double integral of k_x and k_y). However, the term $\exp(\pm ik_z|z|)$ remains as is.

Equation (3.32) was originally derived to hold true in the range $z > 0$, but it is shown in the literature[3] that if it is in this form, it also holds true for $z = 0$ (excluding the origin) and $z < 0$.

3.3.2 Spherical wave expansion of plane waves

Let U_1 be the solution to the Helmholtz equation [Eq. (3.3)] for a plane wave traveling in the z direction:

$$U_1 = \exp(-ikz). \tag{3.34}$$

Expand this with a solution in spherical coordinates. This corresponds to finding the coefficient $a_{\mu\nu}$ in Eq. (3.27). Since there is no ϕ dependence, $\mu = 0$ can be set. Also, in the solution of the spherical Bessel differential equation, only j_ν does not diverge at $r = 0$, so Eq. (3.34) can be expanded as follows:

$$U_1 = \exp(-ikz) = \exp(-ikr\cos\theta) = \sum_{\nu=0}^{\infty} a_\nu j_\nu(kr)P_\nu(\cos\theta). \tag{3.34}$$

The detail for the derivation of a_ν is omitted here and only the results are shown:[1]

$$\exp(-ikz) = \sum_{\nu=0}^{\infty} (-i)^\nu (2\nu + 1) j_\nu(kr) P_\nu(\cos\theta). \tag{3.35}$$

This is the spherical wave expansion of a plane wave.

3.4 Electric Dipole Radiation

When there is an electric dipole $p(t)$ oscillating in the z direction at the origin, the electromagnetic potential (scalar potential ψ and vector potential $\boldsymbol{A}$) at position r is expressed by the following equations:[8–10]

$$\psi(\boldsymbol{r}, t) = \frac{\cos\theta}{4\pi\varepsilon}\left\{\frac{1}{r^2}p(t_1) + \frac{1}{cr}\dot{p}(t_1)\right\}, \tag{3.37}$$

$$A_z(\boldsymbol{r}, t) = \frac{\mu}{4\pi r}\dot{p}(t_1), A_x = A_y = 0. \tag{3.38}$$

These formulas are derived from the retarded potential.[11] In the above equations, $\dot{p}(t_1)$ represents the time derivative of $p(t_1)$. The time t_1 includes a time delay that depends on the distance from the origin and is given by the following equation:

$$t_1 = t - \frac{n_r}{c}r, \tag{3.39}$$

where n_r is the refractive index. The z differential of Eq. (3.38) is expressed by

$$\begin{aligned}\frac{\partial A_z}{\partial z} &= \frac{\mu}{4\pi\varepsilon}\left\{-\frac{1}{r}\frac{\partial r}{\partial z}\dot{p}(t_1) + \frac{\partial t_1}{\partial r}\frac{\partial r}{\partial z}\ddot{p}(t_1)\right\}\\ &= -\frac{\mu\cos\theta}{4\pi\varepsilon}\left\{\frac{1}{r}\dot{p}(t_1) + \frac{1}{c}\ddot{p}(t_1)\right\}.\end{aligned} \tag{3.40}$$

From Eqs. (3.37) and (3.40), it can be seen that ψ and $\boldsymbol{A}$ satisfy the Lorenz condition [Eq. (3.A12) in Appendix 3A].

The r, θ, and ϕ components of vector potential $\boldsymbol{A}$ are represented by the following equations:

$$A_r = \frac{\mu}{4\pi}\frac{\cos\theta}{r}\dot{p}(t_1), \tag{3.41a}$$

$$A_\theta = -\frac{\mu}{4\pi}\frac{\sin\theta}{r}\dot{p}(t_1), \tag{3.41b}$$

$$A_\phi = 0. \tag{3.41c}$$

Since $\boldsymbol{E}$ and $\boldsymbol{H}$ are given by Eqs. (3.18) and (3.19), the r, θ, and ϕ components of $\boldsymbol{E}$ and $\boldsymbol{H}$ are derived as follows:

$$E_r = -\frac{\partial\psi}{\partial r} - \frac{\partial A_r}{\partial t} = \frac{\cos\theta}{2\pi\varepsilon r^2}\left\{\frac{1}{r}p(t_1) + \frac{1}{c}\dot{p}(t_1)\right\}, \tag{3.42a}$$

$$E_\theta = -\frac{1}{r}\frac{\partial\psi}{\partial\theta} - \frac{\partial A_\theta}{\partial t} = \frac{\sin\theta}{4\pi\varepsilon r}\left\{\frac{1}{r^2}p(t_1) + \frac{1}{cr}\dot{p}(t_1) + \frac{1}{c^2}\ddot{p}(t_1)\right\}, \tag{3.42b}$$

$$E_\phi = 0, \tag{3.42c}$$

$$H_r = \frac{1}{\mu}\frac{1}{r\sin\theta}\left\{\frac{\partial}{\partial\theta}(\sin\theta A_\phi) - \frac{\partial A_\theta}{\partial\phi}\right\} = 0, \tag{3.43a}$$

$$H_\theta = \frac{1}{\mu r}\left\{\frac{1}{\sin\theta}\frac{\partial A_r}{\partial\phi} - \frac{\partial}{\partial r}(rA_\phi)\right\} = 0, \tag{3.43b}$$

$$H_\phi = \frac{1}{\mu r}\left\{\frac{\partial}{\partial r}(rA_\theta) - \frac{\partial A_r}{\partial\theta}\right\} = \frac{\sin\theta}{4\pi r}\left\{\frac{1}{r}\dot{p}(t_1) + \frac{1}{c}\ddot{p}(t_1)\right\}. \tag{3.43c}$$

Here, let the time dependence of $p(t_1)$ be $\exp(i\omega t_1)$ and express it as follows:

$$p(t_1) = p_0\exp(i\omega t_1) = p_0\exp(-ikr + i\omega t), \tag{3.44}$$

where the relationship $\omega = (c/n_r)k$ was used.

Substituting Eq. (3.44) into Eqs. (3.42) and (3.43), the following equations are obtained:

$$E_r = p_0\frac{k\cos\theta}{2\pi\varepsilon r^2}\left\{\frac{1}{kr} + i\right\}\exp(-ikr + i\omega t), \tag{3.45a}$$

$$E_\theta = p_0\frac{k^2\sin\theta}{4\pi\varepsilon r}\left\{\frac{1}{k^2r^2} + \frac{i}{kr} - 1\right\}\exp(-ikr + i\omega t), \tag{3.45b}$$

$$E_\phi = 0, \tag{3.45c}$$

$$H_r = H_\theta = 0, \tag{3.46a}$$

$$H_\phi = p_0\frac{\omega k\sin\theta}{4\pi r}\left\{\frac{i}{kr} - 1\right\}\exp(-ikr + i\omega t). \tag{3.46b}$$

Let's draw the electric flux lines in the same way as in the previous chapter. At first, from Eqs. (3.45a) and (3.45b),

$$\frac{\mathrm{Re}(E_\theta)}{\mathrm{Re}(E_r)} = \frac{\tan\theta\left\{\left(\frac{1}{kr} - kr\right)\cos(kr - \omega t) + \sin(kr - \omega t)\right\}}{2\left\{\frac{\cos(kr - \omega t)}{kr} + \sin(kr - \omega t)\right\}}$$
$$= \frac{r\tan\theta\frac{d}{dr}\left\{\frac{\cos(kr - \omega t)}{kr} + \sin(kr - \omega t)\right\}}{2\left\{\frac{\cos(kr - \omega t)}{kr} + \sin(kr - \omega t)\right\}}. \tag{3.47}$$

From Eq. (2.41) in the previous chapter, the equation representing the electric flux lines is given by

$$r\frac{d\theta}{dr} = \frac{\mathrm{Re}(E_\theta)}{\mathrm{Re}(E_r)}. \tag{3.48}$$

Substituting Eq. (3.47) and integrating, the following equation is obtained:

$$\ln(\sin\theta) + C_1 = \frac{1}{2}\ln\left[\frac{\cos(kr - \omega t)}{kr} + \sin(kr - \omega t)\right]. \tag{3.49}$$

From this,

$$\frac{C_2}{\sin^2\theta} = \sin(kr - \omega t) + \frac{\cos(kr - \omega t)}{kr}. \tag{3.50}$$

Equation (3.50) gives a parameter θ as a function of r and enables the drawing of the electric flux lines.

Figure 3.4 shows an example of drawing electric flux lines in electric dipole radiation. In Fig. 3.4(a), the electric flux lines and magnetic flux lines are shown. Since the magnetic flux lines are curves wherein Eq. (3.46b) is constant, they are concentric circles centered on the z-axis. Figure 3.4(b) shows a group of electric flux lines with different values of C_2.

3.5 Conclusion

Spherical waves obtained as a solution to the wave equation in a spherical coordinate system have been introduced. For spherical waves as scalar waves, there is a simple solution with spherical symmetry, but this does not satisfy Maxwell's equations. For a vector solution that satisfies Maxwell's equations, we arrive at electric dipole radiation.

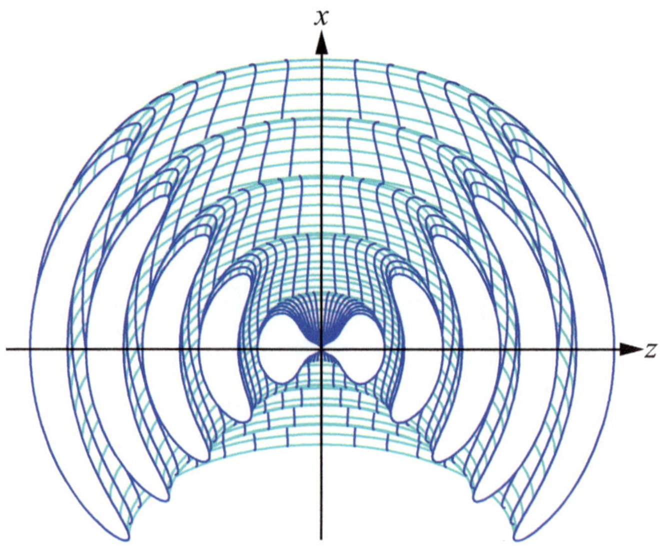

(a)

(b)

Figure 3.4 Electric dipole radiation: (a) electric flux lines (dark lines) and magnetic flux lines (light lines) and (b) a group of electric flux lines plotted by changing the value of C_2.

References

1. M. Abramowitz and I. A. Stegun, eds., *Handbook of Mathematical Functions: with Formulas, Graphs, and Mathematical Tables, 9th Rev. Ed.*, Dover Publications (1965).
2. T. Ishizu, M. Sato, and N. Kaneko, eds., *Applied Mathematics Exercise 2*, Baifukan, Tokyo (1969) [in Japanese].

3. E. W. Hobson, *The Theory of Spherical and Elliptical Harmonics*, Cambridge University Press (1931).
4. J. R. Schulenberger, "The Debye potential: A scalar factorization for Maxwell's equations," *J. Mathematical and Applications* **63**(2), 502–520 (1978).
5. Y. Honguh, *Landscape of Wave Optics, Mie scattering (1)*, Advanced Communication Media, Tokyo (2014) [in Japanese].
6. L. Mandel and E. Wolf, "3.2.4 The Weyl representation of spherical wave," *Optical Coherence and Quantum Optics*, Cambridge University Press (1995).
7. Y. Honguh, *Landscape of Wave Optics, Diffraction (1)*, Advanced Communication Media, Tokyo (2014) [in Japanese].
8. K. Shimoda, "Why are radio waves emitted from an oscillating dipole?" *Physics Education Soc. Japan* **26**(2), 141–145 (1978) [in Japanese].
9. K. Ishiguro, "Drawing electric flux lines—Tracking temporal changes in electric flux lines around an electric dipole," *Physics Education Soc. Japan* **26**(2), 141–145 (1977) [in Japanese].
10. K. Iga and G. Hatakoshi, Chapter 5, *Treasure Microbox of Optoelectronics I*, Advanced Communication Media, Tokyo (2018) [in Japanese].
11. M. Born and E. Wolf, *Principles of Optics, 7th Ed.*, Cambridge University Press (1999).
12. J. C. Maxwell, "A dynamical theory of the electromagnetic field," *Phil. Trans. R. Soc. Lond.* **155**, 459–512 (1865).

Appendix 3A Electromagnetic Potential

The Maxwell's equations that appeared in Chapter 1 are listed again in the following:

$$\nabla \times \boldsymbol{E} = -\frac{\partial \boldsymbol{B}}{\partial t}, \tag{3.A1}$$

$$\nabla \times \boldsymbol{H} = \boldsymbol{j} + \frac{\partial \boldsymbol{D}}{\partial t}, \tag{3.A2}$$

$$\nabla \cdot \boldsymbol{D} = \rho, \tag{3.A3}$$

$$\nabla \cdot \boldsymbol{B} = 0. \tag{3.A4}$$

Equation (3.A4) holds true if magnetic flux density $\boldsymbol{B}$ satisfies the following equation:

$$\boldsymbol{B} = \nabla \times \boldsymbol{A}. \tag{3.A5}$$

Equation (3.A1) holds if the electric field $\boldsymbol{E}$ is represented, by using this vector $\boldsymbol{A}$ and an arbitrary scalar function ψ, in the following form:

$$\boldsymbol{E} = -\nabla\psi - \frac{\partial \boldsymbol{A}}{\partial t}. \tag{3.A6}$$

In order for the remaining Eqs. (3.A2) and (3.A3) to hold, the following should hold:

$$\nabla^2\psi + \nabla \cdot \left(\frac{\partial \boldsymbol{A}}{\partial t}\right) = -\frac{\rho}{\varepsilon}, \tag{3.A7}$$

$$\nabla\left(\varepsilon\mu\frac{\partial\psi}{\partial t} + \nabla \cdot \boldsymbol{A}\right) + \left(\varepsilon\mu\frac{\partial^2}{\partial t^2} - \nabla^2\right)\boldsymbol{A} = \mu\boldsymbol{j}. \tag{3.A8}$$

In Eqs. (3.A5) to (3.A8), $\boldsymbol{A}$ is called a vector potential and ψ is called a scalar potential. In fact, Maxwell's original paper[12] uses this electromagnetic potential to describe the electromagnetic field.

As shown in the following section, by imposing certain conditions on $\boldsymbol{A}$ and ψ, Eqs. (3.A7) and (3.A8) become relatively simple.

3A.1 Coulomb gauge

Assuming that

$$\nabla \cdot \boldsymbol{A} = 0, \tag{3.A9}$$

Eqs. (3.A7) and (3.A8) become the following equations:

$$\nabla^2\psi = -\frac{\rho}{\varepsilon}, \tag{3.A10}$$

$$\nabla^2\boldsymbol{A} - \varepsilon\mu\frac{\partial^2 \boldsymbol{A}}{\partial t^2} - \varepsilon\mu\frac{\partial}{\partial t}(\nabla\psi) = -\mu\boldsymbol{j}. \tag{3.A11}$$

Equation (3.A10) is nothing but Poisson's equation.

3A.2 Lorenz gauge

Consider the following condition:

$$\nabla \cdot \boldsymbol{A} + \varepsilon\mu\frac{\partial\psi}{\partial t} = 0. \tag{3.A12}$$

In this case, Eqs. (3.A7) and (3.A8) become as follows:

$$\nabla^2\psi - \varepsilon\mu\frac{\partial^2\psi}{\partial t^2} = -\frac{\rho}{\varepsilon}, \tag{3.A13}$$

$$\nabla^2\boldsymbol{A} - \varepsilon\mu\frac{\partial^2 \boldsymbol{A}}{\partial t^2} = -\mu\boldsymbol{j}. \tag{3.A14}$$

Equation (3.A12) is called the Lorenz condition or Lorenz gauge condition. When there is no charge or current, that is, when $\rho = 0$ and $\boldsymbol{j} = 0$, Eqs. (3.A13) and (3.A14) become wave equations of the same form.

Appendix 3B Excel File

Appendix I-A contains the following Excel file. Each tab corresponds to a figure in the text. Refer to the operation instructions for running the animation.

03_Spherical wave.xlsm

Instruction	
Spherical wave	Fig. 3.2
Electric dipole radiation	Fig. 3.4

Chapter 4
Reflection and Refraction

4.1 Introduction

Transmission and reflection of light at interfaces have already been discussed in Chapter 1, "Propagation of Light: Plane Waves," where the case of normal incidence was treated. Reflection and refraction[1–3] are covered in various books as the basis of optics, so the principles and detailed derivation of formulas will be omitted here. From a simulation perspective, this chapter focuses on familiar diagrams for reflection and refraction. Some parts are easier to understand by utilizing Excel animation, which will also be introduced.

4.2 Reflection and Refraction at Media Interfaces

4.2.1 Snell's law

As shown in Fig. 4.1, reflection and refraction of light occur at the boundary between media with different refractive indices. If the angles that the incident ray, reflected ray, and transmitted ray make with the normal to the interface are θ_i, θ_r, and θ_t, respectively, then the following equations hold in the case of an isotropic medium:

$$\theta_r = \theta_i, \tag{4.1}$$

$$n_1 \sin\theta_i = n_2 \sin\theta_t. \tag{4.2}$$

Equation (4.2) is Snell's law.

Figure 4.1 shows the case of linearly polarized light where the electric field vector $\boldsymbol{E}$ is within the plane of incidence (xz plane), and this polarized light is called p-polarized light. On the other hand, when $\boldsymbol{E}$ is perpendicular to the plane of incidence, that is, perpendicular to the plane of the paper, it is called s-polarized light. The letters p and s are based on the initial letters of the German words parallel (same as in English) and senkrecht (perpendicular). As shown below, the symbols $\|$ and $\perp$ may also be used.

Another symbol convention is that $\otimes$ represents a vector that goes from the front of the page to the other side of the page, and $\odot$ represents a vector

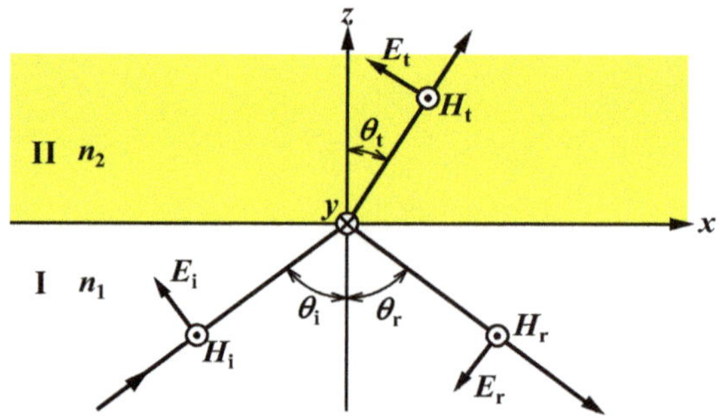

Figure 4.1 Reflection and refraction at the interface in the case for $n_1 < n_2$.

that goes from the back side of the page to the front of the page. This convention comes from the shape of an arrow seen from each direction. For example, in Fig. 4.1, the y-axis is represented by ⊗. Also, since the magnetic field vector is ⊙, it points in the $-y$ direction.

Figure 4.2 is similar to Fig. 4.1 and shows the wavefronts of the incident wave, reflected wave, and transmitted wave. In Fig. 4.1, the electric field vector and magnetic field vector were indicated by arrows, while in Fig. 4.2(a), the

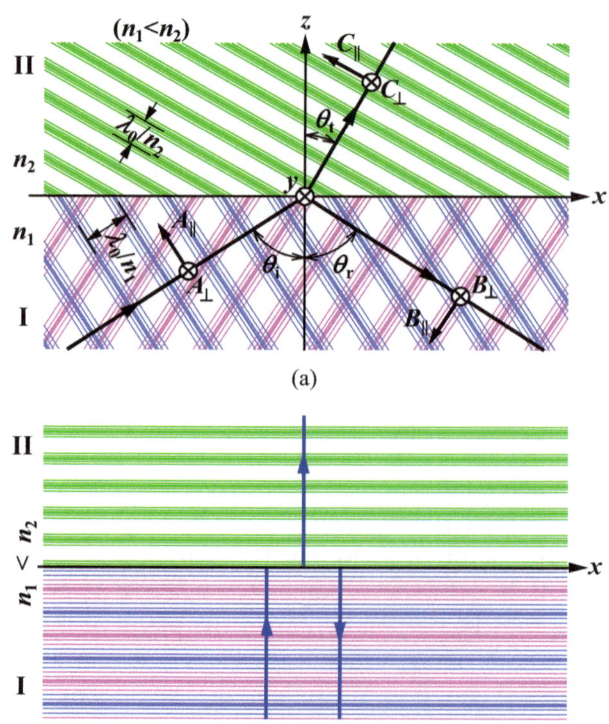

Figure 4.2 Rays and wavefronts in reflection and refraction. (a) Oblique incidence and (b) normal incidence.

p-polarized component ($A_{\parallel}$, $B_{\parallel}$, and $C_{\parallel}$) and s-polarized component ($A_{\perp}$, $B_{\perp}$, and $C_{\perp}$) of the electric field vector are shown respectively.† This will be used in the explanation of the reflection coefficient and transmission coefficient in Section 4.3.

Assuming that the wavelength in vacuum is λ_0, the wavelengths in a medium with refractive indices n_1 and n_2 are λ_0/n_1 and λ_0/n_2, respectively, as shown in Fig. 4.2(a). Also, the speed of light is c/n_1 and c/n_2, respectively. This difference in speed can be seen in the Excel animation in the appendix. When looking at Fig. 4.2(b), it may seem strange that the wavelength and velocity suddenly change at the medium interface. However, if the angle of incidence in Fig. 4.2(a) is gradually changed until it reaches the state shown in Fig. 4.2(b), this change in wavelength and velocity can be understood.

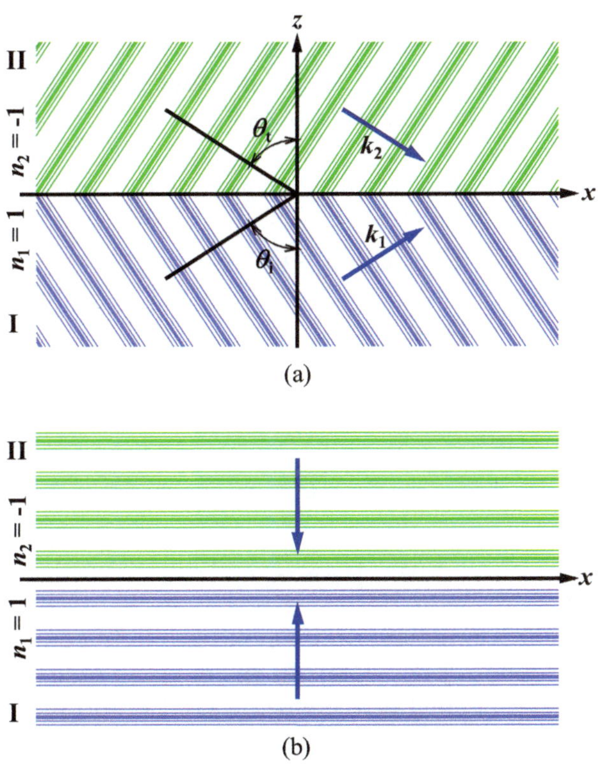

Figure 4.3 Wavefront at air/metamaterial interface. (a) Oblique incidence and (b) normal incidence.

†Note: The symbols were taken such that $A_{\parallel} = B_{\parallel}$ and $A_{\perp} = B_{\perp}$ in the case of normal incidence. When $n_1 < n_2$ and $\theta < \theta_B$, the phase deviation of the reflection is π, so the actual directions of both $B_{\parallel}$ and $B_{\perp}$ will be opposite to their directions in this figure. For example, $B_{\parallel}$ will be as shown in Fig. 4.2(a). This will be explained in Fig. 4.4 below.

4.2.2 Metamaterial

If Snell's law in Eq. (4.2) holds even for metamaterials with a negative refractive index, the refraction of light at the interface between air and metamaterials will be as shown, for example, in Fig. 4.3.

In Fig. 4.3(b), the directions of the propagation vectors are reversed on both sides of the interface, which looks strange. However, similarly to Fig. 4.2, the state of Fig. 4.3(a) can be shifted to the state of Fig. 4.3(b) by gradually changing the angle of incidence. Therefore, the state of Fig. 4.3(a) is not that strange.

The refractive index and relative magnetic permeability in Fig. 4.3 were set as $n_1 = 1$, $\mu_{r1} = 1$ for air, and $n_2 = -1$, $\mu_{r2} = -1$ for metamaterial. In this case, $n_1/\mu_{r1} = n_2/\mu_{r2}$, so from Eqs. (4.8) and (4.9) described in the following section, the reflection coefficient becomes 0 and no reflected wave appears.

4.3 The Reflection Coefficient and Transmission Coefficient

In Fig. 4.2(a), the electric fields of the incident wave, reflected wave, and transmitted wave are expressed as follows:†

$$\boldsymbol{E}_{\rm i} = \boldsymbol{A}\exp(-i\boldsymbol{k}_{\rm i}\cdot\boldsymbol{r} + i\omega t), \tag{4.3}$$

$$\boldsymbol{E}_{\rm r} = \boldsymbol{B}\exp(-i\boldsymbol{k}_{\rm r}\cdot\boldsymbol{r} + i\omega t), \tag{4.4}$$

$$\boldsymbol{E}_{t} = \boldsymbol{C}\exp(-i\boldsymbol{k}_{t}\cdot\boldsymbol{r} + i\omega t). \tag{4.5}$$

The magnitude of each propagation vector is given by the following formula, assuming that $k_0 = 2\pi/\lambda_0$:

$$|\boldsymbol{k}_{\rm i}| = |\boldsymbol{k}_{\rm r}| = k_0 n_1, \tag{4.6}$$

$$|\boldsymbol{k}_{\rm t}| = k_0 n_2. \tag{4.7}$$

The p-polarized light components and s-polarized light components of vectors $\boldsymbol{A}$, $\boldsymbol{B}$, and $\boldsymbol{C}$ are shown in Fig. 4.2(a). At the medium interface, the tangential components (xy components) of the electric field $\boldsymbol{E}$ and magnetic field $\boldsymbol{H}$ are continuous, and the normal components (z component) of the electric flux density $\boldsymbol{D}$ and magnetic flux density $\boldsymbol{B}$ are continuous.[1] Using these boundary conditions, the amplitude reflection coefficient and amplitude transmission coefficient are determined and are represented by the following equations:[2]

†Note: Symbols $\boldsymbol{A}$, $\boldsymbol{B}$, and $\boldsymbol{C}$ are used for convenience, but $\boldsymbol{B}$ in this case is different from the magnetic flux density $\boldsymbol{B}$. Symbols $\boldsymbol{a}$, $\boldsymbol{b}$, and $\boldsymbol{c}$ can be used, but in this case, $\boldsymbol{c}$ can be confused with the speed of light. Similarly, when adopting $\boldsymbol{A}$, $\boldsymbol{R}$, and $\boldsymbol{T}$, the symbols $\boldsymbol{R}$ and $\boldsymbol{T}$ can be confused with the reflectivity and the transmittance. In any case, there are not enough symbols.

$$r_p \equiv \frac{B_\parallel}{A_\parallel} = \frac{(n_2/\mu_{r2})\cos\theta_i - (n_1/\mu_{r1})\cos\theta_t}{(n_2/\mu_{r2})\cos\theta_i + (n_1/\mu_{r1})\cos\theta_t}, \tag{4.8}$$

$$r_s \equiv \frac{B_\perp}{A_\perp} = \frac{(n_1/\mu_{r1})\cos\theta_i - (n_2/\mu_{r2})\cos\theta_t}{(n_1/\mu_{r1})\cos\theta_i + (n_2/\mu_{r2})\cos\theta_t}, \tag{4.9}$$

$$t_p \equiv \frac{C_\parallel}{A_\parallel} = \frac{2(n_1/\mu_{r1})\cos\theta_i}{(n_2/\mu_{r2})\cos\theta_i + (n_1/\mu_{r1})\cos\theta_t}, \tag{4.10}$$

$$t_s \equiv \frac{C_\perp}{A_\perp} = \frac{2(n_1/\mu_{r1})\cos\theta_i}{(n_1/\mu_{r1})\cos\theta_i + (n_2/\mu_{r2})\cos\theta_t}, \tag{4.11}$$

where μ_{ri} is the relative magnetic permeability of medium i.

For many materials such as dielectrics and semiconductors, the relative magnetic permeability can be set to 1, so by setting $\mu_{r1} = \mu_{r2} = 1$ and using Snell's law, the following can be obtained from Eqs. (4.8) to (4.11):

$$r_p = \frac{\tan(\theta_i - \theta_t)}{\tan(\theta_i + \theta_t)}, \tag{4.12}$$

$$r_s = -\frac{\sin(\theta_i - \theta_t)}{\sin(\theta_i + \theta_t)}, \tag{4.13}$$

$$t_p = \frac{2\cos\theta_i \sin\theta_t}{\sin(\theta_i + \theta_t)\cos(\theta_i - \theta_t)}, \tag{4.14}$$

$$t_s = \frac{2\cos\theta_i \sin\theta_t}{\sin(\theta_i + \theta_t)}. \tag{4.15}$$

Equations (4.12) to (4.15) are called Fresnel equations.

In the case of normal incidence, that is, $\theta_i = \theta_r = \theta_t = 0$ in Eqs. (4.8) to (4.11), the amplitude coefficients are given by the following equations when $\mu_{r1} = \mu_{r2} = 1$:

$$r_p = \frac{n_2 - n_1}{n_2 + n_1}, \tag{4.16}$$

$$r_s = \frac{n_1 - n_2}{n_1 + n_2}, \tag{4.17}$$

$$t_p = \frac{2n_1}{n_1 + n_2}, \tag{4.18}$$

$$t_s = \frac{2n_1}{n_1 + n_2}. \tag{4.19}$$

At normal incidence, there is no distinction between p-polarized light and s-polarized light, so the reflection coefficient and transmission coefficient should be the same equation, but Eqs. (4.16) and (4.17) have opposite signs. This is due to the direction of $B_{||}$ in Fig. 4.2(a). If $B_{||}$ is taken in the opposite direction, the sign of Eq. (4.8) is reversed, and the expressions for r_p and r_s at normal incidence coincide with each other.

Figure 4.4 shows the dependence of the amplitude reflection coefficient on the angle of incidence expressed by Eqs. (4.8) and (4.9) or Eqs. (4.12) and (4.13). Figures 4.4(a) and (b) are essentially the same. Depending on the direction of $B_{||}$, the diagram representing the r-θ_i relationship will be either (a) or (b).

As shown in Fig. 4.4, for p-polarized light, there is an angle of incidence θ_B at which the reflectivity is 0. This is Brewster's angle and is given by the following formula:

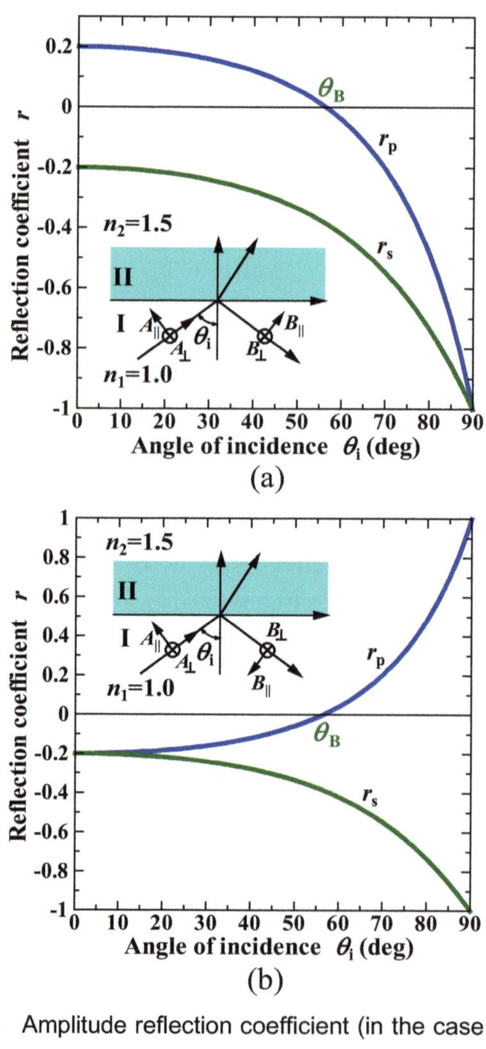

Figure 4.4 Amplitude reflection coefficient (in the case of $n_1 < n_2$).

$$\theta_B = \tan^{-1}\left(\frac{n_2}{n_1}\right). \tag{4.20}$$

4.4 Amplitude Distribution of Plane Waves in Reflection and Refraction

An example of transmission and reflection at an interface in the case of normal incidence was given in Chapter 1 (see Fig. 1.4). Let's plot the amplitude distribution of a plane wave in the case of oblique incidence.

Figure 4.5 shows the real part amplitude of the electric field given by Eqs. (4.3) to (4.5). The coordinate system in this figure is that for Fig. 4.2(a) rotated by 90 deg. The diagram looks similar to Fig. 4.2, but it takes some effort to create. This figure does not plot the wavefront as in Fig. 4.2, but what corresponds to the wavefront can be seen as the density of the lines. This animation can be executed by an Excel file in Appendix I-A, where you can see how the plane wave travels.

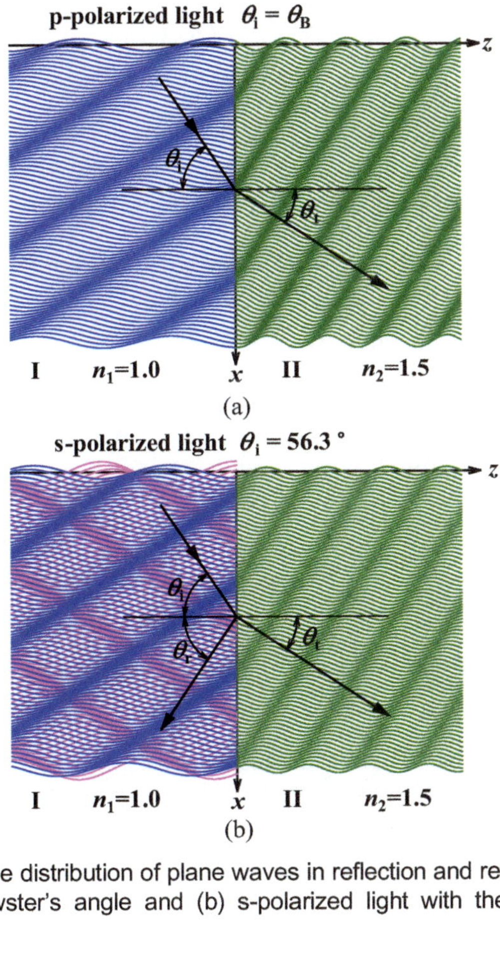

Figure 4.5 Amplitude distribution of plane waves in reflection and refraction. (a) p-polarized light incident at Brewster's angle and (b) s-polarized light with the same incident angle as in (a).

Figure 4.5(a) shows the case for p-polarized light incident at Brewster's angle. The amplitude distribution for p-polarized light shows a continuous component in media I and II, that is, a normal component of $\boldsymbol{D}$ (proportional to the tangential component of $\boldsymbol{E}$). The Excel file has a version that displays the sum of the incident wave and reflected wave in medium I, and a version that plots each separately. If you change the incident angle little by little, you can see how the reflected wave disappears at Brewster's angle.

Figure 4.5(b) shows the case for s-polarized light incident at the same angle as in (a). There is a reflected wave here; the amplitude of the reflected wave is multiplied by a certain coefficient to increase it in order to make it easier to see.

4.5 Total Internal Reflection

When $n_1 > n_2$, if θ_i exceeds the value given by the following equation, there is no θ_t that satisfies Eq. (4.2), and total internal reflection occurs:

$$\theta_c = \sin^{-1}\left(\frac{n_2}{n_1}\right), \tag{4.21}$$

where θ_c is the critical angle for total reflection, and the absolute value of the amplitude reflection coefficient is unity when $\theta_i > \theta_c$. In this case, the electric field in region II is expressed by the following equation:

$$\boldsymbol{E}_t = \boldsymbol{C}\exp(-\gamma z)\exp(-ik_0 n_1 x \sin\theta_i + i\omega t), \tag{4.22}$$

$$\gamma = k_0\sqrt{n_1^2\sin^2\theta_i - n_2^2}. \tag{4.23}$$

The amplitude reflection coefficient and the amplitude transmission coefficient are expressed by complex numbers as shown below, where the relative permeabilities were assumed as $\mu_{r1} = \mu_{r2} = 1$:

$$r_p \equiv \frac{B_\parallel}{A_\parallel} = \frac{(n_2/n_1)^2\cos\theta_i + i\sqrt{\sin^2\theta_i - (n_2/n_1)^2}}{(n_2/n_1)^2\cos\theta_i - i\sqrt{\sin^2\theta_i - (n_2/n_1)^2}}, \tag{4.24}$$

$$r_s \equiv \frac{B_\perp}{A_\perp} = \frac{\cos\theta_i + i\sqrt{\sin^2\theta_i - (n_2/n_1)^2}}{\cos\theta_i - i\sqrt{\sin^2\theta_i - (n_2/n_1)^2}}, \tag{4.25}$$

$$t_p \equiv \frac{C_\parallel}{A_\parallel} = \frac{2(n_2/n_1)\cos\theta_i}{(n_2/n_1)^2\cos\theta_i - i\sqrt{\sin^2\theta_i - (n_2/n_1)^2}}, \tag{4.26}$$

$$t_s \equiv \frac{C_\perp}{A_\perp} = \frac{2\cos\theta_i}{\cos\theta_i - i\sqrt{\sin^2\theta_i - (n_2/n_1)^2}}. \tag{4.27}$$

The reflection coefficients r_p and r_s can be expressed in the following form:

$$r_p = r_{p0} \exp(-i\delta_p), \tag{4.28}$$

$$r_s = r_{s0} \exp(-i\delta_s). \tag{4.29}$$

In the total reflection region where $\theta_i > \theta_c$, $r_{p0} = r_{s0} = 1$ and the phase discontinuity δ of the reflected wave with respect to the incident wave is given below:

$$\tan\frac{\delta_p}{2} = -\frac{\sqrt{\sin^2\theta_i - (n_2/n_1)^2}}{(n_2/n_1)^2 \cos\theta_i}, \tag{4.30}$$

$$\tan\frac{\delta_s}{2} = -\frac{\sqrt{\sin^2\theta_i - (n_2/n_1)^2}}{\cos\theta_i}. \tag{4.31}$$

Figure 4.6 shows the dependence of the amplitude reflection coefficient on the angle of incidence, including the region of $\theta_i \leq \theta_c$. In the region of $\theta < \theta_B$ for p-polarized light, $r_{p0} < 0$ and $\delta_p = 0$ are plotted.

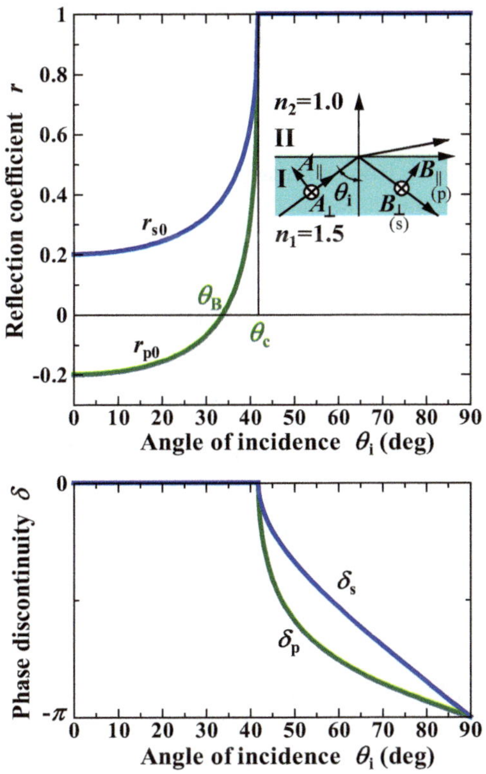

Figure 4.6 Amplitude reflection coefficient and phase discontinuity when $n_1 > n_2$.

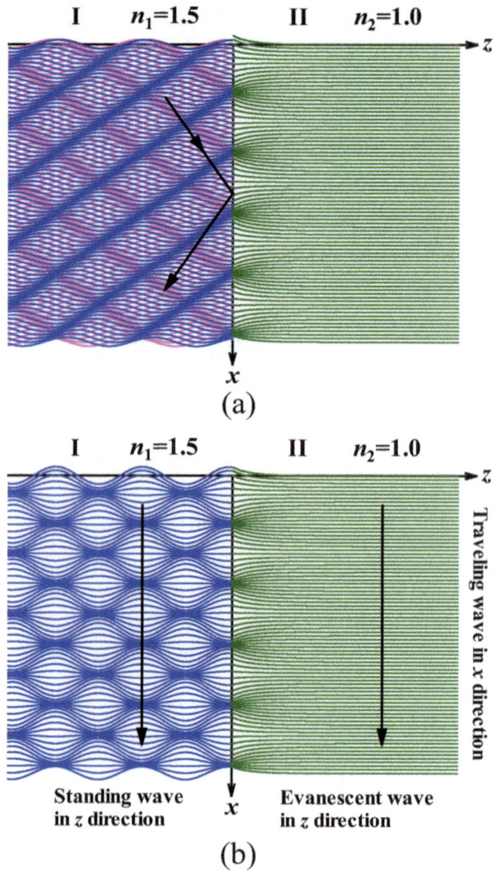

Figure 4.7 Amplitude distribution of a plane wave for the case of total internal reflection. (a) Amplitudes of E_i and E_r are plotted separately, and (b) the amplitude of $E_i + E_r$ is plotted in region I.

Figure 4.7(a) shows the amplitude distribution of $\boldsymbol{E}_i$, $\boldsymbol{E}_r$, and $\boldsymbol{E}_t$ when total internal reflection is occurring. Figure 4.7(b) is a plot of $\boldsymbol{E}_i + \boldsymbol{E}_r$ in region I, which is continuous with $\boldsymbol{E}_t$ in region II at the boundary. Since the absolute value of the amplitude reflection coefficient is unity, a standing wave is formed in the z direction in region I. In region II, the wave becomes an evanescent wave that decays exponentially, as shown by Eq. (4.22). In the x direction, the wave is a traveling wave propagating in the $+x$ direction in both regions I and II.

4.6 Conclusion

A simulation example of electric field amplitude for the reflection and refraction at a medium interface was introduced. The appearance of reflected waves and transmitted waves is relatively easy to understand by executing the animation.

Side Topic: Pulse propagation in metamaterials

It is well known that, assuming that Snell's law and the laws of reflection and refraction hold for metamaterials with a negative refractive index, imaging beyond the diffraction limit, such as Pendry's perfect lens,[4] will be realized. Negative refraction is shown in Figs. 4.3(a) and (b).

Now, how does pulse propagation occur in such a metamaterial? It is common sense to simply think that the pulse propagates as shown in Fig. 4.S1(a). In the metamaterial, the propagation vector is reversed, but the energy must flow from left to right. In other words, even if the phase velocity is in the direction to the left, the group velocity should be in the direction to the right.

However, based on the formula the pulse propagation is not like that in Fig. 4.S1(a). Let's consider each wavenumber component by Fourier expansion of the pulse. The amplitude distribution of the electric and magnetic fields in each wavenumber component should be symmetrical with $z = 0$ as the boundary.

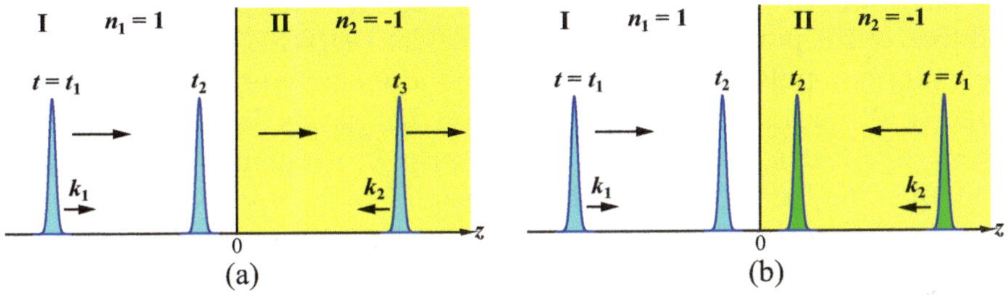

Figure 1.S1 Pulse propagation from air to metamaterial.

The synthesized pulse also appears at a symmetrical position with respect to the original pulse. Therefore, as shown in Fig. 4.S1(b), if there is a pulse at a certain position in air at $t = t_1$, there must be a pulse whose propagation vector is in the opposite direction at a symmetrical position due to the boundary conditions. The two pulses travel in opposite directions and disappear at $z = 0$. In this case, it seems that the law of causality and the law of conservation of energy do not hold.

Figure 4.S1(b) shows that the pulse exists both in air and in the metamaterial at the same time of $t = t_1$. This is the origin of the strange phenomenon, but considering the image formation in Pendry's perfect lens shown in Fig. 4.S2(a), it seems to be correct.

In Fig. 4.S2(a), point A is imaged onto point B in the metamaterial. As with image formation with a normal lens, all optical path lengths from point A to point B should be equal. Calculating the optical path length, we obtain $1 \times \overline{AP} + (-1) \times \overline{PB} = 0$.[5] In other words, since the optical path lengths from point A to point B are all 0, the original image and the formed image exist at point A and point B at the same time. The optical path length from point B to point C is also 0.

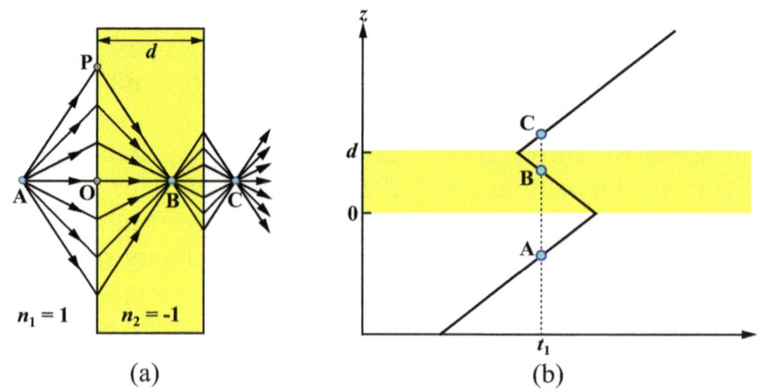

Figure 1.S2 (a) Image formation in Pendry's perfect lens and (b) the position and time of the pulse passing through the perfect lens. (Is this true?).

When a pulse passes through Pendry's perfect lens, plotting the position and time of the pulse results in something like Fig. 4.S2(b). From the above explanation, a pulse exists at point A and at its imaging positions B and C at the same time of t_1. From this, the pulse that passes through a perfect lens travels $2d/c$ faster than when it passes through vacuum. Pulses propagate faster than the speed of light, but is this true? Pendry's perfect lens is famous in the metamaterial field, so it is a topic that has already been discussed and explained in the literature.

References

1. M. Born and E. Wolf, *Principles of Optics, 7th Ed.*, Cambridge University Press (1999).
2. K. Kuroda, *Physical Optics—Propagation of light waves in a medium*, Asakura Publishing (2011) [in Japanese].
3. K. Iga and G. Hatakoshi, Chapter 6, *Treasure Microbox of Optoelectronics I*, Advanced Communication Media (2018) [in Japanese].
4. J. B. Pendry, "Negative refraction makes a perfect lens," *Phys. Rev. Lett.* **85**(18), 3966–3969 (2000).
5. K. Iga and G. Hatakoshi, Chapter 32, *Treasure Microbox of Optoelectronics I*, Advanced Communication Media (2018) [in Japanese].

Appendix 4A Excel File

Appendix I-A contains the following Excel file. Each tab corresponds to a figure/s in the main text. Refer to the operation instructions for executing the animation.

04_Reflection Refraction.xlsm

Instruction	
Wave front	Fig. 4.2, Fig. 4.3
p-polarized light (1)	Fig. 4.5(a)
p-polarized light (2)	
s-polarized light (1)	Fig. 4.7(b)
s-polarized light (2)	Fig. 4.5(b), Fig. 4.7(a)

In the above file, reflection and refraction for p-polarized light (1) and reflection and refraction for s-polarized light (1) are versions that display the sum of the incident wave and reflected wave in medium I. Reflection and refraction for p-polarized light (2) and reflection and refraction for s-polarized light (2) are versions that plot the incident wave and reflected wave separately.

Chapter 5
Refractive Index Part 1

5.1 Introduction

In the previous chapter, simulation examples of reflection and refraction at the interface of media with different refractive indices were introduced. When performing such calculations, the refractive index value of each medium is required. How is the refractive index determined in the first place? Is it something that can be calculated through simulation? This chapter describes the model and approximate formula for calculating the refractive index.

5.2 What Is the Refractive Index?

Starting from Maxwell's equations, the refractive index n_r of the medium is expressed by the following equation using permittivity ε and permeability μ:

$$n_r = \sqrt{\varepsilon_r}\sqrt{\mu_r} = \sqrt{\varepsilon}\sqrt{\mu}/\sqrt{\varepsilon_0\mu_0}, \tag{5.1}$$

where ε_0 and μ_0 are the permittivity and permeability in vacuum, and ε_r and μ_r are the relative permittivity and relative permeability, respectively.

As mentioned in Chapter 1, the electric flux density $\boldsymbol{D}$ and the electric field $\boldsymbol{E}$, and the magnetic flux density $\boldsymbol{B}$ and the magnetic field $\boldsymbol{H}$ are connected by the following relationships using ε and μ, respectively:

$$\boldsymbol{D} = \varepsilon \boldsymbol{E}, \tag{5.2}$$

$$\boldsymbol{B} = \mu \boldsymbol{H}. \tag{5.3}$$

Equation (5.2) expresses the macroscopic response of the medium to the electric field and shows that the degree of response is defined as the permittivity ε. Similarly, Eq. (5.3) expresses the macroscopic response of the medium to the magnetic field, and its degree is the permeability μ.

Since the electrons in the dielectric are bound to atoms, Eq. (5.2) in that case represents the macroscopic response of the bound electrons. The electric flux density in vacuum is $\varepsilon_0 \boldsymbol{E}$, and it becomes $\boldsymbol{D}$ in Eq. (5.2) due to the presence of the dielectric. The difference between the two is defined as the (dielectric) polarization $\boldsymbol{P}$.[1,2] That is,

$$\boldsymbol{P} = \boldsymbol{D} - \varepsilon_0 \boldsymbol{E} = (\varepsilon - \varepsilon_0)\boldsymbol{E}. \tag{5.4}$$

Polarization is the density of electric dipoles induced by an electric field. The degree of polarization induced by an electric field is defined as electric susceptibility χ as follows:

$$\boldsymbol{P} = \chi \varepsilon_0 \boldsymbol{E}. \tag{5.5}$$

From Eqs. (5.4) and (5.5),

$$\boldsymbol{D} = (1 + \chi)\varepsilon_0 \boldsymbol{E}. \tag{5.6}$$

Therefore, the relative permittivity ε_{r} is expressed by the following formula:

$$\varepsilon_{\mathrm{r}} = 1 + \chi. \tag{5.7}$$

The above represents the response of electrons bound to atoms. On the other hand, the response of conduction electrons in metals is based on the well-known Ohm's law and is represented by the following equation:

$$\boldsymbol{J} = \sigma \boldsymbol{E}, \tag{5.8}$$

where $\boldsymbol{J}$ is the current density and σ is the electrical conductivity.

In Chapter 1, it was shown that the wave equation can be derived from Maxwell's equations, where the electric charge (corresponding to polarization here) and the current density were assumed to be 0. The wave equation for the case wherein polarization $\boldsymbol{P}$ and current density $\boldsymbol{J}$ are not 0 can be expressed as follows:

$$\nabla \times (\nabla \times \boldsymbol{E}) = -\varepsilon\mu \frac{\partial^2 \boldsymbol{E}}{\partial t^2} - \mu \frac{\partial \boldsymbol{J}}{\partial t} = -\varepsilon_0 \mu \frac{\partial^2 \boldsymbol{E}}{\partial t^2} - \frac{\partial^2 \boldsymbol{P}}{\partial t^2} - \mu \frac{\partial \boldsymbol{J}}{\partial t}. \tag{5.9}$$

It is assumed that $\nabla \cdot \boldsymbol{E} = 0$ and $\mu = \mu_0$ because the relative permeability of most media is unity. Then Eq. (5.9) becomes

$$\nabla^2 \boldsymbol{E} - \frac{1}{c^2}\frac{\partial^2 \boldsymbol{E}}{\partial t^2} = \mu_0 \frac{\partial^2 \boldsymbol{P}}{\partial t^2} + \mu_0 \frac{\partial \boldsymbol{J}}{\partial t}. \tag{5.10}$$

The first term on the right side of Eq. (5.10) results from polarization due to bound electrons, and the second term results from current density due to conduction electrons. In vacuum, these terms are 0, resulting in the commonly seen wave equation where the left side of Eq. (5.10) equals 0.

Suppose that, in Eq. (5.10), the time dependence of $\boldsymbol{E}$, $\boldsymbol{P}$, and $\boldsymbol{J}$ is set as $\exp(i\omega t)$, and that Eq. (5.10) can be converted into the following form:

$$\nabla^2 \boldsymbol{E} = -\frac{\omega^2}{c^2} {n_c}^2 \boldsymbol{E}, \tag{5.11}$$

where the parameter n_c is the refractive index of the medium, which is generally a complex number.

5.3 Refractive Index of Dielectrics

5.3.1 Model for a dielectric refractive index

Assuming that the electric field felt by the electrons inside the dielectric is $\boldsymbol{E}_l$, the electric dipole moment $\boldsymbol{p}$ induced by this electric field is given by the following equation:

$$\boldsymbol{p} = -e\boldsymbol{r} = \alpha \boldsymbol{E}_l, \tag{5.12}$$

where e is the elementary charge, $\boldsymbol{r}$ is the position vector, and α is the polarizability. If the electron density is N, the polarization $\boldsymbol{P}$ is represented by

$$\boldsymbol{P} = N\boldsymbol{p} = -Ne\boldsymbol{r} = N\alpha \boldsymbol{E}_l. \tag{5.13}$$

When $N\alpha/\varepsilon_r \ll 1$, the electric field $\boldsymbol{E}_l$ approximately equals the macroscopic electric field $\boldsymbol{E}$ in the medium, but strictly speaking, it is different. In the Lorentz local electric field model, $\boldsymbol{E}_l$ is given as follows:[1,2]

$$\boldsymbol{E}_l = \boldsymbol{E} + \frac{\boldsymbol{P}}{3\varepsilon_0}. \tag{5.14}$$

From Eqs. (5.13) and (5.14),

$$\boldsymbol{E}_l = \frac{\boldsymbol{E}}{1 - N\alpha/(3\varepsilon_0)}. \tag{5.15}$$

From Eqs. (5.4), (5.13), and (5.15), the vacuum permittivity ε_0 is expressed as follows:

$$\varepsilon = \varepsilon_0 + \frac{N\alpha}{1 - N\alpha/(3\varepsilon_0)}. \tag{5.16}$$

Since $\varepsilon/\varepsilon_0 = \varepsilon_r$, the following equation is derived from Eq. (5.16):

$$\frac{\varepsilon_r - 1}{\varepsilon_r + 2} = \frac{N\alpha}{3\varepsilon_0}. \tag{5.17}$$

Equation (5.17) is called the Clausius–Mossotti relation. By setting

$$\varepsilon_r = n_c^2, \tag{5.18}$$

then the following equation is derived:

$$\frac{n_c^2 - 1}{n_c^2 + 2} = \frac{N\alpha}{3\varepsilon_0}. \tag{5.19}$$

Equation (5.19) is called the Lorentz–Lorenz equation.

In Eq. (5.12), $\boldsymbol{r}$ represents the position coordinate of the electron. Using this, the equation of motion for electrons bound to atoms in a dielectric can be represented as follows:

$$m_e \frac{d^2\boldsymbol{r}}{dt^2} + m_e \gamma \frac{d\boldsymbol{r}}{dt} + K\boldsymbol{r} = -e\boldsymbol{E}_1, \tag{5.20}$$

where m_e is the electron mass, K is the spring constant between the atom and the bound electron, and γ is the damping coefficient. Since $\boldsymbol{P} = -Ne\boldsymbol{r}$ from Eq. (5.13), Eq. (5.20) can be converted into an equation for $\boldsymbol{P}$ and $\boldsymbol{E}_1$. If the time dependence of $\boldsymbol{E}$, $\boldsymbol{E}_1$, and $\boldsymbol{P}$ is set as $\exp(i\omega t)$, which is the same as in the case of Eq. (5.11), the following equation is obtained from Eqs. (5.13) and (5.20):

$$N\alpha = \frac{Ne^2/m_e}{\omega_0^2 - \omega^2 + i\gamma\omega}. \tag{5.21}$$

From Eqs. (5.16) and (5.21),

$$\varepsilon_r = \frac{\varepsilon}{\varepsilon_0} = 1 + \frac{Ne^2/(m_e\varepsilon_0)}{\omega_0^2 - \omega^2 + i\gamma\omega}, \tag{5.22}$$

where ω_0 is the effective resonance angular frequency and is given by the following equation:

$$\omega_0 \equiv \sqrt{\frac{K}{m_e} - \frac{Ne^2}{3m_e\varepsilon_0}}. \tag{5.23}$$

The real and imaginary parts of Eq. (5.22) are respectively expressed by the following equations:

$$\varepsilon_1 \equiv \mathrm{Re}(\varepsilon_r) = 1 + \frac{Ne^2}{m_e\varepsilon_0} \frac{\omega_0^2 - \omega^2}{(\omega_0^2 - \omega^2)^2 + \gamma^2\omega^2}, \tag{5.24}$$

$$\varepsilon_2 \equiv \mathrm{Im}(\varepsilon_r) = -\frac{Ne^2}{m_e\varepsilon_0} \frac{\gamma\omega}{(\omega_0^2 - \omega^2)^2 + \gamma^2\omega^2}. \tag{5.25}$$

The complex refractive index n_c can be represented as follows:[†]

$$n_c = n_r - i\kappa, \tag{5.26}$$

where the imaginary part κ corresponds to the extinction coefficient.

From Eq. (5.18),

$$\varepsilon_1 = n_r^2 - \kappa^2, \tag{5.27}$$

$$\varepsilon_2 = -2n_r\kappa. \tag{5.28}$$

Therefore, n_r and κ can be determined using the following formulas:

$$n_r = \sqrt{\frac{\sqrt{\varepsilon_1^2 + \varepsilon_2^2} + \varepsilon_1}{2}}, \tag{5.29}$$

$$\kappa = \sqrt{\frac{\sqrt{\varepsilon_1^2 + \varepsilon_2^2} - \varepsilon_1}{2}}. \tag{5.30}$$

Figure 5.1 shows the ω dependence of the real and imaginary parts of the refractive index given by Eqs. (5.24) to (5.30). As can be seen from the figure, there is large absorption ($\kappa > 0$) near the resonance angular frequency ω_0, and the real part of the refractive index exhibits anomalous dispersion; that is, n_r decreases as ω increases. In the region away from ω_0, the refractive index shows normal dispersion.

Equation (5.22) is based on the assumption that all electrons are bound in the same way (bound with the same spring constant). When there are several kinds of bound electrons, that is, when there are several resonance angular frequencies, Eq. (5.22) is extended as follows:

$$\varepsilon_r = 1 + \frac{Ne^2}{m_e\varepsilon_0}\sum_j \frac{f_j}{\omega_j^2 - \omega^2 + i\gamma_j\omega}, \tag{5.31}$$

where ω_j and γ_j are the jth resonance angular frequency and the jth damping coefficient, respectively, and f_j is the ratio of electrons with ω_j to the electrons with γ_j. This variable f_j corresponds to the oscillator strength.

[†]Note: As explained in Chapter 1, if the time dependence of $\boldsymbol{E}$ is set as $\exp(-i\omega t)$ instead of $\exp(i\omega t)$, the complex refractive index is represented as $n_c = n_r + i\kappa$. Similarly, the sign of the imaginary part of the permittivity is reversed. It should be noted that, in most literature, the imaginary part of the permittivity has the opposite sign from the formula shown here. In the refractive index model for semiconductors introduced in the next chapter, the sign of the imaginary part of the permittivity is adjusted to match the notation in the literature to avoid confusion. Therefore, note in advance that the sign is reversed from this chapter.

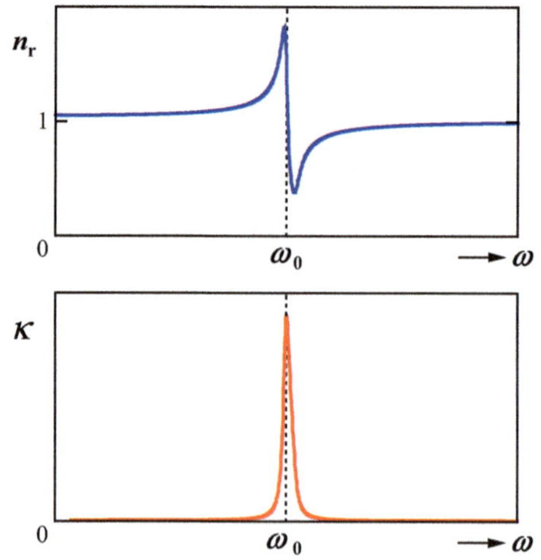

Figure 5.1 Real and imaginary parts of the refractive index of dielectrics.

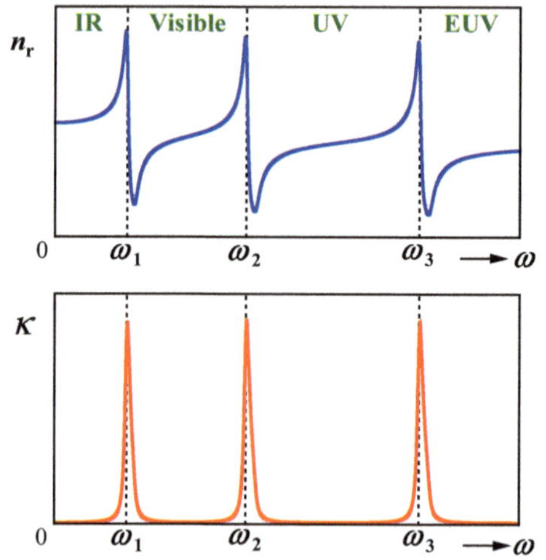

Figure 5.2 Real and imaginary parts of the refractive index of dielectrics in the case of multiple ω_j.

According to the model of Eq. (5.31), the refractive index of the dielectric has a ω dependence, for example, as shown in Fig. 5.2.

5.3.2 The Sellmeier equation

Next, consider approximation in a region away from the resonance angular frequency ω_j. For simplicity, the damping coefficient γ_j is approximated to 0. Then Eq. (5.31) becomes a real number, and the real part of the refractive index is given by the following equation:

$$n_r{}^2 = 1 + \frac{Ne^2}{m_e \varepsilon_0} \sum_j \frac{f_j}{\omega_j{}^2 - \omega^2}. \tag{5.32}$$

Expressing ω by the wavelength λ,

$$\omega = \frac{2\pi c}{\lambda}. \tag{5.33}$$

Then Eq. (5.32) can be represented in the following form:

$$n_r{}^2 = 1 + \sum_j \frac{B_j \lambda^2}{\lambda^2 - C_j}. \tag{5.34}$$

This is called the Sellmeier equation or Sellmeier's dispersion formula and was proposed by Wilhelm Sellmeier in 1871.

As an approximation formula for the refractive index in the normal dispersion region, a Cauchy equation was published by Augustin-Louis Cauchy in 1836. This is a form in which the refractive index is expanded to a power of $1/\lambda^2$, and its general form is expressed as follows:

$$n_r = A_0 + \frac{A_1}{\lambda^2} + \frac{A_2}{\lambda^4} + \cdots. \tag{5.35}$$

If the wavelength range is limited, the refractive index can be expressed with a relatively good approximation by Eq. (5.35) even up to the term $1/\lambda^2$.

The Sellmeier equation usually uses three sets of (B_j, C_j), resulting in a formula that fits the measured values well over a wide wavelength range. The Cauchy equation gives an equivalent fit by six parameters, A_0 to A_5.

Table 5.1 shows the parameters of the Sellmeier equation for various oxides.[3–9] Figure 5.3 shows the wavelength dependence of the refractive index based on the data in Table 5.1.

5.4 The Refractive Index of Metal

In metals, it is necessary to consider the current density due to conduction electrons. Current density $\boldsymbol{J}$ can be expressed as follows using electron density N and electron velocity v:[2]

$$\boldsymbol{J} = -Nev. \tag{5.36}$$

Similar to Eq. (5.20), the equation of motion is as follows:

$$m_e \frac{dv}{dt} + \frac{m_e}{\tau} v = -e\boldsymbol{E}, \tag{5.37}$$

Table 5.1 Parameters of the Sellmeier equation for various oxides.[6–10] Reprinted with permission from Refs. 6–10 © Optica Publishing Group.

	B_j	C_j (μm^2)	Ref.
SiO_2	0.6961663	$(0.0684043)^2$	6
	0.4079426	$(0.1162414)^2$	
	0.8974794	$(9.896161)^2$	
Al_2O_3	1.4313493	$(0.0726631)^2$	7
	0.6505471	$(0.1193242)^2$	
	5.3414021	$(18.028251)^2$	
ZrO_2	1.347091	$(0.062543)^2$	8
	2.117788	$(0.166739)^2$	
	9.452943	$(24.32057)^2$	
GeO_2	0.8068664	$(0.0689726)^2$	9
	0.7181585	$(0.1539661)^2$	
	0.8541683	$(11.841931)^2$	
HfO_2	1.345	$(0.0634)^2$	10
	1.9558	$(0.15494)^2$	
	10.41	$(27.12)^2$	

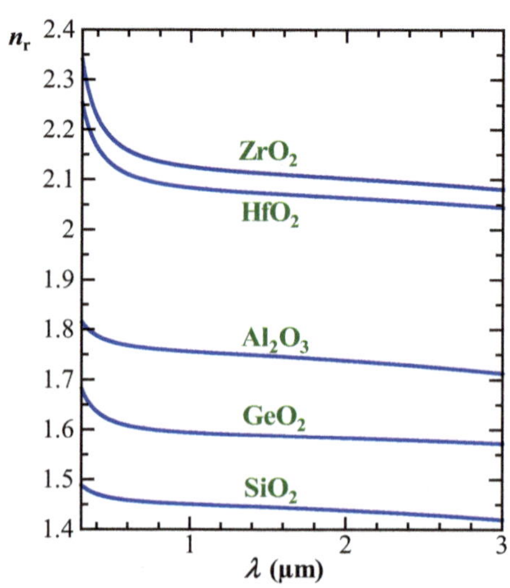

Figure 5.3 Refractive index of various oxides according to the Sellmeier equation. Plotted with permission based on data from Refs. 6–10 © Optica Publishing Group.

where τ is the relaxation time. Since conduction electrons are not bound to atoms, the spring constant K is 0. From Eqs. (5.36) and (5.37), the following equation is obtained:

$$\frac{d\boldsymbol{J}}{dt} + \frac{1}{\tau}\boldsymbol{J} = \frac{Ne^2}{m_e}\boldsymbol{E}. \tag{5.38}$$

Ohm's law in Eq. (5.8) holds for a steady state ($dJ/dt = 0$); therefore, from Eq. (5.38), the conductivity σ is given by the following equation:

$$\sigma = \frac{Ne^2}{m_e}\tau. \tag{5.39}$$

On the other hand, for the angular frequency ω, the following equation is obtained by setting the time dependence of $\boldsymbol{E}$ and $\boldsymbol{J}$ to $\exp(i\omega t)$ in Eq. (5.38):

$$\boldsymbol{J} = \frac{\sigma}{1 + i\omega\tau}\boldsymbol{E}. \tag{5.40}$$

Substituting this into Eq. (5.10),

$$\nabla^2 \boldsymbol{E} = -\frac{\omega^2}{c^2}\left(1 - \frac{\mu_0 \sigma c^2/\tau}{\omega^2 - i\omega/\tau}\right)\boldsymbol{E}. \tag{5.41}$$

From Eq. (5.11),

$$\varepsilon_r = n_c{}^2 = 1 - \frac{\omega_p{}^2}{\omega^2 - i\omega_\tau \omega}, \tag{5.42}$$

where

$$\omega_p = \sqrt{\frac{\mu_0 \sigma c^2}{\tau}} = \sqrt{\frac{Ne^2}{m_e \varepsilon_0}}, \tag{5.43}$$

$$\omega_\tau = \frac{1}{\tau}, \tag{5.44}$$

is the plasma angular frequency. Equation (5.42) represents the dielectric function of the Drude model. Here, the effect of conduction electrons is treated as the current density according to Eq. (5.36), but the same equation can be obtained if it is treated as polarization (electric polarization) in the same way as Eq. (5.13).

The real and imaginary parts of Eq. (5.42) are given by the following equations, respectively:

$$\varepsilon_1 \equiv \mathrm{Re}(\varepsilon_r) = n_r{}^2 - \kappa^2 = 1 - \frac{\omega_p{}^2}{\omega^2 + \omega_\tau{}^2}, \tag{5.45}$$

$$\varepsilon_2 \equiv \mathrm{Im}(\varepsilon_r) = -2n_r\kappa = -\frac{\omega_p{}^2}{\omega^2 + \omega_\tau}\left(\frac{\omega_\tau}{\omega}\right). \tag{5.46}$$

Figure 5.4 shows the ω dependence of the real and imaginary parts of the metal refractive index expressed by Eqs. (5.45) and (5.46). As can be seen from

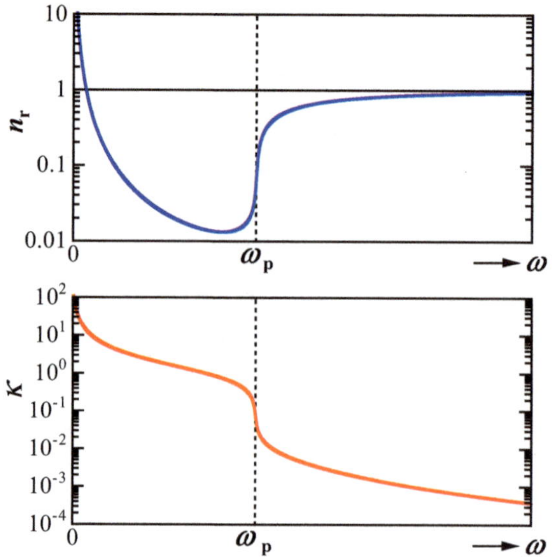

Figure 5.4 Real and imaginary parts of the refractive index for metal.

the figure, n_r takes a value smaller than unity in a wide region including the plasma angular frequency ω_p. In the region where ω is small (long wavelength region), the extinction coefficient κ becomes very large. In other words, the longer the wavelength, the greater the absorption. On the other hand, in the region $\omega > \omega_p$, κ becomes small. Therefore, metals become transparent in the short-wavelength region.

Figure 5.5 shows the real and imaginary parts of the metal refractive index reported as measured values.[11–17] In actual metals, the refractive index is not determined by a simple model such as the Drude model but has complex characteristics that include factors that reflect the band structure. However, for alkali metals, Au, Ag, Cu, etc., it can be partially approximated by the Drude model. There are several reported examples of fitting the refractive index of metals using the Drude model expressed by Eq. (5.42).[18–20]

Figure 5.6 shows the calculated examples for the real and imaginary parts of the metal refractive index according to the Drude model. If the plasma frequency ω_p is fixed (8 eV in this case), as shown in the figure, the smaller ω_p is, the smaller the real part n_r of the refractive index in the visible region becomes. The Drude parameter ω_τ for Au, Ag, and Cu is around 0.01–0.02 eV.[18] Therefore, the real parts n_r of the refractive indices for these metals take small values less than unity.

As shown in Fig. 5.5, metals have large absorption coefficients in the visible range. The absorption coefficient is so large that it seems that all visible radiation is absorbed. However, this is not true. Assuming that the refractive index of air is $n_0 = 1$ and the complex refractive index of metal is $n_c = n_r - i\kappa$,

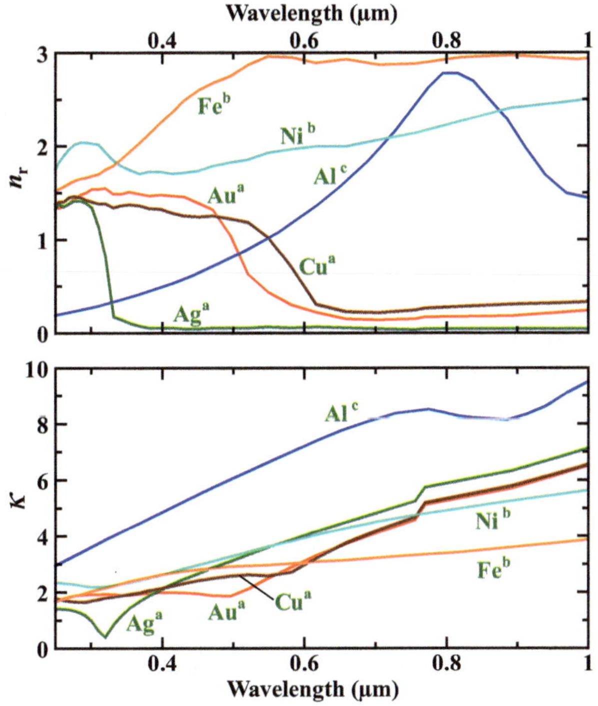

Figure 5.5 Reported example of real and imaginary parts of metal refractive index. [a]Plotted based on data from P. B. Johnson and R. W. Christy[11] © American Physical Society. [b]Plotted based on data from P. B. Johnson and R. W. Christy[12] © American Physical Society. [c]Plotted with permission based on data from A. D. Rakić[13] © Optica Publishing Group.

the reflectance R in the case of normal incidence is expressed by the following equation:

$$R = \frac{(1 - n_r)^2 + \kappa^2}{(1 + n_r)^2 + \kappa^2}. \tag{5.47}$$

As mentioned in Fig. 5.6, Au, Ag, and Cu have small n_r and large κ in the visible range. Therefore, $R \cong 1$ in the visible range, and as a result metals reflect almost all of the visible radiation. This is the reason metals are glossy.[21]

In the actual case as shown in Fig. 5.7, Au, Ag, and Cu have a reflectance close to unity for light with wavelengths above a certain critical wavelength λ_c (indicated by the arrow in Fig. 5.7). For example, λ_c for silver is in the ultraviolet range, so almost all visible radiation is reflected, giving it a silver appearance. On the other hand, the λ_c values of gold and copper are around 520 nm and 590 nm, respectively, and blue light shorter than that is absorbed, so they appear gold and copper, respectively.

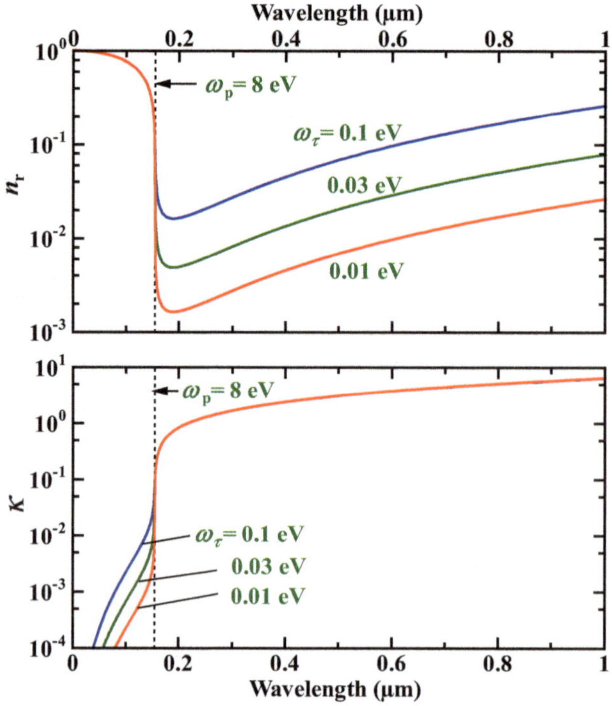

Figure 5.6 Real and imaginary parts of the refractive index of metals using the Drude model.

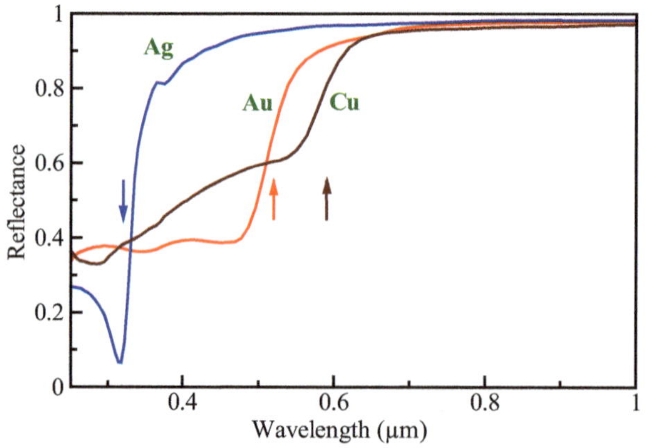

Figure 5.7 Reflectances of gold (Au), silver (Ag), and copper (Cu), calculated based on the data shown in Fig. 5.5.

5.5 Conclusion

A model of the refractive index of dielectrics and metals was introduced. In the next chapter, a model of the refractive index based on the band structure of semiconductors will be introduced.

Side Topic: Reflection on the surface of a perfect conductor

Consider the case where there is no scattering of conduction electrons in the Drude model, that is, the case when $1/\tau = 0$. This case corresponds to a perfect conductor because $\sigma = \infty$ from Eq. (5.39). In this case, in the region $\omega < \omega_p$, the relative permittivity becomes a negative real number according to Eq. (5.42), and the refractive index becomes a pure imaginary number. Since the real part n_r of the refractive index is 0, the reflectance is unity from Eq. (5.46) regardless of the value of κ.

Now when a plane wave is incident on such a perfect conductor from air, the reflectance is unity, so a standing wave is formed on the incident side. Since it is a perfect conductor, one would think that the interface should be a node of standing waves, but this is not actually the case.

In the footnote in Chapter 1, the terms fixed end and open end were explained. As mentioned there, in a standing electromagnetic wave, the nodes of the electric field $\boldsymbol{E}$ are the antinodes of the magnetic field $\boldsymbol{H}$. If the metal surface were a completely fixed end, light would not be able to penetrate the metal at all, but this contradicts the fact that $\boldsymbol{H}$ does not become 0 at the interface.

As in the case of total internal reflection in Chapter 4, there is a phase discontonuity at the interface, and $\boldsymbol{E}$ does not become a node. Using Eq. (5.17), the amplitude reflection coefficient in the case of normal incidence is given as follows:

$$r_s = \frac{1 + i\kappa}{1 - i\kappa} = \exp(-i\delta), \tag{5.S1}$$

$$\delta = -\tan^{-1}\left(\frac{2\kappa}{1 - \kappa^2}\right). \tag{5.S2}$$

In other words, there is always a phase discontinuity δ given by Eq. (5.S2). Therefore, the sum of the incident wave and reflected wave does not become 0, and the interface does not become a node of $\boldsymbol{E}$.

Assuming $n_1 = 1$ (air) and $n_2 = -2i$ (perfect conductor), a diagram similar to Fig. 1.4(b) can be drawn as Fig. 5.S1. Since the reflectance is unity, the wave becomes a perfect standing wave in air. It can be seen that the boundary surface ($z = 0$) does not become a node of the electric field, and both the electric and magnetic fields penetrate the metal.

By substituting the above refractive index parameters into the Excel file [01-2_Plane wave.xlsm: Standing wave (3)] in Chapter 1 and running the animation, it will be easier to understand the state of the standing wave.

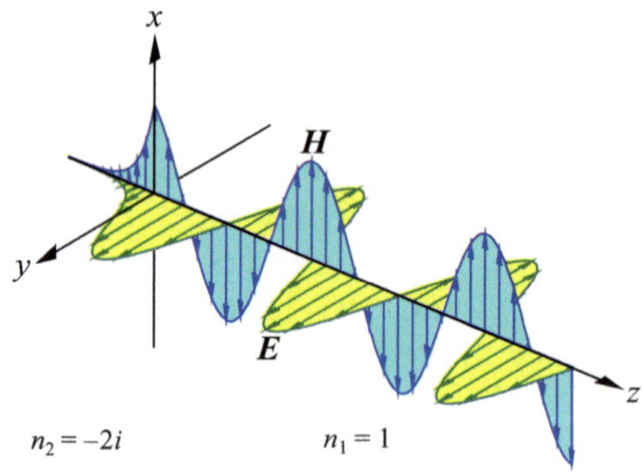

Figure 5.S1 Reflection of light by a perfect conductor.

References

1. M. Born and E. Wolf, *Principles of Optics, 7th Ed.*, Cambridge University Press (1999).
2. G. R. Fowles, *Introduction to Modern Optics, 2nd Ed.*, Dover Publications, New York (1975).
3. Refractive Index Info, Refractive index database (2025), https://refractiveindex.info/
4. "Downloads for Optical Glass," Schott, https://www.schott.com/en-us/products/optical-glass-p1000267/downloads
5. Hoya, Hoya Group Optics Division (2025), https://www.hoya-opticalworld.com/english/
6. I. H. Malitson, "Interspecimen comparison of the refractive index of fused silica," *J. Opt. Soc. Am.* **55**(10), 1205–1209 (1965).
7. I. H. Malitson, "Refraction and dispersion of synthetic sapphire," *J. Opt. Soc. Am.* **52**(12), 1377–1379 (1962).
8. D. L. Wood and K. Nassau, "Refractive index of cubic zirconia stabilized with yttria," *Appl. Opt.* **21**(16), 2978–2981 (1982).
9. J. W. Fleming, "Dispersion in GeO_2-SiO_2 glasses," *Appl. Opt.* **23**(24), 4486–4493 (1984).
10. D. L. Wood, K. Nassau, T. Y. Kometani, and D. L. Nash, "Optical properties of cubic hafnia stabilized with yttria," *Appl. Opt.* **29**(4), 604–607 (1990).
11. P. B. Johnson and R. W. Christy, "Optical constants of the noble metals," *Phys. Rev. B* **6**(12), 4370–4379 (1972).
12. P. B. Johnson and R. W. Christy, "Optical constants of transition metals: Ti, V, Cr, Mn, Fe, Co, Ni, and Pd," *Phys. Rev. B* **9**(12), 5056–5076 (1974).

13. A. D. Rakić, "Algorithm for the determination of intrinsic optical constants of metal films: application to aluminum," *Appl. Opt.* **34**(22), 4755–4767 (1995).
14. W. S. M. Werner, K. Glantschnig, and C. Ambrosch-Draxl, "Optical Constants and Inelastic Electron-Scattering Data for 17 Elemental Metals," *J. Phys. Chem. Ref. Data* **38**(4), 1013–1092 (2009).
15. G. Hass, "Optical constants of metals," *American Institute of Physics Handbook*, D. E. Gray, ed., McGraw Hill, New York (1957).
16. K. Kudo, *Basic physical properties chart*, Kyoritsu Shuppan, Tokypo (1972) [in Japanese].
17. K.-H. Hellwege and J. L. Olsen, eds., "Metals: Electronic Transport Phenomena—Electrical Resistivity, Thermoelectrical Power and Optical Properties," *Landolt–Börnstein: Numerical Data and Functional Relationships in Science and Technology—New Series, Group III Condensed Matter, Vol. 15B*, Springer (1985).
18. M. A. Ordal, R. J. Bell, R. W. Alexander, L. L. Long, and M. R. Querry, "Optical properties of fourteen metals in the infrared and far infrared: Al, Co, Cu, Au, Fe, Pb, Mo, Ni, Pd, Pt, Ag, Ti, V, and W," *Appl. Opt.* **24**(24), 4493–4499 (1985).
19. M. G. Blaber, M. D. Arnold, and M. J. Ford, "Search for the ideal plasmonic nanoshell: The effects of surface scattering and alternatives to gold and silver," *J. Phys. Chem. C* **113**(8), 3041–3045 (2009).
20. E. J. Zeman and G. C. Schatz, "An accurate electromagnetic theory study of surface enhancement factors for Ag, Au, Cu, Li, Na, Al, Ga, In, Zn, and Cd," *J. Phys. Chem.* **91**(3), 634–643 (1987).
21. K. Iga and G. Hatakoshi, Chapter 1, *Treasure Microbox of Optoelectronics I*, Advanced Communication Media (2018) [in Japanese].

Chapter 6
Refractive Index Part 2

6.1 Introduction

In the previous chapter, the refractive index models for dielectrics and metals were introduced, wherein the models were based on the equation of motion for bound electrons and conduction electrons. On the other hand, light absorption in semiconductors is caused by interband transitions and exhibits complex behavior that reflects the band structure. This chapter introduces models of permittivity and refractive index in semiconductors.

6.2 Band Structures and the Refractive Index Model for Semiconductors

6.2.1 Permittivity and refractive index

As mentioned in the previous chapter, the relationship between the complex refractive index n_c and the relative permittivity ε_r is given by the following equation:

$$\varepsilon_r = {n_c}^2. \tag{6.1}$$

In this book, the time dependence of the electric field is expressed as $\exp(i\omega t)$, so the relationship between the complex refractive index n_c, the real part n_r, and the extinction coefficient κ is given as follows:

$$n_c = n_r - i\kappa. \tag{6.2}$$

The sign of the imaginary part of ε_r is also determined by Eq. (6.2), but many papers discussing refractive index use the opposite sign. In this chapter, the sign of the imaginary part of the permittivity is selected so as to match the notation in the literature, in order to avoid confusion. Therefore, the relative permittivity ε_r described in this chapter is defined as follows:

$$\varepsilon_r = n_c^{*2}, \tag{6.3}$$

$$n_c^* = n_r + i\kappa. \tag{6.4}$$

The relationship between the real part ε_1 and the imaginary part ε_2 of ε_r, n_r, and κ is the same as Eqs. (5.29) and (5.30), which are shown again as follows:

$$n_r = \sqrt{\frac{\sqrt{\varepsilon_1{}^2 + \varepsilon_2{}^2} + \varepsilon_1}{2}}, \tag{6.5}$$

$$\kappa = \sqrt{\frac{\sqrt{\varepsilon_1{}^2 + \varepsilon_2{}^2} - \varepsilon_1}{2}}. \tag{6.6}$$

The real part ε_1 and the imaginary part ε_2 are connected by the well-known Kramers–Kronig relations. That is,

$$\varepsilon_1(\omega) = \varepsilon_\infty + \frac{2}{\pi}\int_0^\infty \frac{\omega'\varepsilon_2(\omega')}{\omega'^2 - \omega^2} d\omega', \tag{6.7a}$$

$$\varepsilon_2(\omega) = -\frac{2\omega}{\pi}\int_0^\infty \frac{\varepsilon_1(\omega') - \varepsilon_\infty}{\omega'^2 - \omega^2} d\omega'. \tag{6.7b}$$

These relations can also be expressed in terms of photon energy E,

$$\varepsilon_1(E) = \varepsilon_\infty + \frac{2}{\pi}\int_0^\infty \frac{E'\varepsilon_2(E')}{E'^2 - E^2} dE', \tag{6.8a}$$

$$\varepsilon_2(E) = -\frac{2E}{\pi}\int_0^\infty \frac{\varepsilon_1(E') - \varepsilon_\infty}{E'^2 - E^2} dE'. \tag{6.8b}$$

From this, for example, if $\varepsilon_2(E)$ is known over the entire region, $\varepsilon_1(E)$ can be calculated using Eq. (6.8a). The transformation using Eqs. (6.7a) and (b) or Eqs. (6.8a) and (b) is called the Kramers–Kronig transformation (K–K transformation).

6.2.2 Band structures and light absorption

Since the imaginary part of the refractive index is closely related to interband transition due to optical absorption, it is important to understand the band structure of semiconductors. Figure 6.1 shows the band structure of GaAs, where E_0, which corresponds to the band gap of GaAs, is the fundamental absorption edge energy. In the figure, Γ, X, and L are critical points with high symmetry in the reciprocal lattice space (k space), which will be described in Chapter 7 (k-Space and Reciprocal Lattice) of Volume 2: Electrons and Photons. In the region where the bands are nearly parallel or near the stationary point [near the critical point (CP)], an absorption transition occurs near the corresponding k and the absorption coefficient increases. As shown in Fig. 6.1, such transitions include E_0, $E_0 + \Delta_0$, E_0', E_1, $E_1 + \Delta_1$, and E_2.

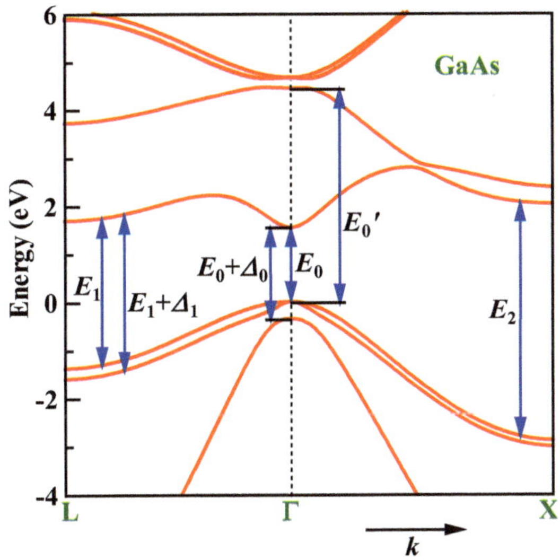

Figure 6.1 Band structure and optical absorption transition for GaAs.

The band structure of semiconductors will be covered in Chapter 7 of Volume 2, so a detailed explanation will be omitted here, but the parabolic approximation of the band structure will be briefly explained. Centering around a highly symmetrical critical point[1] in the reciprocal lattice space (k space), it is approximated that the interband energy E is expressed by the following parabolic function with respect to angular wavenumber k:

$$E = E_g + \beta_1 {k_1}^2 + \beta_2 {k_2}^2 + \beta_3 {k_3}^2, \tag{6.9}$$

where E_g is the minimum interband energy (for example E_0 in Fig. 6.1). Generally, the interband energy is expressed in a three-dimensional k-space as represented by Eq. (6.9), but if the density of state of the bonding state is localized in a certain direction, the dimensions can be reduced and expressed in two or one dimensions. If it is localized in all directions, it is zero-dimensional. Excitons are an example.

6.3 A Model of the Permittivity and Refractive Index of Semiconductors

To calculate the real and imaginary parts of the permittivity, expressions are proposed that represent the respective contributions from several optical transitions,[2–9] as shown in Fig. 6.1. In the following four subsections, the model dielectric function (MDF) model based on Refs. 2–5 will be introduced.

6.3.1 E_0 transition and $E_0 + \Delta_0$ transition

The contribution to the relative permittivity by the fundamental absorption edges E_0 and $E_0 + \Delta_0$ is expressed by the following formula:[2–5]

$$\varepsilon(E) = A\left\{E_0^{-3/2} f(\chi_0) + \frac{1}{2}(E_0 + \Delta_0)^{-3/2} f(\chi_{s0})\right\}, \tag{6.10}$$

$$\chi_0 = \frac{E + i\Gamma_0}{E_0}, \tag{6.11}$$

$$\chi_{s0} = \frac{E + i\Gamma_0}{E_0 + \Delta_0}, \tag{6.12}$$

$$f(\chi_0) = \chi_0^{-2}\{2 - (1 + \chi_0)^{1/2} - (1 - \chi_0)^{1/2}\}. \tag{6.13}$$

χ_0 and χ_{s0} are used for defining Eqs. (6.11) and (6.12), respectively. Note that they are not the electric susceptibility defined in Chapter 5. Equation (6.10) is a complex function, and its real part $\varepsilon_1(E)$ and imaginary part $\varepsilon_2(E)$ satisfy the Kramers–Kronig relations given by Eq. (6.8).

As an example of $\varepsilon(E)$ for the GaAs case, the contribution of E_0, the contribution of $E_0 + \Delta_0$, and the sum of both [Eq. (6.10)] are shown in Fig. 6.2.

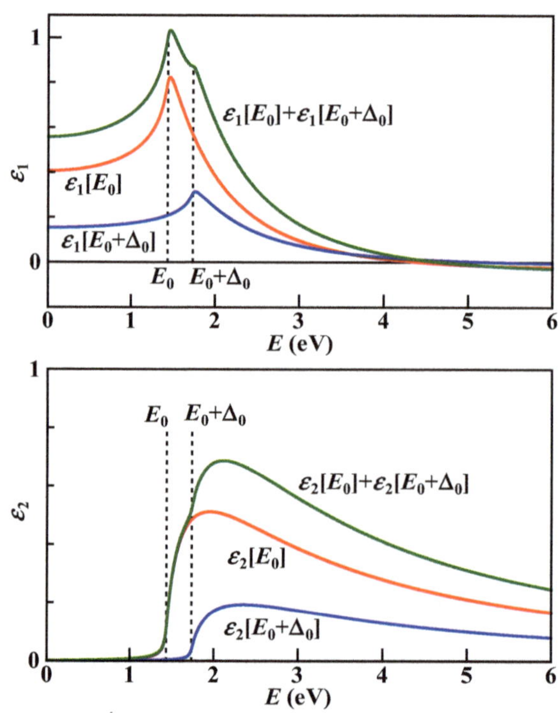

Figure 6.2 $\varepsilon_1(E)$ and $\varepsilon_2(E)$ due to the contribution of the E_0 transition and $E_0 + \Delta_0$ transition.

For GaN-based materials, exciton absorption should be considered. Assuming that the exciton binding energy is G_0, the contribution from the exciton in the bound state is given by the following equation:[2]

$$\varepsilon(E) = \sum_{n=1}^{\infty} \frac{1}{n^3} \frac{A^{ex}}{[E_0 - (G_0/n^2)]^2 - E^2 - i2E\Gamma}, \tag{6.14}$$

where, A^{ex} is the exciton contribution factor and Γ is the damping factor, respectively. The contribution from continuous state excitons is given by the following equation[2] using the parameters F^C and G:

$$\varepsilon(E) = \frac{F^C E_0}{G(E + i\Gamma)^2} \ln \frac{{E_0}^2}{{E_0}^2 - (E + i\Gamma)^2}. \tag{6.15}$$

Equations (6.14) and (6.15) are both complex functions, and the real part $\varepsilon_1(E)$ and the imaginary part $\varepsilon_2(E)$ satisfy the Kramers–Kronig relations. Note that the logarithm of a complex number is defined below:

$$\ln[r \exp(i\theta)] = \ln(r) + i\theta. \tag{6.16}$$

Therefore, by setting $z = r\exp(i\theta)$, the following equation is obtained:

$$\ln(z) = \ln|z| + i \arg(z). \tag{6.17}$$

Side Topic: Sign of the imaginary part of the refractive index

As mentioned in the text, the time dependence of the electric field is expressed as $\exp(+i\omega t)$ in this book, so the refractive index is expressed as Eq. (6.2), but Eq. (6.4) was adopted in this chapter only. This is to match the fact that the imaginary part of the permittivity is expressed as positive in most literature. In ordinary materials, the extinction coefficient κ is never negative, so in the field of optics, the complex refractive index is expressed as Eq. (6.4); therefore, the expression of $\exp(ik_0 n_c z - i\omega t)$ represents an electromagnetic wave that decays in the z direction.

However, in the electrical and electronic fields, we deal with materials that have gain, such as semiconductor lasers, where we would like to express gain as positive and loss as negative. In that case, the refractive index is represented as $n_c = n_r - i\kappa$, and the wave propagation is expressed as $\exp(-ik_0 n_c z + i\omega t)$.

Since both are solutions to the wave equation, there should be no problem as long as the sign is clearly defined. However, care must be taken as this also affects the sign of the exponential part of the Fourier transform.

6.3.2 E_1 transition and $E_1 + \Delta_1$ transition

The contribution to the relative permittivity by E_1 and $E_1 + \Delta_1$ in Fig. 6.1 is expressed by the following formula:[2–5]

$$\varepsilon(E) = -\frac{B_{11}}{\chi_{11}^2}\ln(1-\chi_{11}^2) - \frac{B_{12}}{\chi_{12}^2}\ln(1-\chi_{12}^2), \tag{6.18}$$

$$\chi_{11} = \frac{E + i\Gamma_1}{E_1}, \tag{6.19}$$

$$\chi_{12} = \frac{E + i\Gamma_1}{E_1 + \Delta_1}. \tag{6.20}$$

As an example of $\varepsilon(E)$ for the GaAs case, the contribution of E_1, the contribution of $E_1 + \Delta_1$, and the sum of both [Eq. (6.18)] are shown in Fig. 6.3. Equations (6.10) and (6.18) have different functional forms, but as can be seen by comparing Figs. 6.2 and 6.3, $\varepsilon_1(E)$ and $\varepsilon_2(E)$ exhibit similar behavior.

Similar to Eq. (6.13), there is the following equation as a contribution from bound excitons:[2]

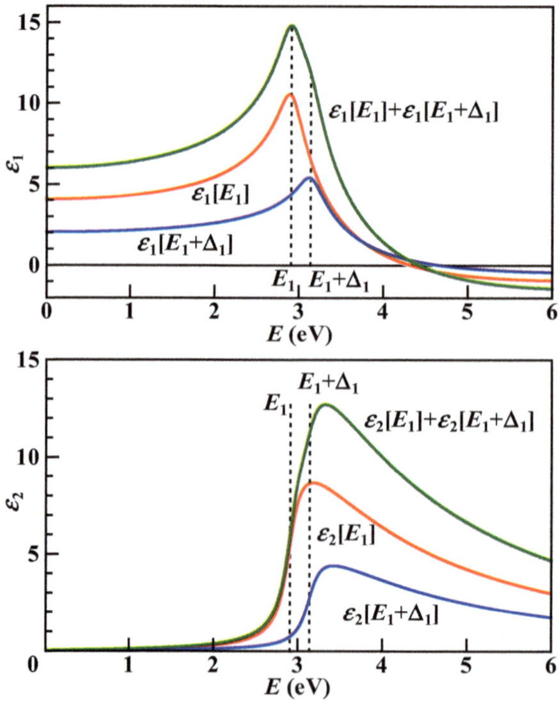

Figure 6.3 $\varepsilon_1(E)$ and $\varepsilon_2(E)$ due to the contribution of the E_1 transition and $E_1 + \Delta_1$ transition.

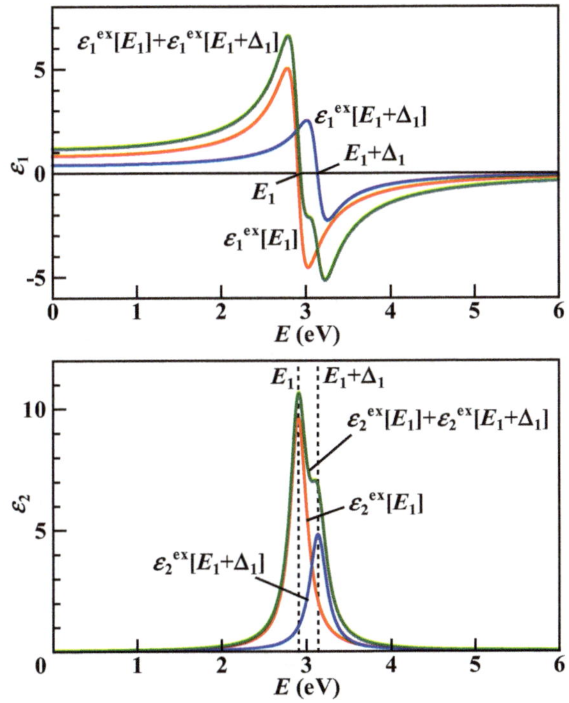

Figure 6.4 $\varepsilon_1(E)$ and $\varepsilon_2(E)$ due to the contribution of bound excitons associated with the E_1 transition and $E_1+\Delta_1$ transition.

$$\varepsilon(E)=\sum_{i=1,2}\sum_{n=1}^{\infty}\frac{B_{ix}}{(2n-1)^3}\frac{1}{\{E_{1i}-4G_{1i}^{2D}/(2n-1)^2\}^2-E^2-i2E\Gamma}, \tag{6.21}$$

$$E_{11}=E_1, E_{12}=E_1+\Delta_1, \tag{6.22}$$

where B_{ix} is the contribution factor of ith bound exciton.

Figure 6.4 shows the photon energy dependence of $\varepsilon_1(E)$ and $\varepsilon_2(E)$ for the GaAs case, due to the contribution of bound excitons associated with the E_1 transition and $E_1+\Delta_1$ transition (example for GaAs).

6.3.3 E_0' transition, E_2 transition, and E_1' transition

The contribution to the relative permittivity due to the transitions of E_0' and E_2 in Fig. 6.1 is expressed by the following formula:[2]

$$\varepsilon(E)=\frac{C}{(1-\chi_2{}^2)-i\chi_2\gamma}, \tag{6.23}$$

$$\chi_2=\frac{E}{E_2},\frac{E}{E_0'}, \tag{6.24}$$

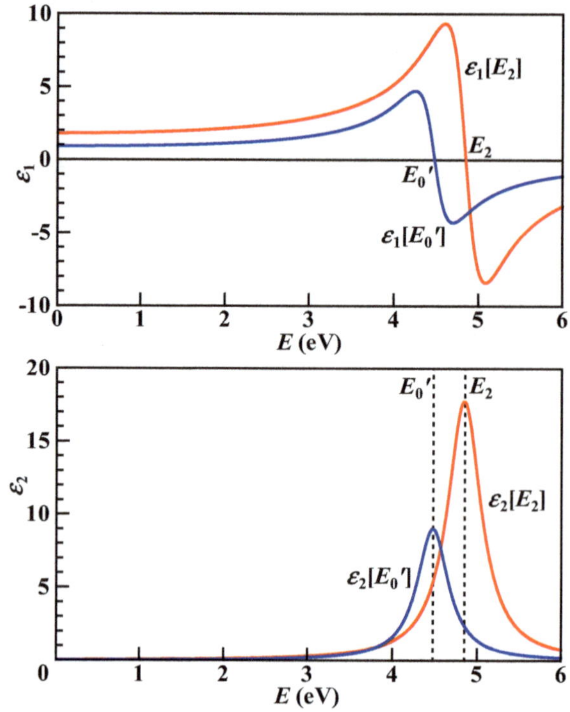

Figure 6.5 $\varepsilon_1(E)$ and $\varepsilon_2(E)$ due to the contribution of the E_0' transition and E_2 transition.

where C is the contribution factor. Equation (6.23) is based on the damped harmonic oscillator (DHO) model, which is the same as the equation based on the Lorentz model [Eq. (5.22) in the previous chapter].†

Figure 6.5 shows $\varepsilon_1(E)$ and $\varepsilon_2(E)$ due to the contribution of the E_0' transition and E_2 transition expressed by Eq. (6.23) (example for GaAs).

The DHO model based on Eq. (6.23) provides a relatively good fit for III-V compound semiconductors.[2] On the other hand, for group IV materials such as Si, fitting using Eq. (6.23) alone is difficult, so a mixed model with the following equation is used:[2]

$$\varepsilon(E) = -\frac{F}{{\chi_{2\mathrm{m}}}^2}\ln\frac{1-{\chi_{\mathrm{c}1}}^2}{1-{\chi_{2\mathrm{m}}}^2}, \tag{6.25}$$

$$\chi_{2\mathrm{m}} = \frac{E+i\Gamma}{E_2}, \tag{6.26}$$

$$\chi_{\mathrm{c}1} = \frac{E+i\Gamma}{E_{\mathrm{c}1}}. \tag{6.27}$$

†Note: The sign of the imaginary part is reversed from the previous chapter due to the situations explained at the beginning.

Note that E_{c1} in Eq. (6.27) is the cutoff energy on the low energy side, which is approximately given by

$$E_{c1} \cong E_1. \tag{6.28}$$

6.3.4 Indirect transition

The contribution of the indirect optical transitions to $\varepsilon_2(E)$ is represented as[2,4]

$$\varepsilon_2(E) = \begin{cases} \dfrac{D}{E^2} n_q (E - E_g^{ID} + E_q)^2 & E > E_g^{ID} - E_q \\ 0 & E < E_g^{ID} - E_q \end{cases}, \tag{6.29}$$

$$n_q = \frac{1}{\exp\left(\dfrac{E_q}{k_B T}\right) - 1}, \tag{6.30}$$

where E_q is the phonon energy and k_B is the Boltzmann constant.

6.4 The Critical Point Parabolic Band (CPPB) Model

The MDF model described in the previous section describes the permittivity at a relatively small number of critical points, and all of them satisfy the Kramers–Kronig relations. Therefore, it is based on physical principles, although fitting work is required.

The degree of freedom in fitting can be increased by increasing the number of CPs by a very similar method. The critical point parabolic band (CPPB) model provides multiple CPs and fits the permittivity over a wide energy range. The permittivity function of semiconductors using this model has been reported.[10]

In this model, the imaginary part of the permittivity is given by the following formula for photon energies above the fundamental absorption edge energy E_0. Note that 0D, 1D, 2D, and 3D below represent the dimensions of the parabolic band explained in Eq. (6.9).

(0D)

$$\varepsilon_2(E) = \mathrm{Im}\left[\sum_j \frac{A_j \exp(i\phi_j)\Gamma_j/2}{E_j - E - i\Gamma_j/2}\right]. \tag{6.31}$$

(1D)

$$\varepsilon_2(E) = \mathrm{Im}\left[\sum_j \frac{A_j \exp(i\phi_j)\Gamma_j/2}{(E_j - E - i\Gamma_j/2)^{1/2}}\right]. \tag{6.32}$$

(2D)

$$\varepsilon_2(E) = \mathrm{Im}\left[\sum_j A_j \exp(i\phi_j)\ln(2E_j - 2E - i\Gamma_j)\right]. \tag{6.33}$$

(3D)

$$\varepsilon_2(E) = \mathrm{Im}\left[\sum_j A_j \exp(i\phi_j)\frac{\Gamma_j}{2}\left(E_j - E - i\frac{\Gamma_j}{2}\right)^{1/2}\right]. \tag{6.34}$$

For $E < E_0$, consider the following exponential absorption tail (called the Urbach tail):[2,10]

$$\varepsilon_2(E) = \varepsilon_2(E_0)\exp\left(\frac{E - E_0}{E_u}\right), \tag{6.35}$$

where, E_u is the Urbach energy.[10]

Using $\varepsilon_2(E)$ obtained by adding up Eqs. (6.31) to (6.35), the real part $\varepsilon_1(E)$ can be found by the K–K transformation of Eq. (6.8a).

6.5 Measurement and Calculation Examples of Permittivity and Refractive Indices of Semiconductors

The wavelength dependence of the refractive index can be measured using spectroscopic ellipsometry,[11] and there are many reports on the measurement for semiconductors.[10,12–17] In areas where measurement is difficult, the real and imaginary parts of the permittivity and refractive index are determined by invoking K–K conversion. Furthermore, many papers have reported parameter extraction using the models described in Sections 6.3 and 6.4 based on these data.[6,18–28] Examples of Si, GaAs, InP, and InGaN are introduced below.

6.5.1 Si

Figure 6.6 shows the real and imaginary parts of the refractive index for Si according to Ref. 23. Equations (6.18), (6.23), (6.25), and (6.29) are used to calculate $\varepsilon_2(E)$. Note that there are no analytical expressions of $\varepsilon_1(E)$ that satisfy the Kramers–Kronig (K–K) relation for formula (6.29). Other formulations for $\varepsilon(E)$ that satisfy the K–K relationship have been reported in Ref. 2. In Fig. 6.6, $\varepsilon_2(E)$ was calculated based on Ref. 23, and $\varepsilon_1(E)$ was derived using the K–K relation. In the figure, the circle indicates data reported by Aspnes et al.[12]

6.5.2 GaAs

Regarding GaAs, diagrams of contributions from each transition have already been introduced in Figs. 6.2 to 6.5. By adding all of these together, the real and imaginary parts of permittivity are obtained.[2,5] Figure 6.7 shows

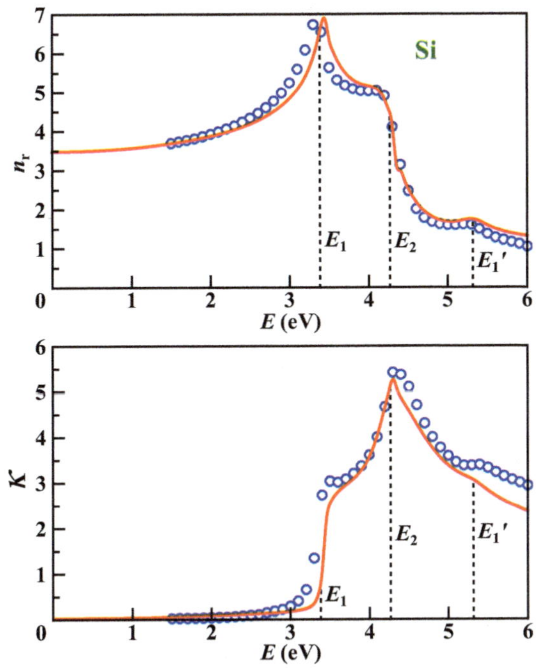

Figure 6.6 Real and imaginary parts of the refractive index for Si.[23] The circle ○ indicates data reported by Aspnes and Studna.[12] Solid lines were plotted based on data from S. Adachi[23] with the permission of the American Institute of Physics. Circles were plotted with permission based on data from D. E. Aspnes and A. A. Studna[12] © American Physical Society.

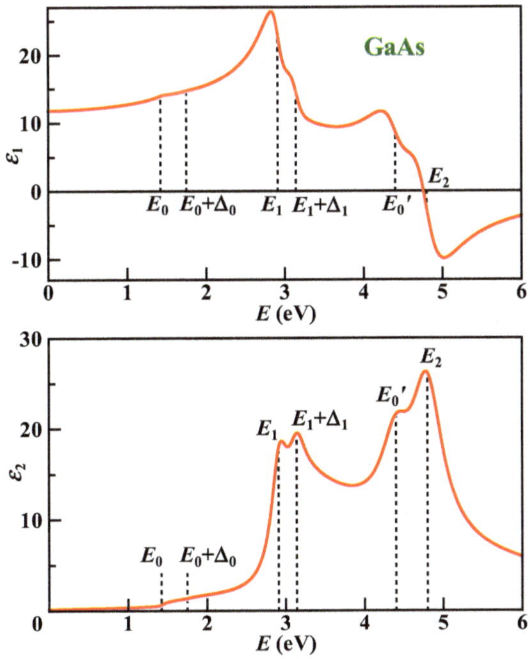

Figure 6.7 Real and imaginary parts of permittivity for GaAs.[5] Plotted based on data from S. Adachi[5] under the Creative Commons license.[29]

$\varepsilon_1(E)$ and $\varepsilon_2(E)$ derived based on the parameters in Ref. 5. Note that the real and imaginary parts of $\varepsilon(E)$ given by Eqs. (6.10), (6.14), (6.15), (6.18), (6.21), and (6.23), satisfy the K–K relations, so the result of adding them also satisfies K–K relations automatically.

Figure 6.8 shows the real and imaginary parts of the refractive index calculated using Eqs. (6.5) and (6.6) based on the parameters in Ref. 5. In the figure, the measured data by Aspnes et al.[13] are plotted together. Although the MDF parameters were not necessarily determined by fitting to this data, they fit well together.

The papers by Aspnes et al.[12,13] are often cited because they have numerical tables and the original data can be seen. Figure 6.9 shows the real and imaginary parts of the refractive index for $Ga_{1-x}Al_xAs$, with Al composition $x = 0, 0.1, 0.2, \ldots, 1.0$, based on GaAlAs measurement data by Aspnes et al. The circle ○ in the figure indicates the refractive index at the bandgap wavelength for each composition.

6.5.3 InP

Figure 6.10 shows the imaginary and real parts of the permittivity for InP according to the CPPB model introduced in Section 6.4. It is fitted using 10

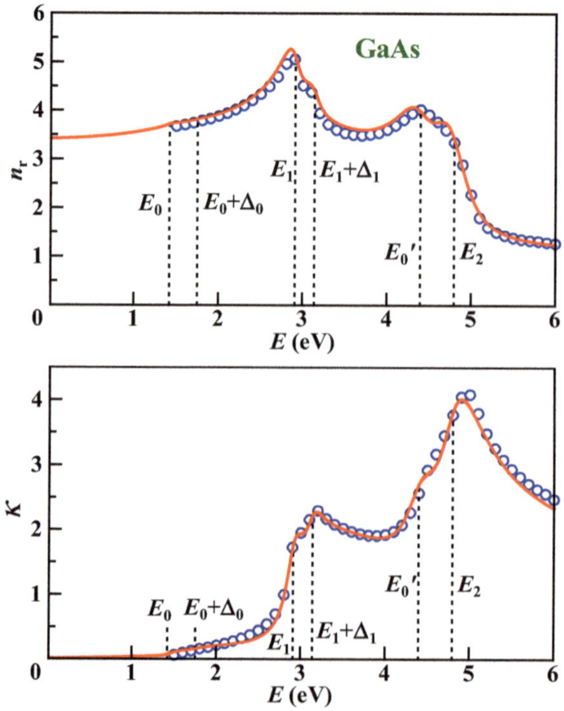

Figure 6.8 Real and imaginary parts of the refractive index for GaAs.[5] The circle ○ indicates data reported by Aspnes et al.[13] Solid lines were plotted based on data from S. Adachi[5] under the Creative Commons license.[29] Circles were plotted with permission based on data from D. E. Aspnes, S. M. Kelso, R. A. Logan, and R. Bhat[13] © American Physical Society.

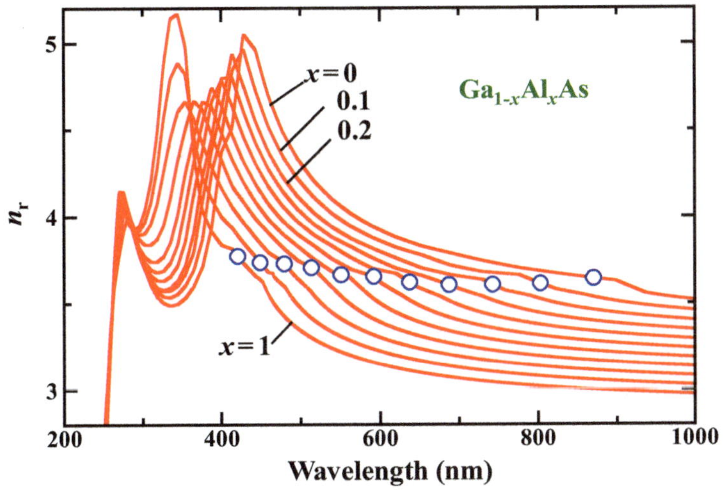

Figure 6.9 Real and imaginary parts of the refractive index for GaAlAs, calculated by interpolation based on Aspnes et al.'s data.[13] Plotted with permission based on data from D. E. Aspnes, S. M. Kelso, R. A. Logan, and R. Bhat[13] © American Physical Society. The circle ○ in the figure indicates the refractive index at the bandgap wavelength for each composition.

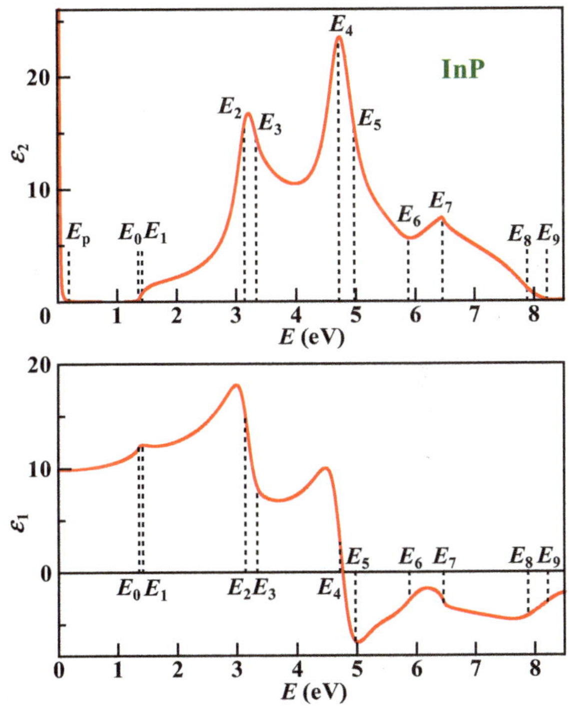

Figure 6.10 Real and imaginary parts of permittivity for InP.[10] Plotted with permission based on data from I. Subedi, M. A. Slocum, D. V. Forbes, S. M. Hubbard, and N. J. Podraza[10] © Elsevier.

CPs including the basic absorption edge E_0, based on Ref. 10. This paper also considers phonon modes and free carrier loss in the low-energy region, and adds equations based on the Lorentz model and Drude model (introduced in the previous chapter) in the region of $E < 0.73$ eV. This is why ε_2 increases in the low-energy region in the upper graph. The curve is fitted by setting the resonance energy of the phonon mode to 0.0378 eV, and the energy E_p corresponding to the plasma frequency in the Drude model to 0.186 eV. The real part ε_1 of the permittivity in the lower figure is obtained by performing K–K conversion on ε_2.

Figure 6.11 shows the results of calculated n_r and κ from these ε_1 and ε_2 using Eqs. (6.5) and (6.6) based on the parameters in Ref. 10. The circles in the figure indicate the data of Aspnes et al.[12]

6.5.4 GaAlN

There are also many reports on GaN-based semiconductors.[9,26–28] Figure 6.12 is an example of calculating the real and imaginary parts of the refractive index for GaAlN using the MDF model in Section 6.3.[28]

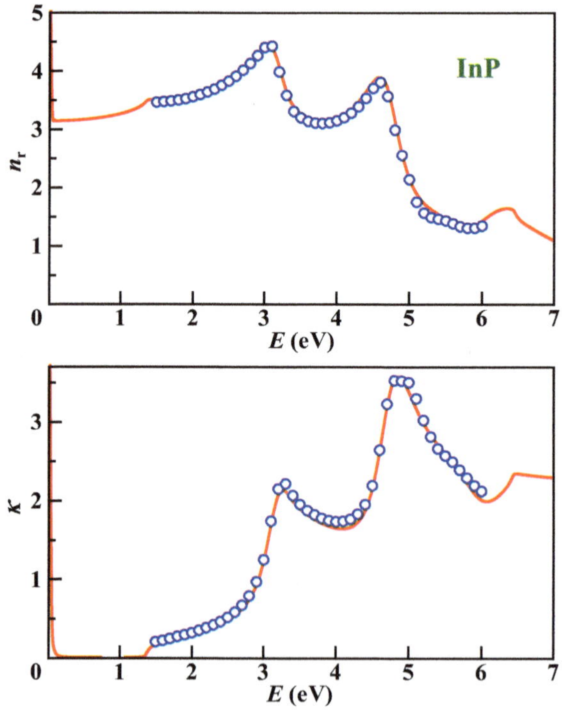

Figure 6.11 Real and imaginary parts of the refractive index for InP, calculated from the permittivity given in Fig. 6.10. The circle ○ indicates data reported by Aspnes and Studna.[12] Solid lines were plotted with permission based on data from I. Subedi, M. A. Slocum, D. V. Forbes, S. M. Hubbard, and N. J. Podraza[10] © Elsevier. Circles were plotted with permission based on data from D. E. Aspnes and A. A. Studna[12] © American Physical Society.

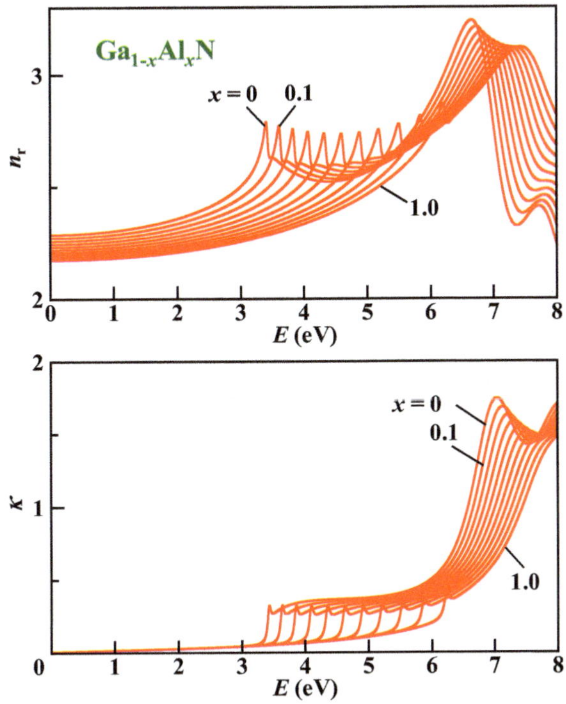

Figure 6.12 Real and imaginary parts of the refractive index for GaAlN.[28] Plotted with permission based on data from K. Takeuchi, S. Adachi, and K. Ohtsuka[28] © American Institute of Physics.

Side Topic: Kramers–Kronig relations and causality

It is well known that the Kramers–Kronig relations are derived from the law of causality. However, since time is not included anywhere in Eq. (6.7) or Eq. (6.8), how is the law of causality involved?

Equation (6.7) shows the frequency response of a material to an electric field. The frequency response is expressed as the Fourier transform of the impulse response function. Impulse response represents the response of a material to a delta function stimulus at time $t = 0$, and the response appears only in the region of $t \geq 0$ according to the law of causality and must be 0 at $t < 0$. This results in the real and imaginary parts of the permittivity not being independent and needing to satisfy the relationship shown in Eq. (6.7).

Now, let's use a permittivity model to confirm that the law of causality holds when the K–K relation is satisfied. To simplify the equation, rewrite Eq. (6.7) using the real part χ_1 and imaginary part χ_2 of the electric susceptibility (see previous chapter), resulting in the following:

$$\chi_1(\omega) = \frac{2}{\pi}\int_0^{\infty} \frac{\omega' \chi_2(\omega')}{\omega'^2 - \omega^2} d\omega', \tag{6.S1}$$

$$\chi_2(\omega) = -\frac{2\omega}{\pi}\int_0^{\infty} \frac{\chi_1(\omega')}{\omega'^2 - \omega^2} d\omega'. \tag{6.S2}$$

Let's consider the Lorentz model, described in the previous chapter, for χ_1 and χ_2, which is represented by

$$\chi_1 + i\chi_2 = \frac{A}{{\omega_0}^2 - \omega^2 - i\gamma\omega}. \tag{6.S3}$$

The real and imaginary parts of Eq. (6.S3) satisfy the K–K relation given by Eqs. (6.S1) and (6.S2). The impulse response function can be derived by the Fourier transform of Eq. (6.S3). That is,

$$f(t) = \frac{1}{\pi}\int_0^{\infty} (\chi_1 + i\chi_2)\exp(-i\omega t) d\omega. \tag{6.S4}$$

This $f(t)$ corresponds to the response of polarization P to the pulsed electric field.

Figure 6.S1 shows χ_1, χ_2, and the real part of $f(t)$. As can be seen, the response appears only at $t \geq 0$, and $f(t) = 0$ for $t < 0$. The vibration at $t \geq 0$ is damped with time, and the larger the term γ, the larger the damping.

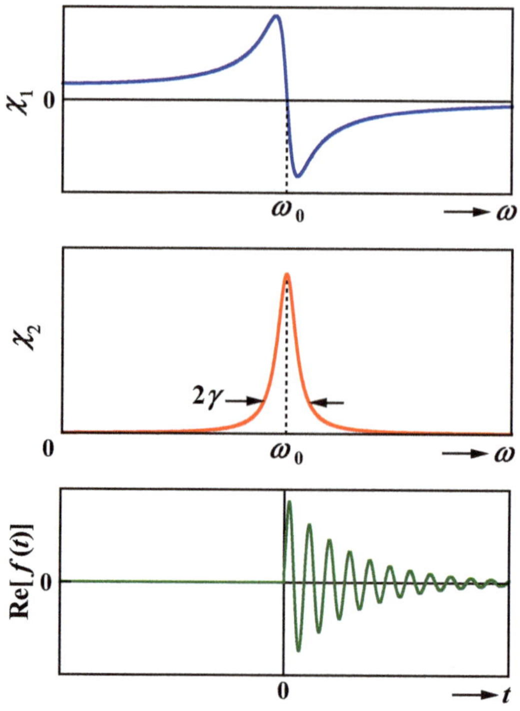

Figure 6.S1 Impulse response of χ_1, χ_2, and polarization $P(t)$ based on the Lorentz model.

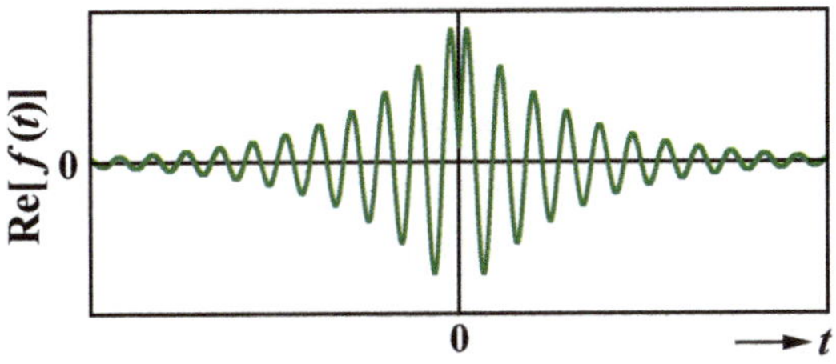

Figure 6.S2 Impulse response when χ_1 and χ_2 do not satisfy the K–K relation.

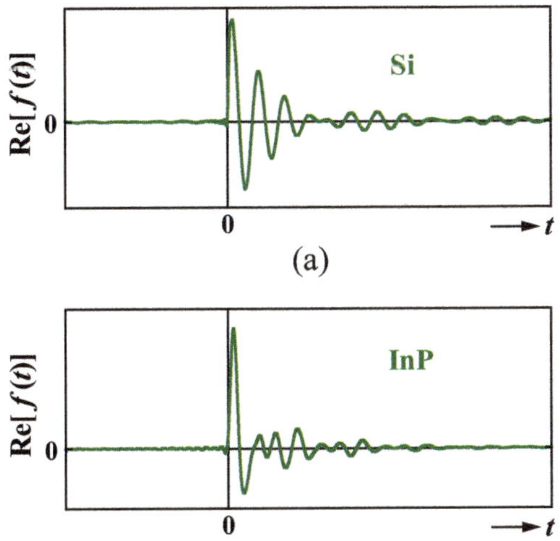

Figure 6.S3 Impulse response of polarization for (a) Si and (b) InP.

Now, what happens if χ_1 and χ_2 do not satisfy the K–K relation? As an example, Fig. 6.S2 shows the impulse response $f(t)$ for the case where $\chi_2 = 0$ in Fig. 6.S1. As can be seen from the figure, a response that is symmetrical to $t \geq 0$ appears at $t < 0$. This, of course, does not satisfy the law of causality. Therefore, there cannot be a material whose real part of electric susceptibility (= real part of refractive index – 1) has the frequency characteristics shown in the upper diagram of Fig. 6.S1 and whose imaginary part is 0 (transparent). Even artificially created metamaterials must satisfy the K–K relation.

Let's check whether the law of causality also holds in the case of a more complicated permittivity function. Figure 6.S3 shows the calculated impulse response of polarization for the permittivity function of Si and InP [$\varepsilon(E)$ for InP is shown in Fig. 6.10]. Since the Fourier transform is performed numerically, there is some noise, but in principle $f(t) = 0$ at $t < 0$.

6.6 Conclusion

Models for the permittivity and refractive index of semiconductors were introduced. The permittivity is expressed by a model based on the band structure of semiconductors. The model is also constructed to satisfy the Kramers–Kronig relations. Finally, parameters are determined by fitting the measured data.

Acknowledgment

The author would like to acknowledge Dr. Sadao Adachi, Professor Emeritus of Gunma University, for the references to his books and papers on the models of the permittivity and refractive index of semiconductors. The author also would like to acknowledge Dr. David Aspnes, Distinguished University Professor Emeritus of North Carolina State University, for the references to his papers on the refractive index data of semiconductors.

References

1. S. Loughin, R. H. French, L. K. De Noyer, W.-Y. Ching, and Y.-N. Xu, "Critical point analysis of the interband transition strength of electrons," *J. Phys. D: Appl. Phys.* **29**(7), 1740–1750 (1996).
2. S. Adachi, *Properties of group-IV, III-V, and II-VI semiconductors, 1st Ed.*, Wiley Series in Materials for Electronic & Optoelectronic Applications, John Wiley & Sons (2005).
3. H. Nagai, S. Adachi, and T. Fukui, *III-V Compound Semiconductors*, Corona Publishing, Tokyo (1988) [in Japanese].
4. S. Adachi, "Model dielectric constants of GaP, GaAs, Gasb, InP, InAs, and InSb," *Phys. Rev. B* **35**(14), 7454–7463 (1987).
5. S. Adachi, "Model dielectric function (MDF) theory for single crystalline materials," *J. Surface Science Soc. Japan* **18**(11), 669–675 (1997) [in Japanese].
6. T. J. Kim, T. H. Ghong, Y. D. Kim, S. J. Kim, D. E. Aspnes, T. Mori, T. Yao, and B. H. Koo, "Dielectric functions of $In_xGa_{1-x}As$ alloys," *Phys. Rev. B* **68**, 115323 (2003).
7. S. G. Choi, C. J. Palmstrøm, Y. D. Kim, S. L. Cooper, and D. E. Aspnes, "Dielectric functions of $Al_xGa_{1-x}Sb$ ($0.00 \leq x \leq 0.39$) alloys from 1.5 to 6.0 eV," *J. Appl. Phys.* **98**, 104108 (2005).
8. Y. W. Jung, T. J. Kim, J. J. Yoon, Y. D. Kim, and D. E. Aspnes, "Model dielectric functions for $Al_xGa_{1-x}As$ alloys of arbitrary compositions," *J. Appl. Phys.* **104**, 013515 (2008).
9. T. J. Kim, S. Y. Hwang, J. S. Byun, D. E. Aspnes, E. H. Lee, J. D. Song, C.-T. Liang, Y.-C. Chang, H. G. Park, J. Choi, J. Y. Kim, Y. R. Kang,

J. C. Park, and Y. D. Kim, "Dielectric functions and interband transitions of $In_xAl_{1-x}P$ alloys," *Cur. Appl. Phys.* **14**(9), 1273–1276 (2014).

10. I. Subedi, M. A. Slocum, D. V. Forbes, S. M. Hubbard, and N. J. Podraza, "Optical properties of InP from infrared to vacuum ultraviolet studied by spectroscopic ellipsometry," *Applied Surface Science* **421**(Part B), 813–818 (2017).
11. D. E. Aspnes, "Spectroscopic ellipsometry—Past, present, and future," *Thin Solid Films* **571**(Part 3), 334–344 (2014).
12. D. E. Aspnes and A. A. Studna, "Dielectric functions and optical parameters of Si, Ge, GaP, GaAs, GaSb, InP, InAs, and InSb from 1.5 to 6.0 eV," *Phys. Rev. B* **27**, 985–1009 (1983).
13. D. E. Aspnes, S. M. Kelso, R. A. Logan, and R. Bhat, "Optical properties of $Al_xGa_{1-x}As$," *J. Appl. Phys.* **60**(2), 754–767 (1986).
14. C. Buchheim, R. Goldhahn, M. Rakel, C. Cobet, N. Esser, U. Rossow, D. Fuhrmann, and A. Hangleiter, "Dielectric function and critical points of the band structure for AlGaN alloys," *Phys. Stat. Sol. (b)* **242**(13), 2610–2616 (2005).
15. H. Lee, M. V. Klein, D. E. Aspnes, C. P. Kuo, M. Peanasky, and M. G. Craford, "Optical study of $(Al_xGa_{1-x})_{0.5}In_{0.5}P/GaAs$ semiconductor alloys by spectroscopic ellipsometry," *J. Appl. Phys.* **73**(1), 400–406 (1993).
16. S. G. Choi, Y. D. Kim, S. D. Yoo, D. E. Aspnes, D. H. Woo, and S. H. Kim, "Optical properties of $Al_xGa_{1-x}P$ ($0 \leq x \leq 0.52$) alloys," *J. Appl. Phys.* **87**(3), 1287–1290 (2000).
17. J. J. Yoon, T. H. Ghong, J. S. Byun, Y. D. Kim, D. E. Aspnes, H. J. Kim, Y. C. Chang, and J. D. Song, "Optical properties of $In_xAl_{1-x}As$ alloy films," *Appl. Phys. Lett.* **92**(15), 151907 (2008).
18. S. Adachi, "Refractive indices of III–V compounds: Key properties of InGaAsP relevant to device design," *J. Appl. Phys.* **53**(8), 5863–5869 (1982).
19. S. Adachi, "Band gaps and refractive indices of AlGaAsSb, GaInAsSb, and InPAsSb: Key properties for a variety of the 2–4 μm optoelectronic device applications," *J. Appl. Phys.* **61**(10), 4869–4876 (1987).
20. S. Adachi, "Model dielectric constants of Si and Ge," *Phys. Rev. B* **38**(18), 12966–12976 (1988).
21. P. Lautenschlager, M. Garriga, L. Vina, and M. Cardona, "Temperature dependence of the dielectric function and interband critical points in silicon," *Phys. Rev. B* **36**(9), 4821–4830 (1987).
22. S. Adachi, "Optical properties of $Al_xGa_{1-x}As$ alloys," *Phys. Rev. B* **38**(17), 12345–12352 (1988).
23. S. Adachi, "Optical dispersion relations for Si and Ge," *J. Appl. Phys.* **66**(7), 3224–3231 (1989).
24. S. Adachi, "Optical dispersion relations for GaP, GaAs, GaSb, InP, InAs, InSb, $Al_xGa_{1-x}As$, and $In_{1-x}Ga_xAs_yP_{1-y}$," *J. Appl. Phys.* **66**(12), 6030–6040 (1989).

25. S. Adachi, "Optical properties of $In_{1-x}Ga_xAs_yP_{1-y}$ alloys," *Phys. Rev. B* **39**(17), 12612–12621 (1989).
26. T. Kawashima, H. Yoshikawa, S. Adachi, S. Fuke, and K. Ohtsuka, "Optical properties of hexagonal GaN," *J. App. Phys.* **82**(7), 3528–3535 (1997).
27. A. B. Djurišić and E. H. Li, "Modeling the optical properties of hexagonal GaN," *Appl. Phys. Lett.* **73**(7), 868–870 (1998).
28. K. Takeuchi, S. Adachi, and K. Ohtsuka, "Optical properties of $Al_xGa_{1-x}N$ alloy," *J. Appl. Phys.* **107**(2), 023306 (2010).
29. CC BY-NC 4.0 Attribution-NonCommercial 4.0 International Deed, Creative Commons, https://creativecommons.org/licenses/by-nc/4.0/deed.en

Chapter 7
Gaussian Beams Part 1

7.1 Introduction

In Chapters 1 to 3, light propagations for plane waves, cylindrical waves, and spherical waves were described. This chapter deals with Gaussian beams. The Gaussian beam formula is generally used in the design and analysis of optical systems.[1–4] It is useful and convenient for representing light propagation through optical components. Although the beam in an actual optical system is not necessarily a Gaussian beam, it is often used as an approximation. First, general Hermite–Gaussian beams will be discussed and the beam diameter and the beam divergence angle of Gaussian beams will be introduced.

7.2 Hermite–Gaussian Beams

7.2.1 Equations of a Hermite–Gaussian beam

Set the electric field of a light wave propagating in the z direction as follows:

$$E = Af(x, y, z)\exp(-ikz + i\omega t), \tag{7.1}$$

$$k = \frac{2\pi}{\lambda} = \frac{2\pi}{\lambda_0} n_r, \tag{7.2}$$

where λ_0 and λ are wavelengths in vacuum and in the medium, respectively, and n_r is the refractive index of the medium. Substituting Eq. (7.1) into the wave equation derived from Maxwell's equations, the following equation for $f(x, y, z)$ is obtained:

$$\frac{\partial^2 f}{\partial x^2} + \frac{\partial^2 f}{\partial y^2} = 2ik\frac{\partial f}{\partial z}. \tag{7.3}$$

This equation is the paraxial Helmholtz equation with the term $\partial^2 f/\partial z^2$ omitted by assuming that $(\partial^2 f/\partial z^2) \ll k(\partial f/\partial z)$. Separating the variables in a Cartesian coordinate system, a cylindrical coordinate system, and an elliptical coordinate system, the solutions for Hermite–Gaussian modes, Laguerre–Gaussian modes, and Ince–Gaussian modes, respectively, are obtained.

Assume that the solution to Eq. (7.3) has the following form:

$$f = f_1(x,z) f_2(y,z). \tag{7.4}$$

Then, $f_1(x, z)$ and $f_2(y, z)$ should satisfy the following equations:

$$\frac{\partial^2 f_1}{\partial x^2} = 2ik\frac{\partial f_1}{\partial z}, \tag{7.5a}$$

$$\frac{\partial^2 f_2}{\partial y^2} = 2ik\frac{\partial f_2}{\partial z}. \tag{7.5b}$$

The solutions to Eqs. (7.5a) and (b) are given by

$$f_1 = a_1\sqrt{\frac{1}{z}}\exp\left(-ik\frac{x^2}{2z}\right), \tag{7.6a}$$

$$f_2 = a_2\sqrt{\frac{1}{z}}\exp\left(-ik\frac{y^2}{2z}\right). \tag{7.6b}$$

In general, the equation represented by Eqs. (7.5a) and (b) and the solutions given by Eqs. (7.6a) and (b) are invariant with respect to the coordinate shift.[5] Kogelnik suggested that the shift direction can be an imaginary number and Gaussian beam solution is obtained by a complex coordinate shift.[6,7] For example, z in Eq. (7.6) can be replaced as follows:

$$\text{For (7.6a)} \quad z \rightarrow z - z_{0x} + iz_{\mathrm{R}x}, \tag{7.7a}$$

$$\text{For (7.6b)} \quad z \rightarrow z - z_{0y} + iz_{\mathrm{R}y}. \tag{7.7b}$$

Then the solutions become as follows, where the coefficients in Eq. (7.6) are set as $a_1 = (iz_{\mathrm{R}x})^{1/2}$ and $a_2 = (iz_{\mathrm{R}y})^{1/2}$:

$$f_1(x,z) = \sqrt{\frac{w_{0x}}{w_x(z - z_{0x})}} \times \exp\left[-\frac{x^2}{\{w_x(z - z_{0x})\}^2} - i\frac{kx^2}{2R_x(z - z_{0x})} + i\varphi_x(z)\right], \tag{7.8a}$$

$$f_2(y,z) = \sqrt{\frac{w_{0y}}{w_y(z - z_{0y})}} \times \exp\left[-\frac{y^2}{\{w_y(z - z_{0y})\}^2} - i\frac{ky^2}{2R_y(z - z_{0y})} + i\varphi_y(z)\right], \tag{7.8b}$$

where

$$w_x(z) = w_{0x}\sqrt{1 + z^2/z_{Rx}^2}, \tag{7.9a}$$

$$w_y(z) = w_{0y}\sqrt{1 + z^2/z_{Ry}^2}, \tag{7.9b}$$

$$R_x(z) = z(1 + z_{Rx}^2/z^2), \tag{7.10a}$$

$$R_y(z) = z(1 + z_{Ry}^2/z^2), \tag{7.10b}$$

$$w_{0x} = \sqrt{2z_{Rx}/k}, \tag{7.11a}$$

$$w_{0y} = \sqrt{2z_{Ry}/k}. \tag{7.11b}$$

$$\varphi_x(z) = \frac{1}{2}\tan^{-1}\left(\frac{z}{z_{Rx}}\right), \tag{7.12a}$$

$$\varphi_y(z) = \frac{1}{2}\tan^{-1}\left(\frac{z}{z_{Ry}}\right). \tag{7.12b}$$

Equations (7.8) to (7.12) are commonly seen Gaussian beam formulas. Note that $w_x(z - z_{0x})$ and $R_x(z - z_{0x})$ in Eq. (7.8a) represent functions with the argument $z - z_{0x}$ and not the product of w_x (or R_x) and $(z - z_{0x})$. The same is true for $w_y(z - z_{0y})$ and $R_y(z - z_{0y})$ in Eq. (7.8b).

In the above equations, z_{0x} and z_{0y} are the z coordinates of the beam waist where the beam diameter is the smallest in the x and y directions, respectively. The functions $w_x(z)$ and $w_y(z)$ are the beam radii in the x and y directions at a position z, and $R_x(z)$ and $R_y(z)$ are the radii of curvature of the wavefront in the x and y directions, respectively. The parameters z_{Rx} and z_{Ry} in Eq. (7.9) are called Rayleigh lengths. The distance between two points away from the beam waist by z_{Rx} or by z_{Ry}, that is, the distance $2z_{Rx}$ or $2z_{Ry}$, is called the confocal parameter or the depth of focus. Note that, in Eqs. (7.7a) and (7.7b), the real parts of the coordinate shift correspond to the beam waist position, and z_{Rx} and z_{Ry}, which are the shift values in the imaginary direction, correspond to the Rayleigh lengths.

More general solutions to Eqs. (7.5a) and (7.5b) are given by the following equations:

$$f_x(x, z) = f_1(x, z)H_m\left(\frac{\sqrt{2}x}{w_x(z - z_{0x})}\right), \tag{7.13a}$$

$$f_y(y, z) = f_2(y, z)H_n\left(\frac{\sqrt{2}y}{w_y(z - z_{0y})}\right), \tag{7.13b}$$

Table 7.1 Hermite polynomials.

m	$H_m(\xi)$
0	1
1	2ξ
2	$4\xi^2 - 2$
3	$8\xi^3 - 12\xi$
4	$16\xi^4 - 48\xi^2 + 12$
5	$32\xi^5 - 160\xi^3 + 120\xi$
6	$64\xi^6 - 480\xi^4 + 720\xi^2 - 120$

where $H_m(\xi)$ is a Hermite polynomial of order m (see Table 7.1). In this case, Eqs. (7.12a) and (7.12b) are replaced by the following equations:

$$\varphi_m(z) = \left(m + \frac{1}{2}\right)\tan^{-1}\left(\frac{z}{z_{Rx}}\right), \tag{7.14a}$$

$$\varphi_n(z) = \left(n + \frac{1}{2}\right)\tan^{-1}\left(\frac{z}{z_{Ry}}\right). \tag{7.14b}$$

The function expressed by Eq. (7.13) is called the Hermite–Gaussian function.

From Eq. (7.1), the general form of the solution for the Hermite–Gaussian beam when $f(x, y, z) = f_x(x, z)\, f_y(y, z)$ is shown below, where the functions $f_x(x, z)$ and $f_y(y, z)$ are given by Eqs. (7.13a) and (7.13b), respectively:[4,8]

$$E(x, y, z) = Af_x(x, z)f_y(y, z)\exp(i\omega t - ikz). \tag{7.15}$$

A so-called Gaussian beam is a Hermite–Gaussian beam in the fundamental mode ($m = n = 0$). Note that $w_x(z)$ and $w_y(z)$ represent the beam radii only when $m = n = 0$, and the beam radii in higher-order modes are different as described in Section 7.2.2.

Let the sum of phase terms given by Eqs. (7.14a) and (7.14b) be $\varphi_{mn}(z)$:

$$\varphi_{mn}(z) = \varphi_m(z) + \varphi_n(z). \tag{7.16}$$

This corresponds to the phase difference with a plane wave propagating in the z direction and is called the Gouy phase. As can be seen from Eqs. (7.14a) and (b), $\varphi_{mn}(z) = (m + n + 1)\pi/2$ at $z \to +\infty$ and $\varphi_{mn}(z) = -(m + n + 1)\pi/2$ at $z \to -\infty$.

Commonly, assuming that the beam waist positions in both the x and y directions are at $z = 0$, $z_{0x} = z_{0y} = 0$ is often set in Eqs. (7.8a) and (7.8b). However, since the beam waist position differs in beams with astigmatism, it is necessary to use a generalized equation such as Eq. (7.8).

Figure 7.1 shows examples of the amplitude distribution of the Hermite–Gaussian function expressed by Eq. (7.13) excluding the phase term. As can

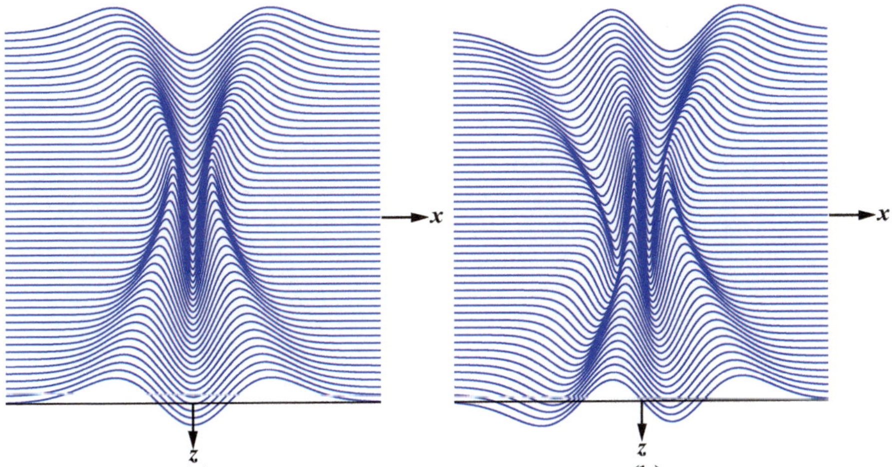

(a) (b)

Figure 7.1 Amplitude distribution change in the z direction for the Hermite–Gaussian beam (excluding the phase term). (a) $m = 2$ and (b) $m = 3$.

be seen from Table 7.1, this is an even function of x when m is an even number, and it is an odd function of x when m is an odd number.

Figure 7.2 shows the z dependence of the intensity distributions ($= |f_x(x,z)|^2$) for the Hermite–Gaussian beams. The highlighted area shows the z dependence of the beam diameter, the definition of which will be described later. As can be seen from the figure, the higher the mode, the larger the beam waist diameter and the larger the beam divergence angle.

7.2.2 Definition of the beam width

The z dependence of the beam diameter was shown in Fig. 7.2 without explanation for the definition of the beam diameter. The beam widths (full widths) or beam diameters $d_{\sigma x}$ and $d_{\sigma y}$ are defined below using the standard deviations σ_x and σ_y:

$$d_{\sigma x}(z) = 4\sigma_x(z), \tag{7.17a}$$

$$d_{\sigma y}(z) = 4\sigma_y(z), \tag{7.17b}$$

$$\sigma_x^2 = \frac{\int_{-\infty}^{\infty} x^2 |f_x(x,z)|^2 dx}{\int_{-\infty}^{\infty} |f_x(x,z)|^2 dx}, \tag{7.18a}$$

$$\sigma_y^2 = \frac{\int_{-\infty}^{\infty} y^2 |f_y(y,z)|^2 dy}{\int_{-\infty}^{\infty} |f_y(y,z)|^2 dy}. \tag{7.18b}$$

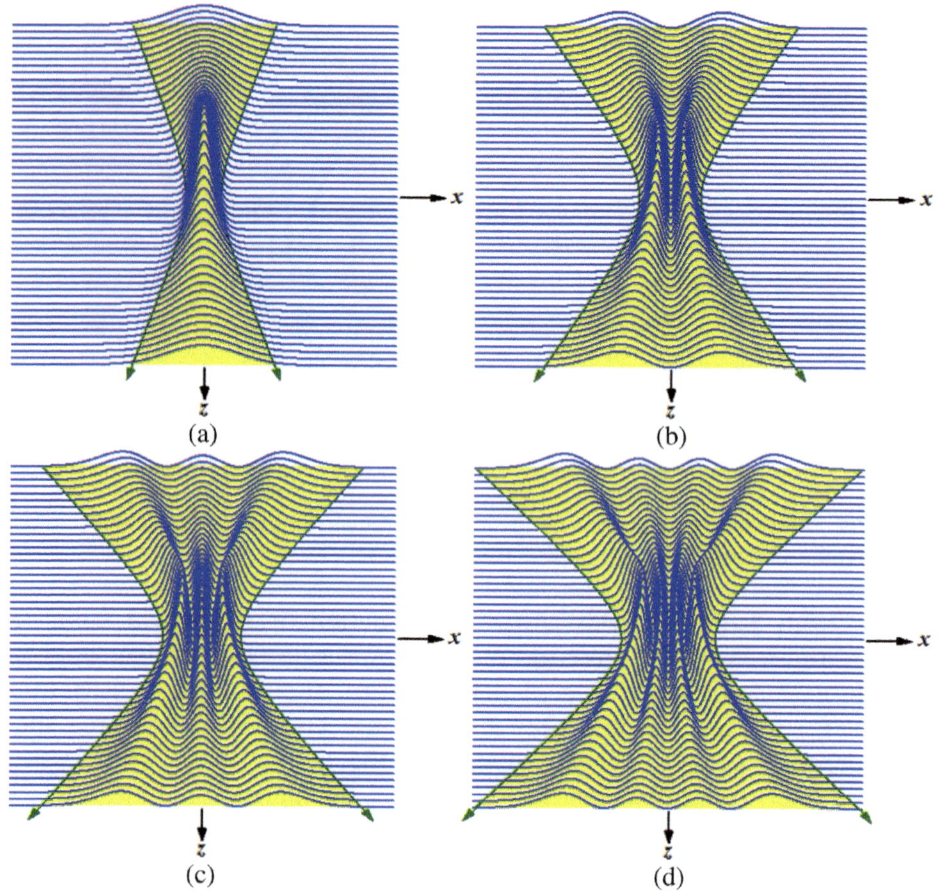

Figure 7.2 The z dependence of the intensity distributions for the Hermite–Gaussian beams. (a) $m=0$, (b) $m=1$, (c) $m=2$, and (d) $m=3$.

To be exact,

$$\sigma_x^2 = \frac{\int_{-\infty}^{\infty} (x-\bar{x})^2 |f_x(x,z)|^2 dx}{\int_{-\infty}^{\infty} |f_x(x,z)|^2 dx}, \ \bar{x} = \frac{\int_{-\infty}^{\infty} x|f_x(x,z)|^2 dx}{\int_{-\infty}^{\infty} |f_x(x,z)|^2 dx}. \tag{7.18c}$$

Since $f_x(x, z)$ is an odd or even function, $\bar{x}=0$. Therefore, σ_x^2 is given by Eq. (7.18a). The same goes for σ_y^2.

Substitute Eqs. (7.13a) and (7.13b) into Eqs. (7.18a) and (7.18b), and apply the orthogonality of the Hermite polynomials given by Eq. (7.19) and the recurrence formula of Eq. (7.20):

$$\int_{-\infty}^{\infty} H_m(\xi)H_n(\xi)\exp(-\xi^2)d\xi = \delta_{mn}\sqrt{\pi}2^m m!, \tag{7.19}$$

$$\xi H_m(\xi) = \frac{1}{2}H_{m+1}(\xi) + mH_{m-1}(\xi). \tag{7.20}$$

Then the following equations are obtained:

$$\sigma_x^2 = \frac{1}{2}\left(m + \frac{1}{2}\right)\{w_x(z)\}^2, \tag{7.21a}$$

$$\sigma_y^2 = \frac{1}{2}\left(n + \frac{1}{2}\right)\{w_y(z)\}^2. \tag{7.21b}$$

Therefore, from Eq, (7.17), the beam widths are expressed by the following equations:

$$d_{\sigma x} = 2\sqrt{2m+1}\,w_x(z), \tag{7.22a}$$

$$d_{\sigma y} = 2\sqrt{2n+1}\,w_y(z). \tag{7.22b}$$

From Eq. (7.22a), the beam width of a Gaussian beam in the fundamental mode ($m = 0$) is given by the following equation:

$$d_{\sigma x} = 2w_x(z). \tag{7.23}$$

This is consistent with the definition of $w_x(z)$ as the beam radius in Eq. (7.9).

On the other hand, as can be seen from Eq. (7.22a), the beam width for the mth-order Hermite–Gaussian beam is $\sqrt{2m+1}$ times the fundamental mode. Therefore, for example, the beam diameter of the third-order Hermite–Gaussian beam in Fig. 7.2(d) is $\sqrt{7}$ times that of the Gaussian beam in Fig. 7.2(a).

7.2.3 Beam divergence angle

In the same way as the beam width, the beam divergence angle also needs to be defined. Since the beam divergence angle is determined from the distribution of the far-field pattern, it can be calculated from the Fourier components expressed by the following equation:

$$k_x = k \sin\theta. \tag{7.24}$$

Let the k_x component of $f_x(x, z)$ be expressed in the following form:

$$f_{kx}(x, z) = g(x, z)\exp(-ik_x x). \tag{7.25}$$

If k_x is expressed as an operator,

$$\hat{k}_x = i\frac{\partial}{\partial x}. \tag{7.26}$$

From this, σ_{kx} is given by the following equation:†

†Note: To be precise, $\bar{k}_x$ must be taken into consideration, but since the average value of $|k_x|$ is 0, σ_{kx}^2 is expressed as Eq. (7.27).

$$\sigma_{kx}^2 = \frac{\int_{-\infty}^{\infty} f_x^*(x,z)\left(i\frac{\partial}{\partial x}\right)^2 f_x(x,z)dx}{\int_{-\infty}^{\infty} |f_x(x,z)|^2 dx} = \frac{\int_{-\infty}^{\infty} \left(i\frac{\partial}{\partial x}\right)^* f_x^*(x,z)\left(i\frac{\partial}{\partial x}\right) f_x(x,z)dx}{\int_{-\infty}^{\infty} |f_x(x,z)|^2 dx} = \frac{\int_{-\infty}^{\infty} \left|\frac{\partial f_x(x,z)}{\partial x}\right|^2 dx}{\int_{-\infty}^{\infty} |f_x(x,z)|^2 dx}. \tag{7.27}$$

The derivative of Eq. (7.13a) is given by

$$\frac{\partial f_x}{\partial x} = \sqrt{\frac{w_{0x}}{w_x}}\left\{\frac{\partial H_m}{\partial x} - 2x\left(\frac{1}{w_x^2} + \frac{ik}{2R_x}\right)H_m\right\} \times \exp\left[-\frac{x^2}{w_x^2} - i\frac{kx^2}{2R_x} + i\varphi_m(z)\right], \tag{7.28}$$

where

$$w_x \equiv w_x(z - z_{0x}),\ R_x \equiv R_x(z - z_{0x}). \tag{7.29}$$

Apply the recurrence formula of Eq. (7.20) and the recurrence formula regarding the differentiation of the Hermite polynomials [given by Eq. (7.30)] to Eq. (7.28):

$$\frac{dH_m(\xi)}{d\xi} = 2mH_{m-1}(\xi). \tag{7.30}$$

Then the following equation is obtained:

$$\left|\frac{\partial f_x}{\partial x}\right|^2 = \frac{2w_{0x}}{w_x}\exp\left(-\frac{2x^2}{w_x^2}\right)\left[\frac{1}{w_x^2}\left\{mH_{m-1}\left(\frac{\sqrt{2}x}{w_x}\right) - \frac{1}{2}H_{m+1}\left(\frac{\sqrt{2}x}{w_x}\right)\right\}^2 + \frac{k^2w_x^2}{4R_x^2}\left\{mH_{m-1}\left(\frac{\sqrt{2}x}{w_x}\right) + \frac{1}{2}H_{m+1}\left(\frac{\sqrt{2}x}{w_x}\right)\right\}^2\right]. \tag{7.31}$$

From Eqs. (7.27) and (7.28), and the orthogonality of Eq. (7.19), the following equation is derived:

$$\sigma_{kx}^2 = 2\left(\frac{1}{w_x^2} + \frac{k^2w_x^2}{4R_x^2}\right)\left(m + \frac{1}{2}\right). \tag{7.32}$$

Using Eqs. (7.9) to (7.11), this becomes

$$\sigma_{kx}^2 = \frac{2}{w_{0x}^2}\left(m + \frac{1}{2}\right). \tag{7.33}$$

From Eq. (7.33), the dispersion σ_{kx}^2 of the angular wave number k_x is constant regardless of the coordinate z in the propagation direction. This is reasonable considering that the angular wave number components do not change due to propagation. The beam divergence angle Θ_x (full width) in the x direction has the following relationship with σ_{kx}:[9–12]

$$\Theta_x = 4\sigma_{kx}/k. \tag{7.34}$$

From Eqs. (7.33) and (7.34),

$$\Theta_x = \frac{4}{kw_{0x}}\sqrt{2m+1} = \frac{2w_{0x}}{z_{\mathrm{R}x}}\sqrt{2m+1}. \tag{7.35}$$

From this, it is understood that the beam divergence angle is proportional to $\sqrt{2m+1}$. Considering this together with the beam width characteristics, both the beam width and the beam divergence angle for the mth-order Hermite–Gaussian beam are $\sqrt{2m+1}$ times those for the fundamental mode (see Fig. 7.2).

7.3 Gaussian Beam Characteristics

Figure 7.3 shows the relationship between the beam radius, Rayleigh length, and wavefront radius of curvature defined by Eqs. (7.9) to (7.11) for a Gaussian beam with $m = 0$. As shown in Fig. 7.3(b), the center of the radius of curvature of the wavefront is not the beam waist position and changes depending on z. As can be seen from Eq. (7.10), $R_x(z) = z$ when $z \to \infty$, and $R_x(z) \to \pm\infty$ at $z \to \pm 0$.

A curve tracing the beam diameter is often used to represent the rays of a Gaussian beam. The curve representing $w_x(z) = x$ in Eq. (7.9a) is a hyperbola given by the following equation:

$$\frac{x^2}{w_{0x}^2} - \frac{z^2}{z_{\mathrm{R}x}^2} = 1. \tag{7.36}$$

This curve is also plotted in Fig. 7.3. Although this hyperbola is commonly used to represent a Gaussian beam, the curve is not a ray perpendicular to the isophase plane.

From Eqs. (7.22) and (7.34), the beam diameter and beam divergence angle (both full width) for $m = 0$ are given by

$$d_{\sigma x} = 2w_x(z), \tag{7.37}$$

$$\Theta_x = \frac{4}{kw_{0x}} = \frac{2w_{0x}}{z_{\mathrm{R}x}}. \tag{7.38}$$

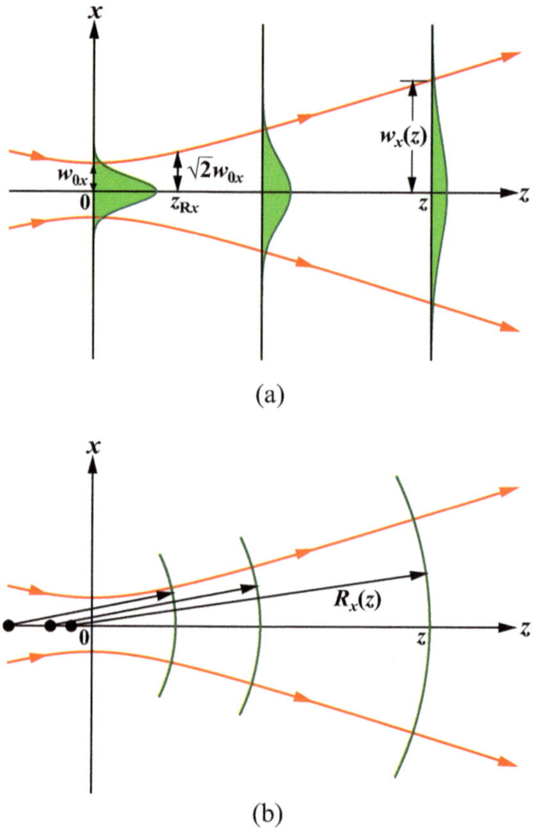

Figure 7.3 Gaussian beam: (a) beam radius and Rayleigh length and (b) radius of curvature of the wavefront.

The beam divergence angle may also be defined as given below:[13]

$$\Theta_x = \lim_{z\to\infty} \frac{d_{\sigma x}}{z}. \tag{7.39}$$

From Eqs. (7.9a), (7.37), and (7.39), Θ_x is represented by

$$\Theta_x = \lim_{z\to\infty} \frac{2w_{0x}\sqrt{1 + z^2/z_{Rx}^2}}{z} = \frac{2w_{0x}}{z_{Rx}}. \tag{7.40}$$

As a matter of course, Θ_x given by Eq. (7.40) agrees with Eq. (7.38).

One method for measuring the beam divergence angle is to use a Fourier transform as shown in Fig. 7.4.[12,13] The amplitude distribution at the focal length f of the lens corresponds to the Fourier component of the incident light amplitude; that is, it represents the far-field image. The intensity distribution of this far-field image does not depend on the beam waist position of the incident light. Therefore, if the beam diameter measured on this plane is $d_{\sigma f}$, the beam divergence angle Θ_σ is given by the following equation:

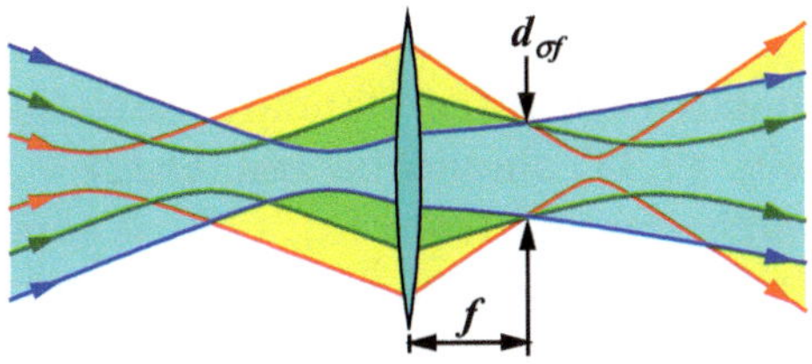

Figure 7.4 Measurement of beam divergence angle.

$$\Theta_\sigma = \frac{d_{\sigma f}}{f}. \tag{7.41}$$

Side Topic: Paraxial approximation and diffraction limit

As described in the main text, the Gaussian beam equation is a solution to the paraxial Helmholtz equation. Therefore, it is expected that as the beam divergence angle increases, the degree of approximation will deteriorate. Figure 7.S1 shows the equiphase of the wavefront expressed by Eq. (7.15). Figure 7.S1(a) is a relatively simple equiphase surface when $w_{0x} = \lambda$. On the other hand, Fig. 7.S1(b) shows the case where $w_{0x} = \lambda/2$, and the equiphase surface becomes distorted in the small z region. Although this is an exact solution to Eq. (7.3), it is thought that the influence of the paraxial approximation appears in such a case of a small beam waist width.

As can be seen from Eq. (7.34), the beam divergence angle increases as the beam waist width w_{0x} is reduced. Note that w_{0x} in Eq. (7.8) can be made as small as desired. Therefore, according to the formula, there is no diffraction limit for the spot diameter. On the other hand, when $m = 0$ in Eq. (7.34), $w_{0x} \cdot \Theta_x = 4/k =$ constant, which is an equation that expresses the diffraction limit. Even though they start from the same equation, why does one have no diffraction limit, while the other one has a diffraction

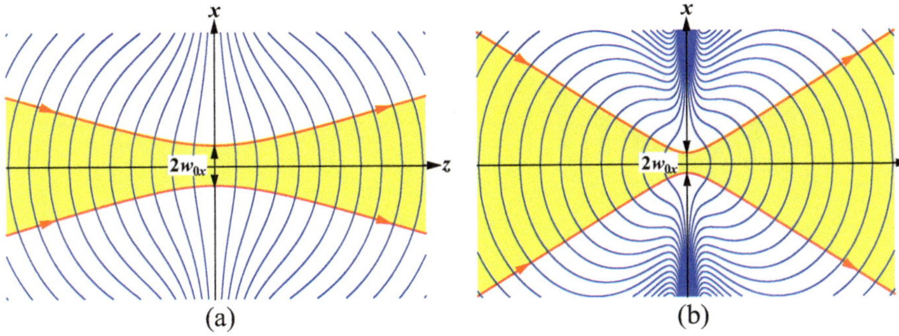

Figure 7.S1 Equiphase surfaces of a Gaussian beam expressed by Eq. (7.15). (a) $w_{0x} = \lambda$ and (b) $w_{0x} = \lambda/2$.

limit? This is due to the paraxial approximation represented by $\theta \cong \sin\theta \cong \tan\theta$. In Eq. (7.32), the standard deviation of $k_x(=k\sin\theta)$ is set to be σ_{kx}, so Eq. (7.33) should be $k\sin(\Theta_x/2) = 2\sigma_{kx}$. However, the beam divergence angle derived from Eq. (7.8) is $\tan(\Theta_x/2) = 2/(kw_{0x})$ as shown in Eq. (7.38). Since it is a [tan] function, it can take any large value, so in the formula, the beam waist width w_{0x} can be made as small as desired. The diffraction limit will be discussed again in Chapter 9 on laser beam quality.

Side Topic: Rigorous solutions of Gaussian beams

As described in the main text, the wave equation is invariant with respect to the coordinate shift.[5–7] A complex coordinate shift given by Eq. (7.7) is also applicable to the rigorous solutions of cylindrical waves in Chapter 2 and spherical waves in Chapter 3. High-divergence-angle Gaussian beam solutions with no singular points are available by this method.[14–17] Analytical expressions of vectorial Gaussian beams as rigorous solutions of Maxwell's equations are obtained by a combined imaginary-shifted-wave method.[15–17] In this method, converging and diverging imaginary-shifted waves are combined, resulting in the elimination of the singular point that existed in the original spherical wave. This method is also applicable to two-dimensional Gaussian beams in optical waveguides.[18]

Figure 7.S2 shows calculated examples of the rigorous solution for Gaussian beams. As shown in Fig. 7.S2(b), this method can also be applied to a Gaussian beam with a small Rayleigh length z_R, i.e., a large divergence angle.

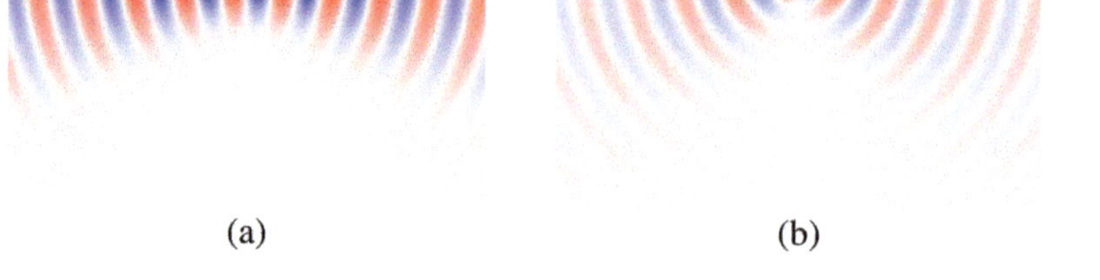

Figure 7.S2 Rigorous solutions of Gaussian beams. (a) $z_R = 2\lambda$ and (b) $z_R = 0.5\lambda$.

7.4 Conclusion

The characteristics of Hermite–Gaussian beams and the definitions of beam width and beam divergence angle for Gaussian beams were described. Note that the equation for Gaussian beams is based on the paraxial approximation. Therefore, it has a limitation for application to high-divergence-angle beams, as described in the first side topic. Rigorous and analytical solutions for Gaussian beams, including high-divergence-angle beams, higher-order modes, and vectorial solutions, have been reported in Refs. 15–18. In the next chapter, Laguerre–Gaussian beams and Gaussian beam imaging by lenses will be discussed.

Acknowledgment

The author would like to acknowledge Dr. Shojiro Kawakami, Professor Emeritus of Tohoku University, for his guidance and discussions on the rigorous solutions of Gaussian beams.

References

1. H. Kogelnik, "On the propagation of Gaussian beams of light through lenslike media including those with a loss or gain variation," *Appl. Opt.* **4**(12), 1562–1569 (1965).
2. H. Kogelnik and T. Li, "Laser beams and resonators," *Appl. Opt.* **5**(10), 1550–1567 (1966).
3. J. Alda, "Laser and Gaussian beam propagation and transformation," in *Encyclopedia of Optical Engineering*, R. Driggers, ed., 999–1013, Marcel Dekker, New York (2003).
4. A. Yariv and P. Yeh, *Photonics: Optical Electronics in Modern Communications, 6th Ed.*, Oxford University Press (2006).
5. H. A. Haus, Chapter 5, *Waves and Fields in Optoelectronics*, Prentice-Hall (1983).
6. H. Kogelnik, *Research Notebook at Bell Telephone Laboratories*, Sept. 1964 (unpublished).
7. H. Kogelnik, Invited talk at 6th International Quantum Electronics Conference, Kyoto (1970).
8. K. Iga and G. Hatakoshi, Chapter 5, *Treasure Microbox of Optoelectronics I*, Advanced Communication Media (2018) [in Japanese].
9. A. E. Siegman, "New developments in laser resonators," *Proc. SPIE* **1224**, 2–14 (1990) [doi: 10.1117/12.18425].
10. T. Taira, "Concept for measuring laser beam-quality parameters," *Rev. Laser Engineering* **26**(10), 723–727 (1998) [in Japanese].

11. G. Hatakoshi, "Analysis of beam quality factor for semiconductor lasers," *Opt. Review* **10**(4), 307–314 (2003).
12. G. Hatakoshi, "Standards for optical properties and measurements of semiconductor lasers," *O plus E* **30**(9), 953–958 (2008) [in Japanese].
13. ISO 11146-1: 2021 "Lasers and laser-related equipment—Test methods for laser beam widths, divergence angles and beam propagation ratios—Part 1: Stigmatic and simple astigmatic beams," ISO (2021).
14. C. J. R. Sheppard and S. Saghafi, "High aperture Gaussian beams," *Proc. SPIE* **3572**, 57–61 (1999) [doi: 10.1117/12.358429].
15. S. Kawakami, "On Gaussian light beams and an extension: magic of complex coordinates," *J. IEICE Japan* **99**, 260–264 (2016) [in Japanese].
16. G. Hatakoshi and S. Kawakami, "Imaginary-shifted expression of Gaussian beams as rigorous solutions of Maxwell equations," *Jpn. J. Appl. Phys.* **63**(3), 03SP30 (2024).
17. S. Kawakami and G. Hatakoshi, "Analytical solution of light beams with its applications," *Proc. 49th Optical Symp. Japan*, 36–39 (2024) [in Japanese].
18. G. Hatakoshi and S. Kawakami, "Rigorous solutions without paraxial approximation for Gaussian beams in planar optical waveguides," *Jpn. J. Appl. Phys.* **64**, 062001 (2025).

Chapter 8
Gaussian Beam Part 2

8.1 Introduction

In the previous chapter, Hermite–Gaussian beams and the beam width and beam divergence angle of Gaussian beams were explained. This chapter describes the solution of the Laguerre–Gaussian mode obtained when variables are separated in a cylindrical coordinate system and the transformation of a Gaussian beam through optical systems.

8.2 Hermite–Gaussian Beams and Laguerre–Gaussian Beams

Set the electric field of a light wave propagating in the z direction as follows:

$$E = Af(x, y, z)\exp(-ikz + i\omega t). \tag{8.1}$$

As mentioned in the previous chapter, the Hermite–Gaussian beam solution was obtained by solving the paraxial Helmholtz equation below:

$$\frac{\partial^2 f}{\partial x^2} + \frac{\partial^2 f}{\partial y^2} = 2ik\frac{\partial f}{\partial z}. \tag{8.2}$$

At first, the Hermite–Gaussian beams are reviewed again, and then the formula for the Laguerre–Gaussian beams using a cylindrical coordinate system is introduced.

8.2.1 Hermite–Gaussian beams

Separate the variables of Eq. (8.1) as follows:

$$E(x, y, z, t) = Af_x(x, z)f_y(y, z)\exp(i\omega t - ikz). \tag{8.3}$$

Since $f_x(x, z)$ and $f_y(y, z)$ are similar formulas, only the formula for $f_x(x, z)$ will be represented again in the following equation. Here, the z coordinate of the beam waist is set as $z_{0x} = 0$.

$$f_x(x,z) = \sqrt{\frac{w_{0x}}{w_x(z)}} H_m\left(\frac{\sqrt{2}x}{w_x(z)}\right) \exp\left[-\frac{x^2}{\{w_x(z)\}^2} - i\frac{kx^2}{2R_x(z)} + i\varphi_m(z)\right], \quad (8.4)$$

$$w_x(z) = w_{0x}\sqrt{1 + z^2/z_{\mathrm{R}x}^2}, \quad (8.5)$$

$$R_x(z) = z(1 + z_{\mathrm{R}x}^2/z^2), \quad (8.6)$$

$$z_{\mathrm{R}x} = kw_{0x}^2/2 = \pi w_{0x}^2/\lambda, \quad (8.7)$$

$$\varphi_m(z) = \left(m + \frac{1}{2}\right)\tan^{-1}\left(\frac{z}{z_{\mathrm{R}x}}\right), \quad (8.8)$$

where $\varphi_m(z)$ is the x-component of the Gouy phase introduced in Chap. 7 [see Eqs. (7.14) and (7.16)].

A new parameter $q_x(z)$ defined by the following equation[1,2] is introduced:

$$\frac{1}{q_x(z)} \equiv \frac{1}{R_x(z)} - i\frac{2}{k\{w_x(z)\}^2}. \quad (8.9)$$

This $q_x(z)$ will be used later when the Gaussian beam is converted by the ray matrix. Using Eq. (8.9), Eq. (8.4) is converted to the following equation:

$$f_x(x,z) = \sqrt{\frac{w_{0x}}{w_x(z)}} H_m\left(\frac{\sqrt{2}x}{w_x(z)}\right) \exp\left[-i\frac{kx^2}{2q_x(z)} + i\varphi_m(z)\right]. \quad (8.10)$$

Using Eqs. (8.5) to (8.7), $q_x(z)$ in Eq. (8.9) becomes a simple formula as shown below:

$$q_x(z) = z + iz_{\mathrm{R}x}. \quad (8.11)$$

Note that this equation is very similar to Eq. (7.7) in the previous chapter.

Figure 7.1 in the previous chapter showed the amplitude distributions of the Hermite–Gaussian beams. Note that the distribution excluded the phase terms. Figure 8.1 shows the amplitude distributions including the phase terms. The function H_n in the y direction is assumed to be the fundamental mode ($n = 0$), and the Gouy phase $\varphi_n(z) = (1/2)\tan^{-1}(z/z_{\mathrm{R}y})$ in the y direction is included in the phase terms in Fig. 8.1. The Hermite–Gaussian mode in which the orders of the Hermite polynomials in the x and y directions are m and n, respectively, is expressed as TEM_{mn}.

Now that the plot for the amplitude distributions including the phase terms is obtained, the next point of interest is the behavior of the propagation. For this purpose, the Excel animations that have been introduced several times (see Appendix 8A) can be used. It is quite interesting to see the animation of the propagation of higher-order modes.

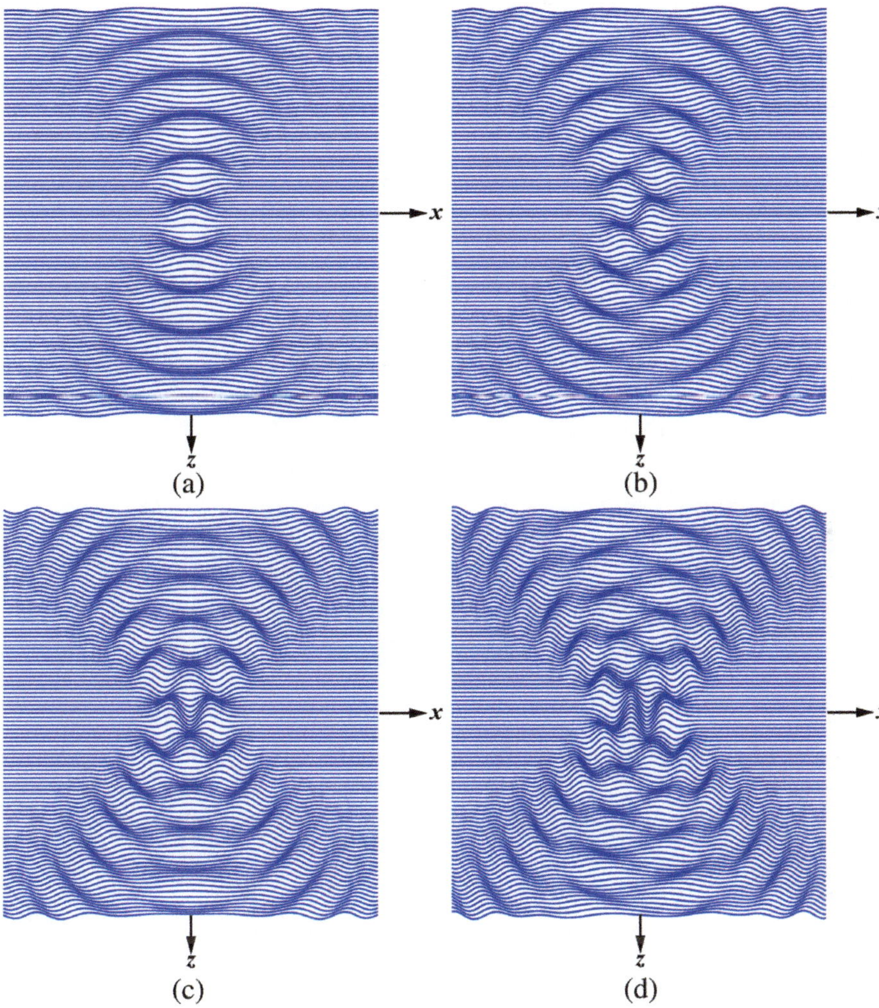

Figure 8.1 z dependence of the amplitude distributions of Hermite–Gaussian beams. (a) $m = 0$ (TEM_{00}), (b) $m = 1$ (TEM_{10}), (c) $m = 2$ (TEM_{20}), and (d) $m = 3$ (TEM_{30}).

Figure 8.2 shows examples of the amplitude distributions in a cross section (xy plane) perpendicular to the propagation direction. The Excel animation for this also allows you to see the time-dependent change in phases of the Gaussian beams. In the region $z > 0$, the isophase plane moves outward from the center. On the other hand, in the region $z < 0$, the equiphase surface moves from the outside toward the center. This is because the sign of the wavefront radius of curvature in Eq. (8.6) corresponds to the sign of z.

Examples of the intensity distributions in the xy cross section are shown in Fig. 8.3. These are commonly seen diagrams. As can be seen from this figure, m and n in TEM_{mn} correspond to the number of nodes in the x and y

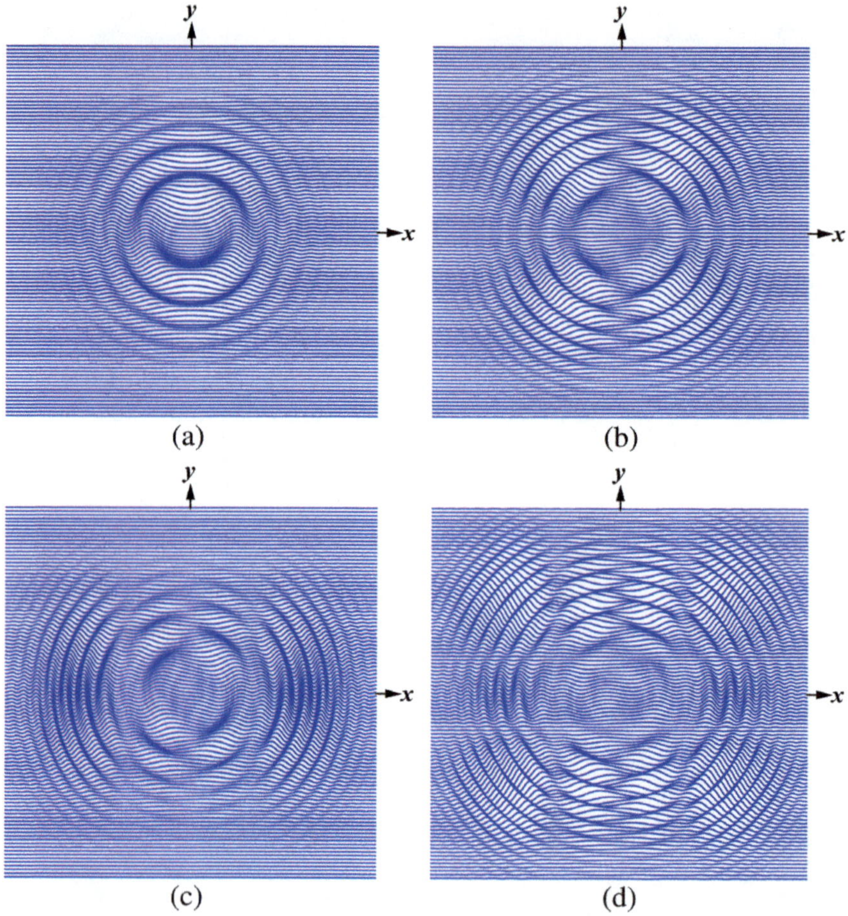

Figure 8.2 Amplitude distributions of Hermite–Gaussian beams in the *xy* cross section. (a) TEM_{00}, (b) TEM_{11}, (c) TEM_{30}, and (d) TEM_{32}.

directions, respectively. Note that in Fig. 8.3, the gradation is not linear with respect to the actual intensity, for ease of viewing.

8.2.2 Laguerre–Gaussian beams

The second differential appeared on the left side of Eq. (8.2) and can be expressed in cylindrical coordinates (r, ϕ, z) as follows:

$$\frac{\partial^2 f}{\partial r^2} + \frac{1}{r}\frac{\partial f}{\partial r} + \frac{1}{r^2}\frac{\partial^2 f}{\partial \phi^2} = 2ik\frac{\partial f}{\partial z}. \tag{8.12}$$

Consider the case where $z_{0x} = z_{0y} = 0$ and $z_{Rx} = z_{Ry}$ in Eqs. (7.8a) and (7.8b) in the previous chapter, and let $f_r(z) = f_1(x, z)\, f_2(y, z)$; then $f_r(z)$ is given by the following equation:

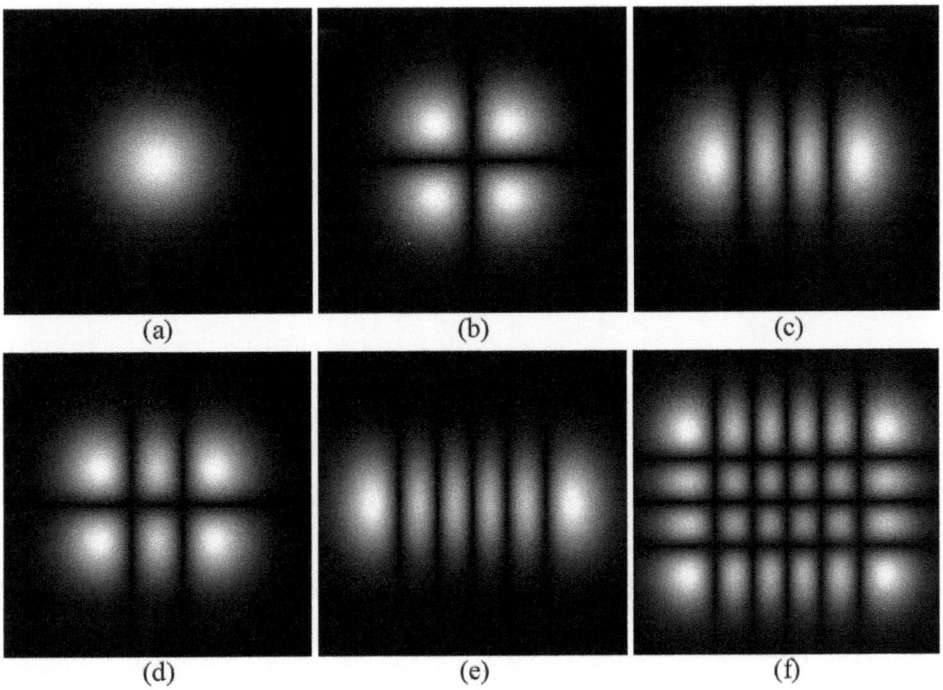

Figure 8.3 Examples of Hermite–Gaussian modes. (a) TEM_{00}, (b) TEM_{11}, (c) TEM_{30}, (d) TEM_{21}, (e) TEM_{50}, and (f) TEM_{53}.

$$f_r(z) = \frac{w_0}{w(z)} \exp\left[-\frac{r^2}{\{w(z)\}^2} - i\frac{kr^2}{2R(z)} + i\varphi(z)\right], \tag{8.13}$$

where

$$w(z) = w_0\sqrt{1 + z^2/z_R^2}, \tag{8.14}$$

$$R(z) = z(1 + z_R^2/z^2), \tag{8.15}$$

$$z_R = kw_0^2/2 = \pi w_0^2/\lambda. \tag{8.16}$$

Naturally, Eq. (8.13) satisfies Eq. (8.12). Note that there is no ϕ dependence; therefore, $\partial^2 f/\partial\phi^2 = 0$. Equation (8.13) multiplied by $\exp(-ikz + i\omega t)$ corresponds to the fundamental mode of the Laguerre–Gaussian mode.[2,3] In other words, the fundamental mode of the Laguerre–Gaussian mode is identical to the fundamental mode of the Hermite–Gaussian mode.

The derivation of the solutions for higher-order Laguerre–Gaussian modes[2,3] is omitted here, and only the results are shown below:

$$E_{nm}(r,\phi,z,t) = A\frac{w_0}{w(z)}\left[\frac{\sqrt{2}r}{w(z)}\right]^{|m|} L_n^{|m|}\left(\frac{2r^2}{w^2(z)}\right)$$
$$\times\,\Phi_m(\phi)\exp\left[-i\frac{kr^2}{2q(z)} - ikz + i\varphi(z) + i\omega t\right], \qquad (8.17)$$

where $L_n^m(\xi)$ are the associated Laguerre polynomials, and $q(z)$ is the same as Eq. (8.9) or Eq. (8.11). That is,

$$\frac{1}{q(z)} \equiv \frac{1}{R(z)} - i\frac{2}{k\{w(z)\}^2}, \qquad (8.18)$$

$$q(z) = z + iz_\mathrm{R}. \qquad (8.19)$$

The function $\Phi_m(\phi)$ is given by the following equation from the periodic boundary condition:

$$\Phi_m(\phi) = \cos(m\phi),\ \sin(m\phi),\ \exp(\pm im\phi),\ m\text{: integer}. \qquad (8.20)$$

The Gouy phase $\varphi(z)$ is given below:

$$\varphi(z) = (2n + |m| + 1)\tan^{-1}\left(\frac{z}{z_\mathrm{R}}\right). \qquad (8.21)$$

The associated Laguerre polynomials are given by

$$L_n^m(\xi) = \frac{1}{n!}\sum_{k=0}^{n}(-1)^k({}_n\mathrm{P}_{n-k})({}_{n+m}\mathrm{C}_{n-k})\xi^k$$
$$= \frac{1}{n!}\sum_{k=0}^{n}(-1)^k\frac{n!}{k!}\frac{(n+m)!}{(n-k)!(m+k)!}\xi^k. \qquad (8.22)$$

Note that the definition of the associated Laguerre polynomials may differ. When $m = 0$, they are called Laguerre polynomials, and $L_n^0(\xi)$ is written as $L_n(\xi)$. The associated Laguerre polynomials are expressed as the m-th derivative of this $L_n(\xi)$ as follows:

$$L_n^m(\xi) = (-1)^m\frac{d^m}{d\xi^m}L_{n+m}(\xi). \qquad (8.22b)$$

The formulas for $L_n^m(\xi)$ with $n = 0$ to 3 are shown below:[2]

$$L_0^m(\xi) = 1, \qquad (8.23)$$

$$L_1^m(\xi) = m + 1 - \xi, \qquad (8.24)$$

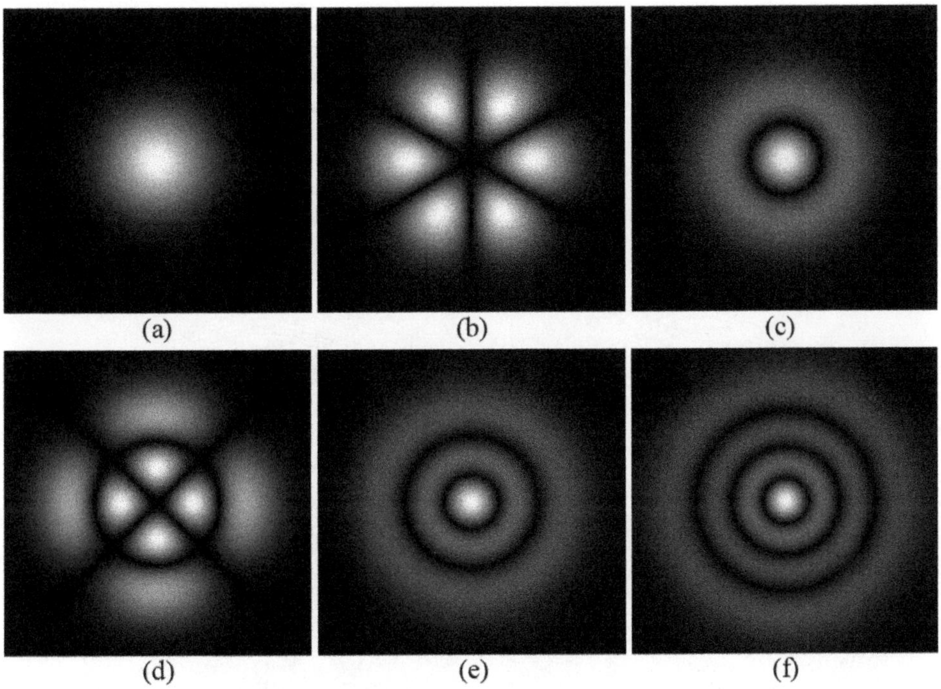

Figure 8.4 Examples of the intensity distributions of the Laguerre–Gaussian modes. (a) LG_{00}, (b) LG_{03}, (c) LG_{10}, (d) LG_{12}, (e) LG_{20}, and (f) LG_{30}.

$$L_2^m(\xi) = \frac{1}{2}\{(m+1)(m+2) - 2(m+2)\xi + \xi^2\}, \tag{8.25}$$

$$L_3^m(\xi) = \frac{1}{6}\{(m+1)(m+2)(m+3) - 3(m+2)(m+3)\xi + 3(m+3)\xi^2 - \xi^3\}. \tag{8.26}$$

Examples of the intensity distributions of the Laguerre–Gaussian modes are shown in Fig. 8.4. Similar to Fig. 8.3, the gradations are not linear with respect to the actual intensity, for ease of viewing. The LG_{00} mode with $n = m = 0$ is the same as the fundamental mode TEM_{00} of the Hermite–Gaussian mode; that is, Figs. 8.3(a) and 8.4(a) have the same distribution.

In Fig. 8.4, $\cos(m\phi)$ in Eq. (8.20) was adopted as $\Phi_m(\phi)$, which is a function of the angle ϕ direction. As shown in Figs. 8.4(b) and (d), when $m > 0$, a node appears in the angular direction, but this node does not appear when the expression $\exp(\pm im\phi)$ is used instead of $\cos(m\phi)$. In Eq. (8.20), the case for $\cos(m\phi)$ or $\sin(m\phi)$ corresponds to a standing wave in the angular direction, and the case for $\exp(\pm im\phi)$ corresponds to a wave whose phase rotates with time.†

†Note: The same is true for the cylindrical wave introduced in Chapter 2 [see Eq. (2.15)]. In Chapter 2, the case for $\exp(i\phi)$ was shown in Fig. 2.7(b).

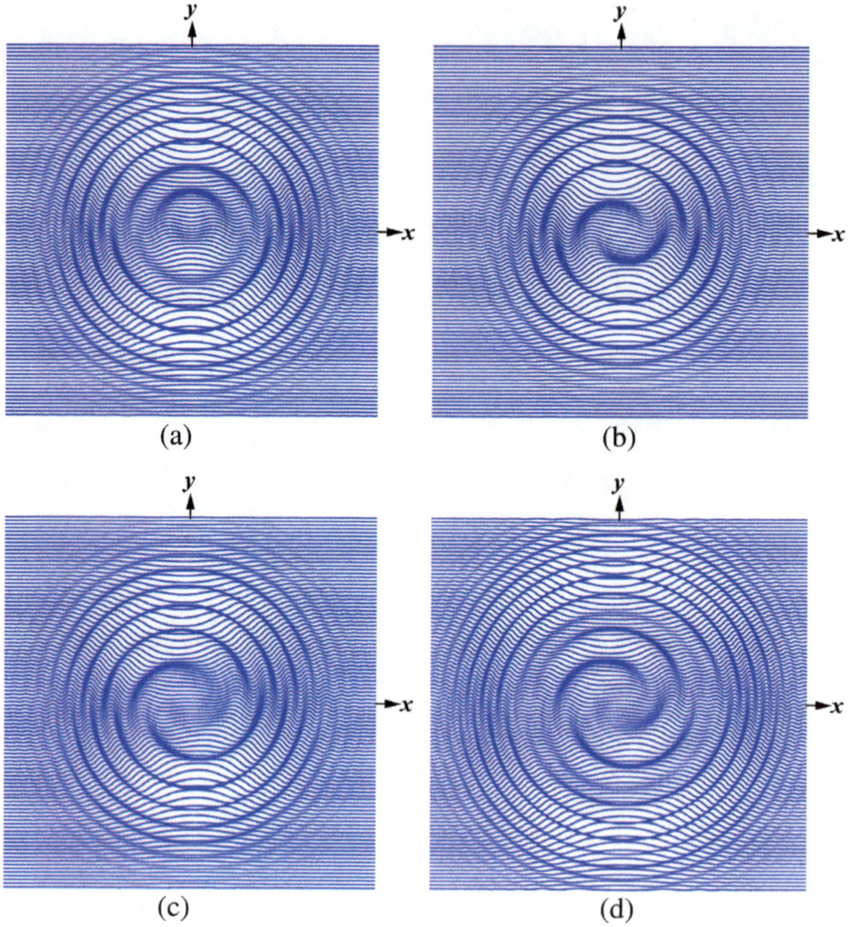

Figure 8.5 Amplitude distributions of Laguerre–Gaussian beams in the *xy* cross section. (a) LG_{10}, (b) LG_{01}, (c) LG_{02}, and (d) LG_{12}.

Similar to Fig. 8.2, Fig. 8.5 shows examples of the amplitude distributions in the *xy* cross section. In this figure, $\exp(\pm im\phi)$ is used as $\Phi_m(\phi)$. In other words, this is a mode in which the phase rotates around the *z*-axis over time.

This phase change can be clearly seen by looking at the animation in Excel. As in the Hermite–Gaussian mode (Fig. 8.2), when $z > 0$, the isophase plane moves outward from the center, and when $z < 0$, it moves from the outside toward the center.

What is interesting is the phase rotation in the modes with $m \geq 1$ [see Figs. 8.5(b), (c), and (d)]. A vortex can be seen spreading outward as it rotates (if $z < 0$, the vortex moves toward the center). It is a so-called optical vortex. As can be seen by operating the Excel animation, the direction of rotation of the vortex is reversed depending on the sign of *m*.

Side Topic: Rotation speed of an optical vortex

The phase rotation of the Laguerre–Gaussian beam is discussed in relation to the orbital angular momentum (OAM) of light. Now, what is this rotation speed? The answer is simple: the angular velocity of ϕ is ω/m. This can be derived from Eqs. (8.17) and (8.20), where, if z and r are constant in Eq. (8.17), the condition for which the phase is constant is represented as $m\phi = \omega t$. Considering the case where $m = 1$, the phase moves at a speed of $v_{\mathrm{p}} = r\omega$ at a location r away from the z-axis. Therefore, if $r > \lambda/(2\pi)$, v_{p} exceeds the speed of light. For example, if $\lambda = 0.6$ μm and $r = 100$ μm, the phase is moving at 1000 times the speed of light. There should not be any problems since this is a phase speed, but it is still worrisome that something is moving at such an incredible speed.

8.3 Gaussian Beams and the Ray Matrix

8.3.1 Ray matrix

The ray matrix is a method for describing the propagation of rays passing through optical elements.[2] The ray matrix is a 2×2 matrix that represents the transformation of the position r and the slope r' of the ray before and after passing through the optical element (see Fig. 8.6), as shown below:

$$\begin{pmatrix} r_2 \\ r_2' \end{pmatrix} = \begin{pmatrix} A & B \\ C & D \end{pmatrix} \begin{pmatrix} r_1 \\ r_1' \end{pmatrix}. \tag{8.27}$$

For example, the ray matrix of a propagation with distance d, an interface with a refractive index $n_1 \to n_2$, and a thin lens with a focal length f is expressed, respectively, as follows:

$$\boldsymbol{M}_d = \begin{pmatrix} 1 & d \\ 0 & 1 \end{pmatrix}, \tag{8.28}$$

$$\boldsymbol{M}_{n_1 \to n_2} = \begin{pmatrix} 1 & 0 \\ 0 & \dfrac{n_1}{n_2} \end{pmatrix}, \tag{8.29}$$

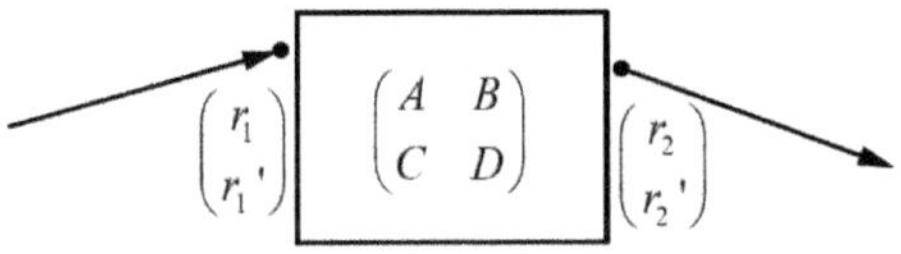

Figure 8.6 Ray matrix.

$$M_f = \begin{pmatrix} 1 & 0 \\ -\frac{1}{f} & 1 \end{pmatrix}. \tag{8.30}$$

Equations (8.29) and (8.30) are based on the paraxial approximations.

8.3.2 Transformation of a Gaussian beam by the ray matrix

Since the ray of light for a Gaussian beam is not a straight line even in a homogeneous medium, Eq. (8.27) cannot be used directly. For Gaussian beams, the following transformation holds true, where $q(z)$ [defined in Eq. (8.9) or Eq. (8.11)] is used:[2]

$$q_2(z) = \frac{Aq_1(z) + B}{Cq_1(z) + D}, \tag{8.31}$$

where, A, B, C, and D are matrix elements in Eq. (8.26). As can be seen from Eqs. (8.9) and (8.11), $q_1(z)$ and $q_2(z)$ are complex numbers. By comparing the real and imaginary parts of Eq. (8.31), it is possible to find out how each parameter of the Gaussian beam is transformed.

Examples of the transformation of Gaussian beam parameters by a ray matrix are shown in Fig. 8.7 and Table 8.1. For example, in a ray matrix for the propagation of distance d, as shown in Table 8.1, d is added to z, and z_R remains unchanged. This is obvious and can be confirmed in the equation by the definition of $q(z)$ in Eq. (8.11).

When a beam passes through two optical elements, calculating the product of the two ray matrices yields the ray matrix corresponding to the combined optical element.

Example 1: Propagation of distance d_1 and propagation of distance d_2

As shown in Fig. 8.7(a), if the propagation of distance d_1 and distance d_2 are performed consecutively, the corresponding ray matrix is given by

$$\begin{pmatrix} 1 & d_2 \\ 0 & 1 \end{pmatrix} \begin{pmatrix} 1 & d_1 \\ 0 & 1 \end{pmatrix} = \begin{pmatrix} 1 & d_1 + d_2 \\ 0 & 1 \end{pmatrix}. \tag{8.32}$$

Therefore, it is confirmed that the propagation is $d_1 + d_2$.

Example 2: Fourier transform with a lens

The ray matrix for a lens with a focal length f and propagation to the Fourier transform plane surface, which is the propagation with distance f (see Fig. 7.4 in the previous chapter), is given by the following ray matrix product:

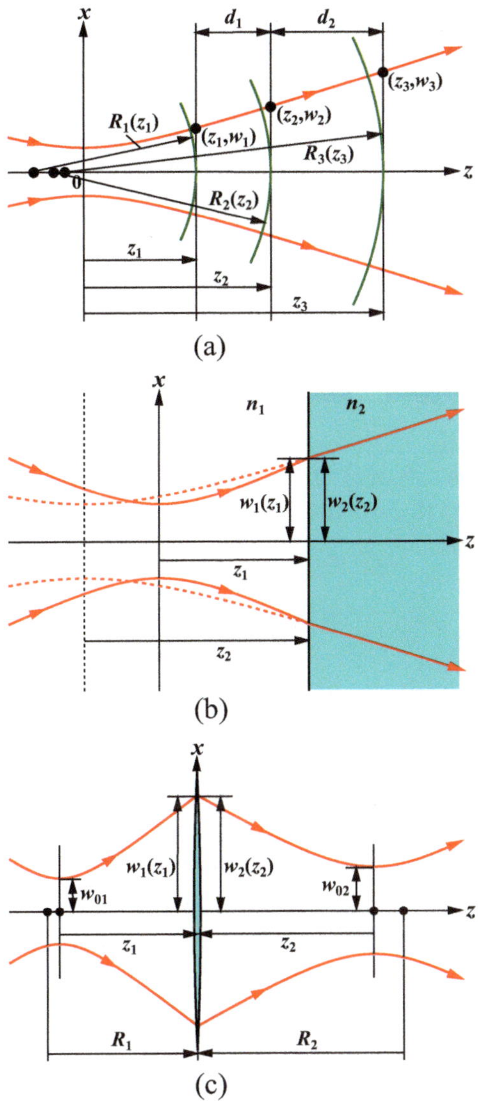

Figure 8.7 Transformation of a Gaussian beam by optical elements: (a) propagation with distance *d*, (b) interface with a refractive index $n_1 \rightarrow n_2$, and (c) a thin lens with a focal length *f*.

$$\begin{pmatrix} 1 & f \\ 0 & 1 \end{pmatrix}\begin{pmatrix} 1 & 0 \\ -1/f & 1 \end{pmatrix} = \begin{pmatrix} 0 & f \\ -1/f & 1 \end{pmatrix}. \tag{8.33}$$

From Eq. (8.31),

$$\frac{1}{q_2} = \frac{Cq_1 + D}{Aq_1 + B} = \frac{-\frac{1}{f}q_1 + 1}{f} = -\frac{1}{f^2}q_1 + \frac{1}{f}. \tag{8.34}$$

Table 8.1 Examples of transformations of Gaussian beam parameters by a ray matrix.

	Propagation with distance d	Interface with a refractive index $n_1 \rightarrow n_2$	Thin lens with a focal length f
Ray matrix	$\begin{pmatrix} 1 & d \\ 0 & 1 \end{pmatrix}$	$\begin{pmatrix} 1 & 0 \\ 0 & \frac{n_1}{n_2} \end{pmatrix}$	$\begin{pmatrix} 1 & 0 \\ -\frac{1}{f} & 1 \end{pmatrix}$
z	$z_2 = z_1 + d$	$z_2 = \frac{n_2}{n_1} z_1$	$z_2 = \frac{z_1 - (z_1^2 + z_{R1}^2)/f}{(1 - z_1/f)^2 + z_{R1}^2/f^2}$
z_R	$z_{R2} = z_{R1}$	$z_{R2} = \frac{n_2}{n_1} z_{R1}$	$z_{R2} = \frac{z_{R1}}{(1 - z_1/f)^2 + z_{R1}^2/f^2}$
$R(z)$	$R_2(z_2) = R_1(z_1 + d)$	$R_2(z_2) = \frac{n_2}{n_1} R_1(z_1)$	$\frac{1}{R_2(z_2)} = \frac{1}{R_1(z_1)} - \frac{1}{f}$
$w(z)$	$w_2(z_2) = w_1(z_1 + d)$	$w_2(z_2) = w_1(z_1)$	$w_2(z_2) = w_1(z_1)$

Using the definitions of $q(z)$ in Eqs. (8.9) and (8.11) and comparing the imaginary part of Eq. (8.34), the following equation is obtained:

$$\frac{2}{kw_2^2} = \frac{z_{R1}}{f^2}. \tag{8.35}$$

That is, the beam width w_2 at the focal plane is determined only by the Rayleigh length z_{R1} of the incident Gaussian beam, and the focal length f of the lens and does not depend on the beam waist position ($=-z_1$) of the incident Gaussian beam.

8.3.3 Thin lens formula and the Gaussian beam

The variable z appeared in the definition of $q(z)$ in Eq. (8.11) and corresponds to the distance from the beam waist position. Assume that the lens position is $z = 0$ for a thin lens. From Table 8.1, the beam waist position of the incident Gaussian beam is $z = -z_1$ [$z_1 > 0$ in Fig. 8.7(c)], and the beam waist position of the transformed Gaussian beam is $z = -z_2$ [$z_2 < 0$ in Fig. 8.7(c)].

Assuming that the beam waist positions of the Gaussian beams before and after the lens are in an imaging relationship with each other, the following equation should hold according to the thin lens formula:

$$\frac{1}{z_1} - \frac{1}{z_2} = \frac{1}{f}. \tag{8.36}$$

However, looking at Table 8.1, this is not the case. The relationship between z_1 and z_2 is expressed by the following equation, as shown in Table 8.1:

$$z_2 = \frac{z_1 - (z_1^2 + z_{R1}^2)/f}{(1 - z_1/f)^2 + z_{R1}^2/f^2}. \tag{8.37}$$

Equation (8.36) holds only when the Rayleigh length is set to $z_{R1} = 0$ in Eq. (8.37).

As can be seen from Table 8.1, the thin lens formula holds between the wavefront radii of curvatures R_1 and R_2 of the two Gaussian beams. That is,

$$\frac{1}{R_1} - \frac{1}{R_2} = \frac{1}{f}. \tag{8.38}$$

Considering that the lens is an optical element that converts the curvature of the wavefront, it is reasonable that Eq. (8.38) holds true.

In any case, as described above, the position where the beam waist of the Gaussian beam is imaged differs depending on the Rayleigh length. In other words, the beam waist position after passing through the lens changes depending on the beam divergence angle. Therefore, even if the astigmatic difference of the incident Gaussian beam is 0, an astigmatic difference will appear in the beam that is being imaged by the lens if the beam divergence angle is different in the x and y directions.[4,5] According to the international standard ISO 11146,[6,7] such beams are classified as "simple astigmatic beams." In other words, even if the astigmatic difference is 0, this is not a stigmatic beam.

8.4 Conclusion

Hermite–Gaussian beams, Laguerre–Gaussian beams, and the transformation of Gaussian beams by ray matrices have been described. The animation of the amplitude distribution (including the phase) changing over time shows the interesting behavior of this distribution.

References

1. H. Kogelnik, "On the propagation of the Gaussian beams of light through lenslike media including those with a loss or gain variation," *Appl. Opt.* **4**(12), 1562–1569 (1965).
2. A. Yariv and P. Yeh, *Photonics: Optical electronics in modern communications, 6th Ed.*, Oxford University Press (2006).
3. G. Goubau and F. Schwering, "On the guided propagation of electromagnetic wave beams," *IRE Trans. Antennas and Propagation* **9**(3), 248–256 (1961).
4. G. Hatakoshi, "Standards for optical properties and measurements of semiconductor lasers," *O plus E* **30**(9), 953–958 (2008) [in Japanese].
5. K. Iga and G. Hatakoshi, *Treasure Microbox of Optoelectronics I*, Chapter 6, Advanced Communication Media (2018) [in Japanese].
6. ISO 11146-1: 2021 "Lasers and laser-related equipment—Test methods for laser beam widths, divergence angles and beam propagation ratios—Part 1: Stigmatic and simple astigmatic beams," ISO (2021).

7. ISO 11146-2: 2021 "Lasers and laser-related equipment—Test methods for laser beam widths, divergence angles and beam propagation ratios—Part 2: General astigmatic beams," ISO (2021).

Appendix 8A Excel File

Appendix I-A contains the following Excel file. Each tab corresponds to a figure in the main text. Refer to the operation instructions for running the animation.

08_HGLG.xlsm

Instruction	
HG_xz	Fig. 8.1
HG_xy	Fig. 8.2
LG_xy	Fig. 8.5

In the above file, VBA macros are used to calculate the Hermite polynomials and the associated Laguerre polynomials. Both are simple programs; the Hermite polynomials use the recurrence formula [Eq. (7.20) in the previous chapter], and the associated Laguerre polynomials use Eq. (8.22) in this chapter.

The input parameters m and n are the orders of the Hermite polynomials or associated Laguerre polynomials. Parameter A is the amplitude coefficient; choose an appropriate value to make the diagram easier to see. The parameters used in Figs. 8.1, 8.2, and 8.5 are shown below. If z is set to a negative value in Figs. 8.2 and 8.5, the phase will move from the outside to the center. In Fig. 8.5, if m is set to a negative value, the direction of rotation of the vortex is reversed.

Fig. 8.1(a) $m = 0$, $A = 1$

Fig. 8.1(b) $m = 1$, $A = 1$

Fig. 8.1(c) $m = 2$, $A = 1$

Fig. 8.1(d) $m = 3$, $A = 1$

Fig. 8.2(a) $m = 0$, $n = 0$, $z = 6$, $A = 1$

Fig. 8.2(b) $m = 1$, $n = 1$, $z = 6$, $A = 1$

Fig. 8.2(c) $m = 3$, $n = 0$, $z = 6$, $A = 1$

Fig. 8.2(d) $m = 2$, $n = 1$, $z = 6$, $A = 1$

Fig. 8.5(a) $n = 1$, $m = 0$, $z = 6$, $A = 1$

Fig. 8.5(b) $n = 0$, $m = 1$, $z = 6$, $A = 1$

Fig. 8.5(c) $n = 0$, $m = 2$, $z = 6$, $A = 0.8$

Fig. 8.5(d) $n = 1$, $m = 2$, $z = 6$, $A = 0.4$

Chapter 9
Beam Quality of Laser Light Part 1

9.1 Introduction

The M^2 factor, which is a parameter representing the beam quality of laser light, is defined by using the formulas for beam width and beam divergence angle introduced in Chapters 7 and 8 on Gaussian beams. The M^2 factor can be applied to arbitrary paraxial beams. In this chapter, it is explained that, in the paraxial region, the beam width or beam diameter for any beam behaves the same as those for a Gaussian beam, and the M^2 factor, which is defined based on this, will be introduced.

9.2 Beam Characteristics and Beam Quality

9.2.1 Definition of beam width and beam diameter

The electric field magnitude of a light wave traveling in the z direction is expressed as the following equation:

$$E = Af(x, y, z) \exp(-ikz + i\omega t). \tag{9.1}$$

The average coordinates $\bar{x}$ and $\bar{y}$ (which are called the first-order moments) of the beam in the x and y directions (referred to as the centroid) and the standard deviations σ_x and σ_y (which are called the second-order moments) corresponding to the beam spread in each direction are expressed, respectively, by the following equations:[1,2]

$$\bar{x}(z) = \frac{\int_{-\infty}^{\infty}\int_{-\infty}^{\infty} x|f(x, y, z)|^2 \, dxdy}{\int_{-\infty}^{\infty}\int_{-\infty}^{\infty} |f(x, y, z)|^2 \, dxdy}, \tag{9.2a}$$

$$\bar{y}(z) = \frac{\int_{-\infty}^{\infty}\int_{-\infty}^{\infty} y|f(x,y,z)|^2\,dxdy}{\int_{-\infty}^{\infty}\int_{-\infty}^{\infty} |f(x,y,z)|^2\,dxdy}, \tag{9.2b}$$

$$\sigma_x^2(z) = \frac{\int_{-\infty}^{\infty}\int_{-\infty}^{\infty} (x-\bar{x})^2|f(x,y,z)|^2\,dxdy}{\int_{-\infty}^{\infty}\int_{-\infty}^{\infty} |f(x,y,z)|^2\,dxdy}, \tag{9.3a}$$

$$\sigma_y^2(z) = \frac{\int_{-\infty}^{\infty}\int_{-\infty}^{\infty} (y-\bar{y})^2|f(x,y,z)|^2\,dxdy}{\int_{-\infty}^{\infty}\int_{-\infty}^{\infty} |f(x,y,z)|^2\,dxdy}. \tag{9.3b}$$

Consider the case where the variables for the function $f(x, y, z)$ in Eq. (9.1) can be separated as follows:

$$f(x,y,z) = f_x(x,z)f_y(x,z). \tag{9.4}$$

Then Eqs. (9.2a) and (9.3a) become integrals of only x as shown below:

$$\bar{x}(z) = \frac{\int_{-\infty}^{\infty} (x-\bar{x})|f_x(x,z)|^2\,dx}{\int_{-\infty}^{\infty} |f_x(x,z)|^2\,dx}, \tag{9.5}$$

$$\sigma_x^2(z) = \frac{\int_{-\infty}^{\infty} (x-\bar{x})^2|f_x(x,z)|^2\,dx}{\int_{-\infty}^{\infty} |f_x(x,z)|^2\,dx}. \tag{9.6}$$

The same applies to Eqs. (9.2b) and (9.3b) for the y direction.

As described in Chapter 7, using σ_x and σ_y in Eq. (9.3), the beam widths $d_{\sigma x}$ and $d_{\sigma y}$ in the x and y directions, respectively, are defined below:

$$d_{\sigma x}(z) = 4\sigma_x(z), \tag{9.7a}$$

$$d_{\sigma y}(z) = 4\sigma_y(z). \tag{9.7b}$$

The beam waist widths $d_{\sigma x0}$ and $d_{\sigma y0}$ are expressed by the following equations:

$$d_{\sigma x0} \equiv d_{\sigma x}(z_{0x}) = 4\sigma_x(z_{0x}), \tag{9.8a}$$

$$d_{\sigma y0} \equiv d_{\sigma y}(z_{0y}) = 4\sigma_y(z_{0y}), \tag{9.8b}$$

where z_{0x} and z_{0y} are the z coordinates of the beam waist in the x and y directions, respectively.

Let the sum of Eqs. (9.3a) and (9.3b) be $\sigma^2(z)$:

$$\sigma^2(z) = \sigma_x^2(z) + \sigma_y^2(z). \tag{9.9}$$

Then the beam radius $w_\sigma(z)$ is given by

$$w_\sigma(z) = \sqrt{2}\sigma(z). \tag{9.10}$$

In this case, the beam diameter $d_\sigma(z)$ is

$$d_\sigma(z) = 2w_\sigma(z) = 2\sqrt{2}\sigma(z). \tag{9.11}$$

When the beam waist positions in the x and y directions coincide with each other, that is, when $z_{0x} = z_{0y} \equiv z_0$, the beam waist diameter $d_{\sigma 0}$ can be defined as follows:

$$d_{\sigma 0} = 2w_{\sigma 0} \equiv 2w_\sigma(z_0) = 2\sqrt{2}\sigma(z_0). \tag{9.12}$$

Equations (9.10) to (9.12) are used especially when the beam is axially symmetric. In the case of an axially symmetric beam, $\bar{x} = \bar{y} = 0$, so from Eqs. (9.3) and (9.9),

$$\sigma^2(z) = \frac{\int_{-\infty}^{\infty}\int_{-\infty}^{\infty}(x^2+y^2)|f(x,y,z)|^2\,dxdy}{\int_{-\infty}^{\infty}\int_{-\infty}^{\infty}|f(x,y,z)|^2\,dxdy} = \frac{\int_0^{\infty} r^3|f(r,z)|^2\,dr}{\int_0^{\infty} r|f(r,z)|^2\,dr}, \tag{9.13}$$

where

$$r = \sqrt{x^2+y^2}. \tag{9.14}$$

9.2.2 Definition of the beam divergence angle

In Chapter 7, two definitions were given for the beam divergence angle Θ_x. One of them is given by

$$\Theta_x = 4\sigma_{kx}/k, \tag{9.15}$$

$$\sigma_{kx}^2(z) = \frac{\int_{-\infty}^{\infty}\int_{-\infty}^{\infty}\left|\frac{\partial f(x,y,z)}{\partial x}\right|^2 dxdy}{\int_{-\infty}^{\infty}\int_{-\infty}^{\infty}|f(x,y,z)|^2\,dxdy}. \tag{9.16}$$

The same goes for the y direction:

$$\Theta_y = 4\sigma_{ky}/k, \tag{9.17}$$

$$\sigma_{ky}^2(z) = \frac{\displaystyle\int_{-\infty}^{\infty}\int_{-\infty}^{\infty}\left|\frac{\partial f(x,y,z)}{\partial y}\right|^2 dxdy}{\displaystyle\int_{-\infty}^{\infty}\int_{-\infty}^{\infty}|f(x,y,z)|^2\,dxdy}. \tag{9.18}$$

Another definition is as follows, which was shown in Eq. (7.33) in Chapter 7:

$$\Theta_x = \lim_{z\to\infty}\frac{d_{\sigma x}}{z - z_{0x}}, \tag{9.19a}$$

$$\Theta_y = \lim_{z\to\infty}\frac{d_{\sigma y}}{z - z_{0y}}. \tag{9.19b}$$

For beams where d_σ can be defined by Eq. (9.11), the following Θ can be used:

$$\Theta = \lim_{z\to\infty}\frac{d_\sigma}{z - z_0}. \tag{9.20}$$

It was shown in Chapter 7 that Eqs. (9.15) and (9.19a) coincide with each other for Gaussian beams.

9.2.3 Any beam is a virtual Gaussian beam

When the beam diameter $d_\sigma(z)$ and the beam divergence angle Θ are defined as in Eqs. (9.7) and (9.19), $d_\sigma(z)$ for an arbitrary beam shows the following dependence on the coordinate z of the beam propagation direction in the paraxial approximation:[3,4]

$$d_\sigma^2(z) = d_{\sigma 0}^2 + \Theta^2(z - z_0)^2. \tag{9.21}$$

This is exactly the equation that shows the z dependence of the beam diameter in a Gaussian beam [see Eq. (7.36) in Chapter 7]. In other words, every beam behaves the same as a Gaussian beam. The derivation of this formula is shown in Appendix 9.1.

In this way, by utilizing the fact that any beam behaves as a "virtual" Gaussian beam as shown in Eq. (9.21), the beam quality factor M^2 shown in Section 9.2.5 is defined.

9.2.4 What is beam quality?

Figure 9.1(a) shows the spot diameter when laser light is focused by a lens. There is a well-known formula for the spot diameter d:

$$d = \alpha\frac{\lambda_0}{NA}, \tag{9.22}$$

where λ_0 is the wavelength in vacuum, and NA is the numerical aperture, which is given by the following equation as a function of θ in Fig. 9.1(a), assuming that the refractive index of the medium is n_r:

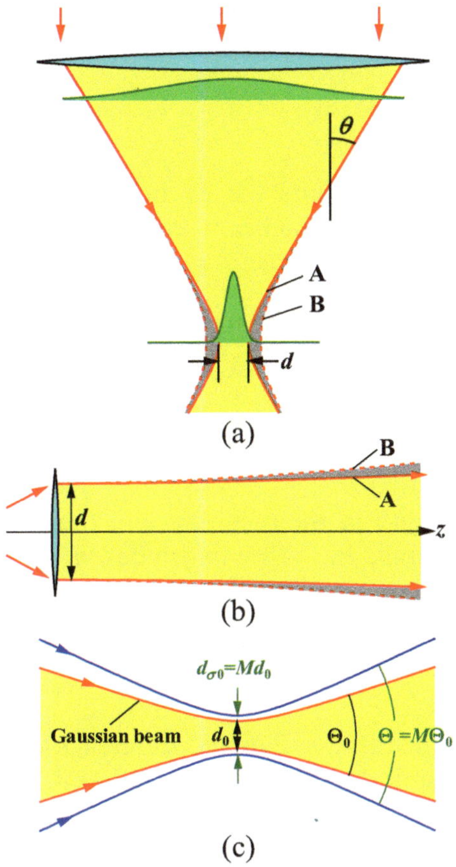

Figure 9.1 Beam quality comparison: (a) focusing of laser light, (b) collimation of laser light, and (c) comparison of Gaussian beam and general beam. The product of $d_{\sigma 0}$ and Θ is $M^2 d_0 \Theta_0$. The coefficient is M^2 factor, which will be described in Section 9.2.5.

$$NA = n_r \sin\theta. \tag{9.23}$$

The coefficient α in Eq. (9.22) differs depending on the type of beam, and in the case of a Gaussian beam, it is given by

$$\alpha = \frac{2}{\pi}. \tag{9.24}$$

Comparing beam A and beam B in Fig. 9.1(a), beam A has a smaller spot diameter. When the wavelength and NA are the same, beam A, which can be focused to a smaller spot diameter, has better beam quality than beam B.

Next, consider the case of collimating laser light as shown in Fig. 9.1(b). Collimated light cannot propagate in parallel forever; it always spreads due to diffraction. It is said that the smaller the spread, the better the beam quality. In Fig. 9.1(b), beam A has better beam quality than beam B.

Figure 9.1(c) shows a comparison between a Gaussian beam and a general beam. As described in Section 9.3, a Gaussian beam has the best beam quality among all beams.

9.2.5 M^2 factor

The M^2 factor, which represents the beam quality, is defined as a factor proportional to the product of the beam waist diameter and beam divergence angle using the following equations:[3–5]

$$M_x^2 = \frac{\pi}{\lambda}\frac{d_{\sigma 0x}\Theta_x}{4}, \tag{9.25a}$$

$$M_y^2 = \frac{\pi}{\lambda}\frac{d_{\sigma 0y}\Theta_y}{4}, \tag{9.25b}$$

where λ is the wavelength in the medium, and the relationship between the wavelength λ_0 in vacuum, the refractive index n_r of the medium, and the angular wave number k is as follows:

$$k = \frac{2\pi}{\lambda} = k_0 n_r = \frac{2\pi}{\lambda_0} n_r. \tag{9.26}$$

For a stigmatic beam with an astigmatic difference of 0, M^2 is given by the following equation, using $d_{\sigma 0}$ in Eq. (9.12) and Θ in Eq. (9.20):

$$M^2 = \frac{\pi}{\lambda}\frac{d_{\sigma 0}\Theta}{4}. \tag{9.27}$$

The coefficients of Eq. (9.25) or Eq. (9.27) are the results of normalization such that $M^2 = 1$ for a Gaussian beam. From Eqs. (9.22) and (9.24), the spot diameter for a Gaussian beam is given by

$$d = \frac{2}{\pi}\frac{\lambda_0}{NA} = \frac{2}{\pi}\frac{\lambda_0}{n_r \sin\theta} = \frac{2}{\pi}\frac{\lambda}{\sin(\Theta/2)}. \tag{9.28}$$

In the paraxial approximation,

$$\sin(\Theta/2) \cong \Theta/2. \tag{9.29}$$

Then Eq. (9.28) becomes

$$\frac{\pi}{\lambda}\frac{d\Theta}{4} = 1, \tag{9.30}$$

indicating that $M^2 = 1$.

9.3 M^2 Factors for a Hermite–Gaussian Beam and Laguerre–Gaussian Beam

Now, let's take a look at the specific value of the M^2 factor of the beam quality index defined above for Hermite–Gaussian beams and Laguerre–Gaussian beams.

9.3.1 M^2 factor for a Hermite–Gaussian beam

For Hermite–Gaussian beams, the beam diameter $d_{\sigma x}(z)$ and beam divergence angle Θ_x are given by the following formulas as indicated by Eqs. (7.22) and (7.35) in Chapter 7:

$$d_{\sigma x}(z) = 2\sqrt{2m+1}\,w_x(z), \tag{9.31}$$

$$\Theta_x = \frac{4}{kw_{0x}}\sqrt{2m+1}. \tag{9.32}$$

As mentioned in Chapter 7, these formulas are derived using the recurrence formula of the Hermite polynomials and orthogonality relations (see Appendix 9.2).

From Eq. (9.31), the beam diameter $d_{\sigma 0x}$ at the beam waist ($z = 0$) is given by

$$d_{\sigma 0x} = 2\sqrt{2m+1}\,w_{x0}. \tag{9.33}$$

Substituting Eqs. (9.32) and (9.33) into Eq. (9.25a) and using the relationship in Eq. (9.26), the following equation is obtained:

$$M_x^2 = 2m+1. \tag{9.34}$$

From Eq. (9.34), the M^2 factor of a Hermite–Gaussian beam has the smallest value of 1 in the fundamental mode ($m = 0$: so-called Gaussian beam) and increases as the mode becomes higher order. Although the proof is omitted here, the value of the M^2 factor of unity in the Gaussian beam is also the theoretical minimum value of Eq. (9.25).[3,6] In other words, a Gaussian beam with $m = 0$ has the best beam quality among all beams.

The relationship between the M^2 factor and the mode order of a Hermite–Gaussian beam is shown in Fig. 9.2.

9.3.2 M^2 factor for a Laguerre–Gaussian beam

The electric field of the Laguerre–Gaussian beam was given by Eq. (8.13) in the previous chapter, which is shown again as follows:

$$E_{nm}(r,\phi,z,t) = A\frac{w_0}{w(z)}\left[\frac{\sqrt{2}r}{w(z)}\right]^{|m|} L_n^{|m|}\left(\frac{2r^2}{w^2(z)}\right) \times \Phi_m(\phi)\exp\left[-i\frac{kr^2}{2q(z)} - ikz + i\varphi(z) + i\omega t\right], \tag{9.35}$$

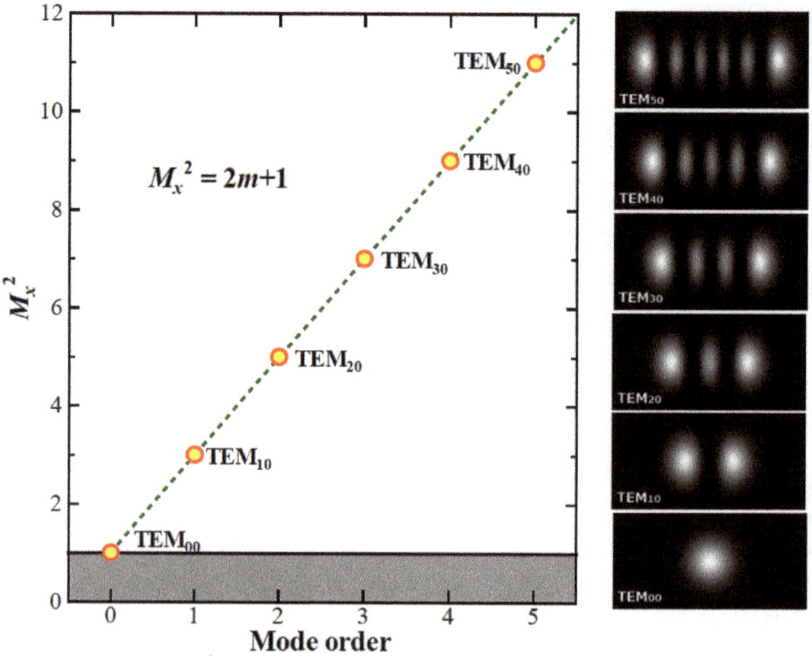

Figure 9.2 The relationship between the M^2 factor and mode order for Hermite–Gaussian beams.

$$\frac{1}{q(z)} \equiv \frac{1}{R(z)} - i\frac{2}{kw^2(z)}. \tag{9.36}$$

Considering the case where $\Phi_m(\phi) = \exp(\pm im\phi)$, $|E_{nm}|^2$ is given as follows:

$$|E_{nm}(r,z)|^2 = A^2\frac{w_0^2}{w^2(z)}\left[\frac{\sqrt{2}r}{w(z)}\right]^{2|m|}\left\{L_n^{|m|}\left(\frac{2r^2}{w^2(z)}\right)\right\}^2. \tag{9.37}$$

From Eq. (9.13), $\sigma^2(z)$ is given by

$$\sigma^2(z) = \frac{\int_0^\infty r^3|E_{nm}(r,z)|^2dr}{\int_0^\infty r|E_{nm}(r,z)|^2dr}. \tag{9.38}$$

From Eq. (9.37), the variable ξ can be defined as follows:

$$\xi \equiv \frac{2r^2}{w^2(z)}. \tag{9.39}$$

Then $|E_{nm}|^2$ is represented by

$$|E_{nm}|^2 = A^2\frac{w_0^2}{w^2(z)}\xi^{|m|}\exp(-\xi)\left\{L_n^{|m|}(\xi)\right\}^2. \tag{9.40}$$

From Eq. (9.39),

$$\frac{4rdr}{w^2(z)} = d\xi. \tag{9.41}$$

From Eqs. (9.38) to (9.41), the following equation is obtained:

$$\sigma^2(z) = \frac{w^2(z)}{2} \frac{\int_0^\infty \xi^{|m|+1} e^{-\xi} \left\{ L_n^{|m|}(\xi) \right\}^2 d\xi}{\int_0^\infty \xi^{|m|} e^{-\xi} \{ L_n^{|m|}(\xi) \}^2 d\xi}. \tag{9.42}$$

Using Eqs. (9.A44) and (9.A45) in Appendix 9C,

$$\sigma^2(z) = \frac{2n + |m| + 1}{2} w^2(z). \tag{9.43}$$

Therefore, from Eq. (9.11), the beam diameter $d_\sigma(z)$ is given by the following equation:

$$d_\sigma(z) = 2\sqrt{2}\sigma(z) = 2\sqrt{2n + |m| + 1}w(z). \tag{9.44}$$

The beam waist diameter $d_{\sigma 0}$ is given by

$$d_{\sigma 0} = 2\sqrt{2}\sigma(0) = 2\sqrt{2n + |m| + 1}w_0. \tag{9.45}$$

If $n = m = 0$ in Eq. (9.45),

$$d_{\sigma 0} = 2w_0. \tag{9.46}$$

This coincides with the beam waist diameter of the Gaussian beam (the TEM_{00} mode of the Hermite–Gaussian beam) given by Eq. (9.32) with $m = 0$.

As shown in Eq. (8.16) in the previous chapter, $w(z)$ in Eq. (9.43) is given by

$$w(z) = w_0\sqrt{1 + z^2/z_R^2}. \tag{9.47}$$

Using this and Eq. (9.20), the beam divergence angle Θ is represented by

$$\Theta = 2\sqrt{2n + |m| + 1}\frac{w_0}{z_R} = \frac{4}{k}\sqrt{2n + |m| + 1}w_0. \tag{9.48}$$

From Eq. (9.45) and Eq. (9.48), using Eq. (9.27), the M^2 factor for the Laguerre–Gaussian beam is given by the following equation:

$$M^2 = \frac{\pi}{\lambda}\frac{d_{\sigma 0}\Theta}{4} = 2n + |m| + 1. \tag{9.49}$$

Of course, when $n = m = 0$, $M^2 = 1$.

The relationship between the M^2 factor and the mode order for Laguerre–Gaussian beams is shown in Fig. 9.3.

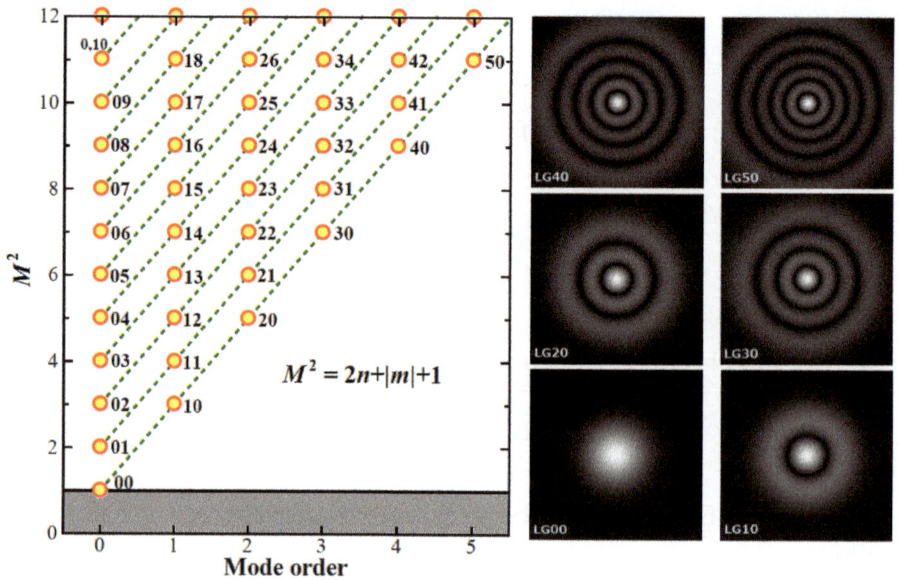

Figure 9.3 The relationship between the M^2 factor and the mode order for Laguerre–Gaussian beams. The numerical values in the figure represent the mode order in *nm*.

Side Topic: Hermite polynomials, associated Laguerre polynomials, and quantum mechanics

The Hermite polynomials and the associated Laguerre polynomials that appear in the formulas for Hermite–Gaussian beams and Laguerre–Gaussian beams are also used in functions representing the modes of optical fibers. Moreover, both are polynomials that are familiar in quantum mechanics.

Associated Laguerre polynomials appear in the function representing the radial component of the electron wave function in the hydrogen atom. The order of the associated Laguerre polynomials determines the principal quantum number and azimuthal quantum number. In addition, the associated Legendre polynomials introduced in Chapter 3 regarding spherical waves appear in the function that represents the angular component of the wave function. This is related to the azimuthal quantum number and the magnetic quantum number. In this way, hydrogen atoms and light seem to have a deep connection.

Hermite polynomials appear in the solution of the Schrödinger equation for harmonic oscillators. The energy eigenvalue of the harmonic oscillator is represented as $E_m = (m + 1/2)\hbar\omega$, and the same coefficient as the value of the M^2 factor in Eq. (9.34) appears. The zero-point energy $E_0 = (1/2)\hbar\omega$ at $m = 0$ is due to the uncertainty principle. Actually, the M^2

factor represents the uncertainty itself in position (beam width) and momentum (beam divergence angle). This is not an accidental coincidence as will be discussed again in the next chapter.

9.4 Conclusion

The M^2 factor, which is a beam quality index, was introduced and an overview of the M^2 factor for Hermite–Gaussian beams and Laguerre–Gaussian beams was described. In the next chapter, the M^2 factor for general beams and related matters will be discussed.

References

1. ISO 11146-1:2021 "Lasers and laser-related equipment—Test methods for laser beam widths, divergence angles and beam propagation ratios—Part 1: Stigmatic and simple astigmatic beams," ISO (2021).
2. G. Hatakoshi, "Standards for optical properties and measurements of semiconductor lasers," *O plus E* **30**(9), 953–958 (2008) [in Japanese].
3. A. E. Siegman, "New developments in laser resonators," *Proc. SPIE* **1224**, 2–14 (1990) [doi: 10.1117/12.18425].
4. G. Hatakoshi, "Analysis of beam quality factor for semiconductor lasers," *Opt. Review* **10**(4), 307–314 (2003).
5. T. Taira, "Concept for measuring laser beam-quality parameters," *Rev. Laser Engineering* **26**, 723–727 (1998) [in Japanese].
6. Y. Honguh, *Landscape of Wave Optics, Optical beam*, Advanced Communication Media, Tokyo (2017) [in Japanese].

Appendix 9A Derivation of Eq. (9.21)

As mentioned in the main text, if the beam width $d_\sigma(z)$ and beam divergence angle Θ are defined as in Eqs. (9.7) and (9.19), then in the paraxial approximation, $d_\sigma(z)$ shows the dependence on the beam propagation direction z as shown in Eq. (9.21) for any beam. The derivation of this equation will be briefly described below.

First, the electric field $E(x, z)$ of a light beam (assuming a paraxial ray) propagating in the z direction is expressed by the following plane wave expansion:

$$E(x,z) = \int_{-\infty}^{\infty} P(s,z)e^{-i2\pi sx}ds, \tag{9.A1}$$

where

$$2\pi s = k_x = k\sin\theta = (2\pi/\lambda)\sin\theta. \tag{9.A2}$$

$P(s, z)$ in Eq. (9.A1) is the Fourier transform of $E(x, z)$ and is expressed by the following equation:

$$P(s,z) = \int_{-\infty}^{\infty} E(x,z)e^{i2\pi sx}dx. \qquad (9.A3)$$

Note that the relationship between k_x, which appears in Eq. (9.15), and s, is as shown in Eq. (9.A2).

Here, the second-order moments for x and s are defined as follows:

$$\sigma^2(z) = \int_{-\infty}^{\infty} x^2|E(x,z)|^2dx, \qquad (9.A4)$$

$$\sigma_s^2(z) = \int_{-\infty}^{\infty} s^2|P(s,z)|^2ds. \qquad (9.A5)$$

It is assumed that the coordinates are selected so that the first-order moments (average values) for x and s are both 0, and $E(x, z)$ and $P(s, z)$ are normalized as follows:

$$\int_{-\infty}^{\infty} |E(x,z)|^2dx = \int_{-\infty}^{\infty} |P(s,z)|^2ds = 1. \qquad (9.A6)$$

Equation (9.A5) is written as a function of z, but actually σ_s^2 does not depend on z. This becomes clear when comparing $P(s, z)$ with the value at $z = z_1$ as shown below:

$$P(s,z) = P(s,z_1)e^{-ik_z(z-z_1)}, \qquad (9.A7)$$

where

$$k_z = k\cos\theta = \sqrt{k^2 - k_x^2} \approx k - \frac{1}{2}\frac{k_x^2}{k} = k - \pi\lambda s^2. \qquad (9.A8)$$

As can be seen from Eq. (9.A7), $P(s, z)$ and $P(s, z_1)$ differ only in phase and have the same amplitude.

Next, the following equations are obtained by differentiating Eqs. (9.A1) and (9.A3):

$$\frac{\partial E(x,z)}{\partial x} = -i2\pi \int_{-\infty}^{\infty} sP(s,z)e^{-i2\pi sx}ds, \qquad (9.A9)$$

$$\frac{\partial P(s,z)}{\partial s} = i2\pi \int_{-\infty}^{\infty} xE(x,z)e^{i2\pi sx}dx. \qquad (9.A10)$$

From the Fourier transform relationship in Eq. (9.A10),

$$
\begin{aligned}
xE(x,z) &= \frac{1}{i2\pi}\int_{-\infty}^{\infty}\frac{\partial P(s,z)}{\partial s}e^{-i2\pi sx}ds \\
&= \frac{1}{i2\pi}\int_{-\infty}^{\infty}\left\{\frac{\partial P(s,z_1)}{\partial s}+i2\pi\lambda s(z-z_1)P(s,z_1)\right\}e^{-ik_z(z-z_1)}e^{-i2\pi sx}ds \\
&= e^{-ik_0(z-z_1)}\left\{xE(x,z_1)+\lambda(z-z_1)\int_{-\infty}^{\infty}sP(s,z_1)e^{-i2\pi sx}ds\right\} \\
&\approx \frac{1}{i2\pi}e^{-ik_0(z-z_1)}\int_{-\infty}^{\infty}\left\{\frac{\partial P(s,z_1)}{\partial s}+i2\pi\lambda s(z-z_1)P(s,z_1)\right\}e^{-i2\pi sx}ds \\
&= e^{-ik_0(z-z_1)}\left\{xE(x,z_1)+\lambda(z-z_1)\int_{-\infty}^{\infty}sP(s,z_1)e^{-i2\pi sx}ds\right\} \\
&= e^{-ik_0(z-z_1)}\left\{xE(x,z_1)-\frac{\lambda(z-z_1)}{i2\pi}\frac{\partial E(x,z_1)}{\partial x}\right\}.
\end{aligned} \tag{9.A11}
$$

From this,

$$
\begin{aligned}
&x^2|E(x,z)|^2 \\
&\quad = x^2|E(x,z_1)|^2+\frac{\lambda(z-z_1)}{i2\pi}x\left\{E(x,z_1)\frac{\partial E^*(x,z_1)}{\partial x}-E^*(x,z_1)\frac{\partial E(x,z_1)}{\partial x}\right\} \\
&\quad + \left(\frac{\lambda}{2\pi}\right)^2(z-z_1)^2\left|\frac{\partial E(x,z_1)}{\partial x}\right|^2,
\end{aligned} \tag{9.A12}
$$

$$
\begin{aligned}
\int_{-\infty}^{\infty}x^2|E(x,z)|^2dx &= \int_{-\infty}^{\infty}x^2|E(x,z_1)|^2dx - A(z_1)(z-z_1) \\
&\quad + \left(\frac{\lambda}{2\pi}\right)^2(z-z_1)^2\int_{-\infty}^{\infty}\left|\frac{\partial E(x,z_1)}{\partial x}\right|^2dx,
\end{aligned} \tag{9.A13}
$$

$$
A(z) = i\frac{\lambda}{2\pi}\int_{-\infty}^{\infty}x\left\{E(x,z)\frac{\partial E^*(x,z)}{\partial x}-E^*(x,z)\frac{\partial E(x,z)}{\partial x}\right\}dx. \tag{9.A14}
$$

$A(z)$ in Eq. (9.A14) can also be expressed as follows using $P(s, z)$:

$$
A(z) = -i\frac{\lambda}{2\pi}\int_{-\infty}^{\infty}s\left\{P(s,z)\frac{\partial P^*(s,z)}{\partial s}-P^*(s,z)\frac{\partial P(s,z)}{\partial s}\right\}ds. \tag{9.A15}
$$

Here, the following relational expression was used:

$$
\begin{aligned}
\int_{-\infty}^{\infty}xE(x,z)\frac{\partial E^*(x,z)}{\partial x}dx &= i2\pi\int_{-\infty}^{\infty}\int_{-\infty}^{\infty}xE(x,z)sP^*(s,z)e^{i2\pi sx}dxds \\
&= \int_{-\infty}^{\infty}sP^*(s,z)\frac{\partial P(s,z)}{\partial s}ds.
\end{aligned} \tag{9.A16}
$$

Similarly,

$$\int_{-\infty}^{\infty} xE^*(x,z)\frac{\partial E(x,z)}{\partial x}dx = \int_{-\infty}^{\infty} sP(s,z)\frac{\partial P^*(s,z)}{\partial s}ds. \tag{9.A17}$$

Transforming the integral of the last term on the right side of Eq. (9.A13), the following equation is obtained:

$$\begin{aligned}\left|\frac{\partial E(x,z_1)}{\partial x}\right|^2 dx &= -i2\pi\int_{-\infty}^{\infty}\int_{-\infty}^{\infty} sP(s,z_1)\frac{\partial E^*(x,z_1)}{\partial x}e^{-i2\pi sx}dxds \\ &= (2\pi)^2\int_{-\infty}^{\infty} s^2P(s,z_1)P^*(s,z_1)ds = (2\pi)^2\sigma_s^2.\end{aligned} \tag{9.A18}$$

From the above,

$$\sigma^2(z) = \sigma^2(z_1) - A(z_1)(z-z_1) + \lambda^2\sigma_s^2(z-z_1)^2. \tag{9.A19}$$

Rewriting this,

$$\sigma^2(z) = \sigma_0^2\left\{1 + \left(\frac{z-z_0}{z_R}\right)^2\right\}, \tag{9.A20}$$

where

$$\sigma_0^2 = \sigma^2(z_1) - \frac{\{A(z_1)\}^2}{4\lambda^2\sigma_s^2}, \tag{9.A21}$$

$$z_0 = z_1 + \frac{A(z_1)}{2\lambda^2\sigma_s^2}, \tag{9.A22}$$

$$z_R = \frac{\sigma_0}{\lambda\sigma_s} = \frac{4\pi\sigma_0^2}{M^2\lambda}. \tag{9.A23}$$

As shown in Eq. (9.7), the beam diameter d and the standard deviation σ are in a proportional relationship, so Eq. (9.A20) is equivalent to Eq. (9.21).

Appendix 9B Hermite Polynomials

The Hermite polynomials were explained in Chapter 7. Here, let's summarize their recurrence formula and orthogonal relationship.

The Hermite polynomials are solutions of the following differential equation:

$$\left(\frac{d^2}{d\xi^2} - 2\xi\frac{d}{d\xi} + 2n\right)H_n(\xi) = 0. \tag{9.A24}$$

$H_n(\xi)$ can be expressed by the following formulas:

$$H_n(\xi) = (-1)^n e^{\xi^2} \frac{d^n}{d\xi^n} e^{-\xi^2}, \tag{9.A25}$$

$$H_n(\xi) = n! \sum_{k=0}^{\lfloor n/2 \rfloor} \frac{(-1)^k}{k!(n-2k)!} (2\xi)^{n-2k}. \tag{9.A26}$$

The recurrence formulas are as follows:

$$H_{n+1}(\xi) = 2\xi H_n(\xi) - 2nH_{n-1}(\xi), \tag{9.A27}$$

$$\frac{d}{d\xi} H_n(\xi) = 2nH_{n-1}(\xi) = 2nH_n(\xi) - H_{n+1}(\xi). \tag{9.A28}$$

The following orthogonal relationship is satisfied:

$$\int_{-\infty}^{\infty} H_n(\xi) H_m(\xi) d\xi = \sqrt{\pi} 2^n n! \delta_{nm}. \tag{9.A29}$$

Appendix 9C Associated Laguerre Polynomials

The associated Laguerre polynomials (also called Sonine polynomials) are solutions of the following differential equation:

$$\left[\xi \frac{d^2}{d\xi^2} + (m+1-\xi) \frac{d}{d\xi} + n\right] L_n^m(\xi) = 0. \tag{9.A30}$$

As mentioned in the previous chapter, $L_n^m(\xi)$ can be expressed by the following formulas:

$$\begin{aligned} L_n^m(\xi) &= \frac{1}{n!} \sum_{k=0}^{n} (-1)^k ({}_nP_{n-k})({}_{n+m}C_{n-k}) \xi^k \\ &= \frac{1}{n!} \sum_{k=0}^{n} (-1)^k \frac{n!}{k!} \frac{(n+m)!}{(n-k)!(m+k)!} \xi^k, \end{aligned} \tag{9.A31}$$

$$L_n^m(\xi) = \frac{\xi^{-m} e^{\xi}}{n!} \frac{d^n}{d\xi^n} (e^{-\xi} \xi^{n+m}). \tag{9.A32}$$

The cases where $m = 0$ are called Laguerre polynomials and are written as follows:

$$L_n(\xi) \equiv L_n^0(\xi). \tag{9.A33}$$

$L_n^m(\xi)$ is expressed by the following formula using these Laguerre polynomials:

$$L_n^m(\xi) = (-1)^m \frac{d^m}{d\xi^m} L_{n+m}(\xi). \tag{9.A34}$$

There is also the following definition:

$$L_n^m(\xi) = \frac{d^m}{d\xi^m} L_n(\xi). \tag{9.A35}$$

Care must be taken because the recurrence formula and orthogonal relationship formula will differ depending on which definition is used.

From Eq. (9.A31) or Eq. (9.A32), the following recurrence formulas are obtained:

$$nL_n^m(\xi) = -\xi L_{n-1}^{m+1}(\xi) + (n+m)L_{n-1}^m(\xi), \tag{9.A36}$$

$$L_n^m(\xi) = L_n^{m+1}(\xi) - L_{n-1}^{m+1}(\xi), \tag{9.A37}$$

$$nL_n^{m+1}(\xi) = (n-\xi)L_{n-1}^{m+1}(\xi) + (n+m)L_{n-1}^m(\xi), \tag{9.A38}$$

$$\xi L_n^{m+1}(\xi) = (n+m)L_{n-1}^m(\xi) - (n-\xi)L_n^m(\xi), \tag{9.A39}$$

$$L_n^m(\xi) = \left(2 + \frac{m-1-\xi}{n}\right) L_{n-1}^m(\xi) - \left(1 + \frac{m-1}{n}\right) L_{n-2}^m(\xi), \tag{9.A40}$$

$$L_n^m(\xi) = \frac{m+1-\xi}{n} L_{n-1}^{m+1}(\xi) - \frac{\xi}{n} L_{n-2}^{m+2}(\xi). \tag{9.A41}$$

In addition, the following recurrence formulas are obtained regarding the differentiation:

$$\frac{d}{d\xi} L_n^m(\xi) = -\frac{m}{\xi} L_n^m(\xi) + \frac{n+m}{\xi} L_n^{m-1}(\xi), \tag{9.A42}$$

$$\frac{d}{d\xi} L_n^m(\xi) = \left(1 - \frac{m}{\xi}\right) L_n^m(\xi) + \frac{n+1}{\xi} L_{n+1}^{m-1}(\xi). \tag{9.A43}$$

$L_n^m(\xi)$ satisfies the following orthogonal relationship:

$$\int_0^\infty \xi^k e^{-\xi} L_n^k(\xi) L_m^k(\xi) d\xi = \frac{(n+k)!}{n!} \delta_{nm}. \tag{9.A44}$$

Here is the formula for the integral relation:

$$\int_0^\infty \xi^{k+1} e^{-\xi} |L_n^k(\xi)|^2 d\xi = \frac{(n+k)!}{n!} (2n+k+1). \tag{9.A45}$$

Chapter 10
Beam Quality of Laser Light Part 2

10.1 Introduction

In the previous chapter, the M^2 factor, which is a beam quality index for laser light, was introduced, and the relationship between the M^2 factor and the mode order in Hermite–Gaussian beams and Laguerre–Gaussian beams was described. This chapter describes the values of the M^2 factors for other beams.

10.2 Beam Quality for the Mode in Optical Waveguides

The characteristic feature of the Hermite–Gaussian beams and the Laguerre–Gaussian beams is the formation of "modes." Other examples where "modes" are formed, are optical waveguides. Since the modes of optical waveguides will be discussed in a separate chapter, the derivation of the equations is omitted here and the mode profiles and the M^2 factors for the waveguide modes will be introduced.

As an example, consider a three-layer semiconductor optical waveguide with a GaAs core and $Ga_{0.6}Al_{0.4}As$ claddings. Assuming that the refractive indices of the core and cladding are 3.6 and 3.3, respectively, and the thickness of the core is 1.5 μm, the zeroth- to third-order modes are formed in this optical waveguide, as shown in Fig. 10.1. There are transverse-electric (TE) modes and transverse-magnetic (TM) modes. Figure 10.1 shows the TE modes.

The distributions shown as "intensity" in the near-field pattern (NFP) in Fig. 10.1 are the intensity distributions observed at the end face of the optical waveguide. The distribution of light emitted from this optical waveguide at a sufficiently far distance is a far-field pattern (FFP), and the figure shows the distributions as a function of the emission angle θ ($-\pi/2 < \theta < \pi/2$). The amplitude of the waveguide mode and the amplitude of the far-field pattern are approximately in a Fourier transform relationship. The word "approximately" refers to the paraxial case in the Fourier transform

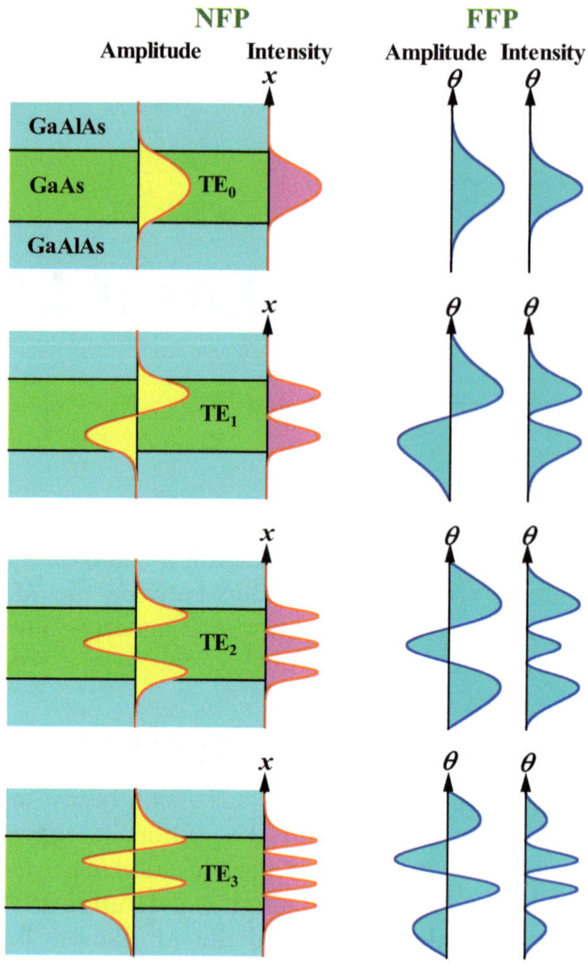

Figure 10.1 Waveguide mode (near-field pattern) and far-field pattern of the output light for a three-layer optical waveguide.

relationship, and in the region where θ is large, the angle correction term is multiplied.

Now, if the intensity distributions of the near-field pattern and the far-field pattern are obtained, the M^2 factor can be calculated using the definition in the previous chapter. The M_x^2 factors for each mode in Fig. 10.1 are shown in Fig. 10.2. The fact that the value of M_x^2 increases as the mode order increases is similar to the case of the Hermite–Gaussian beam shown in the previous chapter. The mode order dependence is similar to Fig. 9.2 in the previous chapter, but it does not take an integer value as given by Eq. (9.34). Furthermore, the value of M_x^2 differs depending on the optical waveguide structure.[1] In Fig. 10.2, M_x^2 of the TE_0 mode is close to 1, but the guided mode

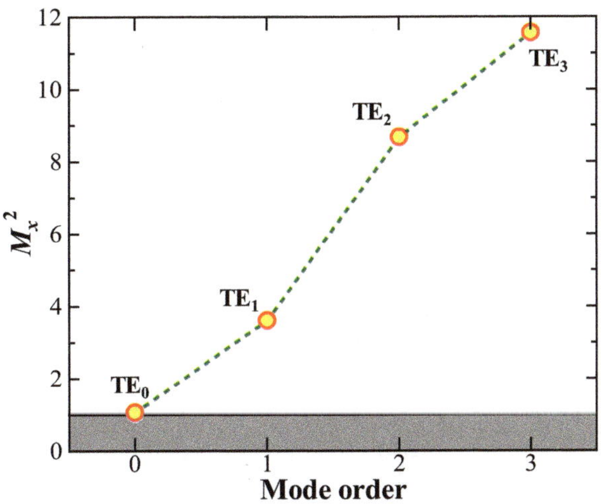

Figure 10.2 M^2 factors for each mode in Fig. 10.1.

is not a Gaussian beam, so $M_x^2 > 1$. In the case of this TE$_0$ mode, $M_x^2 = 1.035$, and the beam quality can be said to be good.

10.3 Beam Quality for Various Beams

As shown in the previous section, the M^2 factor can be calculated if the near-field and far-field patterns are known. Therefore, by using the fact that the amplitude of the near-field pattern and the amplitude of the far-field pattern are in a Fourier transform relationship on the paraxial axis, the M^2 factor for the case where the analytical formula for the Fourier transform is given will be introduced below.

First, if the amplitude distribution of the near-field pattern is $f(x)$, the amplitude $g(k_x)$ of the far-field pattern is given, in the paraxial approximation, by the following Fourier transform:

$$g(k_x) = \frac{1}{\sqrt{2\pi}} \int_{-\infty}^{\infty} f(x) \exp(ik_x x) dx. \tag{10.1}$$

As mentioned in the previous chapter, the standard deviations σ_x and σ_{kx} are defined by the second-order moment as follows:

$$\sigma_x^2 = \frac{\displaystyle\int_{-\infty}^{\infty} x^2 |f(x)|^2 \, dx}{\displaystyle\int_{-\infty}^{\infty} |f(x)|^2 \, dx}, \tag{10.2}$$

Table 10.1 Near-field patterns, far-field patterns, and M^2 factors for various beams.

	Gaussian	sech	tri-sinc^2	rect-sinc
$f(x)$	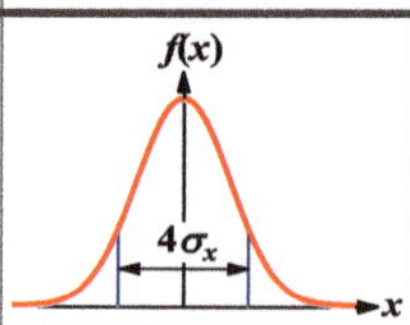	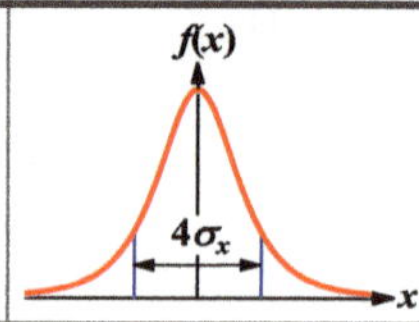	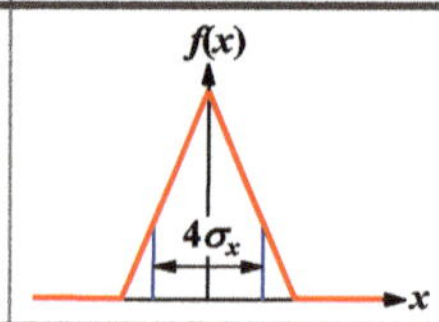	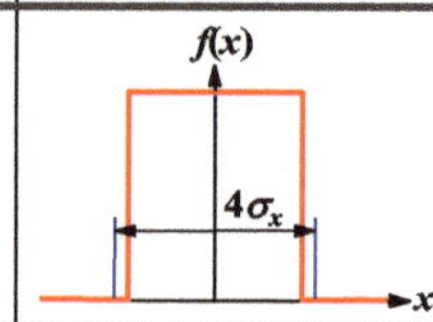
	$\exp\left(-\frac{x^2}{w^2}\right)$	$\mathrm{sech}\left(\frac{x}{a}\right)$	$\mathrm{tri}\left(\frac{x}{a}\right)$	$\mathrm{rect}\left(\frac{x}{a}\right)$
σ_x	$\sigma_x = \frac{w}{2}$	$\sigma_x = 0.907a$	$\sigma_x = \frac{a}{\sqrt{10}}$	$\sigma_x = \frac{a}{2\sqrt{3}}$
$g(k_x)$	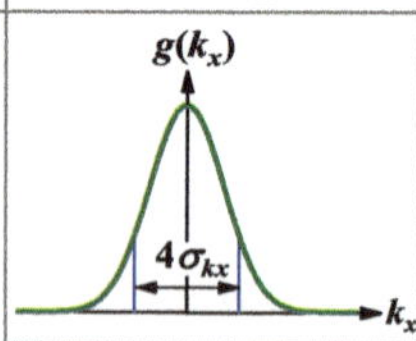	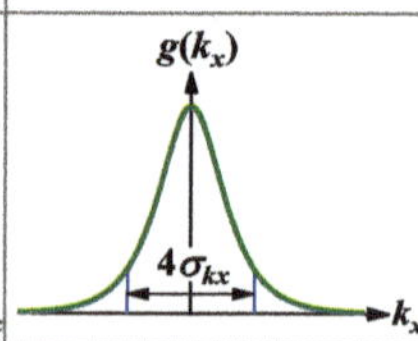	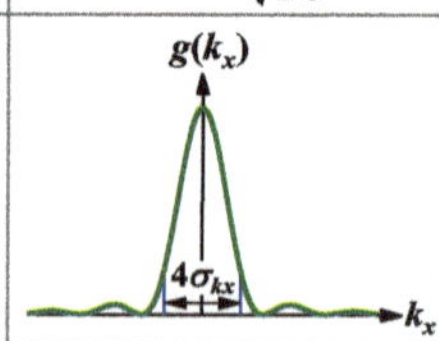	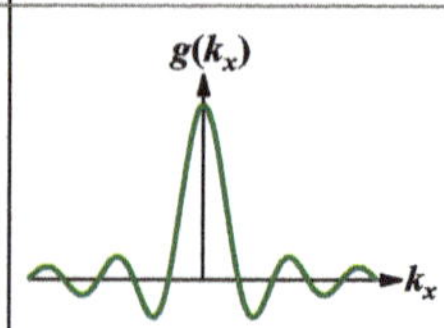
	$\frac{w}{\sqrt{2}}\exp\left(-\frac{w^2k_x^{\,2}}{4}\right)$	$a\sqrt{\frac{\pi}{2}}\,\mathrm{sech}\left(\frac{\pi}{2}ak_x\right)$	$\frac{a}{\sqrt{2\pi}}\mathrm{sinc}^2\left(\frac{ak_x}{2\pi}\right)$	$\frac{a}{\sqrt{2\pi}}\mathrm{sinc}\left(\frac{ak_x}{2\pi}\right)$
σ_{kx}	$\sigma_{kx} = \frac{1}{w}$	$\sigma_{kx} = \frac{0.577}{a}$	$\sigma_{kx} = \frac{1.737}{a}$	$\sigma_{kx} = \infty$
$M_x^{\,2}$	$M_x^{\,2} = 1$	$M_x^{\,2} = 1.047$	$M_x^{\,2} = 1.1$	See Side Topic

$$\sigma_{kx}^2 = \frac{\int_{-\infty}^{\infty} k_x^2 |g(k_x)|^2 \, dk_x}{\int_{-\infty}^{\infty} |g(k_x)|^2 \, dk_x}. \tag{10.3}$$

Using σ_x and σ_{kx}, M_x^2 can be calculated as follows:

$$M_x^2 = 2\sigma_x \sigma_{kx}. \tag{10.4}$$

Table 10.1 shows various beams whose near-field and far-field amplitudes are related by a Fourier transform. A Gaussian beam is also shown for comparison.

Similar to the Gaussian function, there is a hyperbolic secant function "sech" in which the Fourier-transformed function becomes the same as the original function. As shown in Table 10.1, it has a distribution very similar to a Gaussian function. Since it is not a Gaussian beam, the value of M_x^2 is greater than 1, but $M_x^2 = 1.047$ indicates relatively good beam quality.

The triangle function "tri" in Table 10.1 is defined below:

$$\mathrm{tri}(x) = \begin{cases} 1 - |x| & (|x| \le 1) \\ 0 & (|x| \ge 1) \end{cases}. \tag{10.5}$$

The Fourier transform of this function is the square of the sinc function. The sinc function is defined below:

$$\mathrm{sinc}(x) = \frac{\sin(\pi x)}{\pi x}. \tag{10.6}$$

The M^2 factor of this "tri-sinc^2" beam is $M_x^2 = 1.1$, which is not that bad in terms of beam quality.

The fourth example of "rect-sinc" in Table 10.1 corresponds to a beam diverging from a rectangular aperture. However, it is well known that when a rectangular aperture beam is focused by a lens, the distribution at the focal plane is a "sinc" function.

Now, when $g(k_x)$ is given by this "sinc" function as shown in Table 10.1, σ_{kx} can be calculated by using Eq. (10.3) as follows:

$$\begin{aligned} \sigma_{kx}^2 &= \frac{\displaystyle\int_0^\infty k_x^2 \left(\frac{ak_x}{2\pi}\right) dk_x}{\displaystyle\int_0^\infty \mathrm{inc}^2\left(\frac{ak_x}{2\pi}\right) dk_x} = \left(\frac{2}{a}\right)^2 \frac{\displaystyle\int_0^\infty \sin^2 \xi d\xi}{\displaystyle\int_0^\infty \frac{\sin^2 \xi}{\xi^2} d\xi} \\ &= \left(\frac{2}{a}\right)^2 \frac{2}{\pi} \int_0^\infty \sin^2 \xi d\xi = \infty. \end{aligned} \tag{10.7}$$

In this way, σ_{kx} diverges.

The relationship between the near-field pattern and the far-field pattern shown in Table 10.1 holds true in the paraxial range, but it is not strictly accurate. In particular, when high-frequency components with $k_x > k$ $(= 2\pi/\lambda)$ are included, they are evanescent components and do not contribute to the propagating light. Since the M^2 factor is a beam quality index for propagating light, it is necessary to take this situation into account.

If the Fourier component of $f(x,0)$ is given by $g(k_x)$, the amplitude at $z > 0$ can be expressed by the plane wave expansion as follows:

$$\begin{aligned} f(x,z) = \frac{1}{\sqrt{2\pi}} \Bigg\{ &\int_{-k}^{k} g(k_x) \exp(-ik_x x - ik_z z) \\ &+ \int_{-\infty}^{-k} g(k_x) \exp(-ik_x x) \exp(-\gamma_z z) dk_x \\ &+ \int_{k}^{\infty} g(k_x) \exp(-ik_x x) \exp(-\gamma_z z) dk_x \Bigg\}, \end{aligned} \tag{10.8}$$

where

$$k_z = \sqrt{k^2 - k_x^2} \quad (|k_x| < k), \tag{10.9}$$

$$\gamma_z = \sqrt{k_x^2 - k^2} \quad (|k_x| > k). \tag{10.10}$$

In Eq. (10.8), the first term on the right side is the propagating component. The second and third terms are evanescent components that decay exponentially at $z > 0$, so they hardly contribute to the propagating light.

Figure 10.3 shows the z dependence of the light intensity calculated using Eq. (10.8) for the "tri-sinc^2" beam and "rect-sinc" beam shown in Table 10.1.

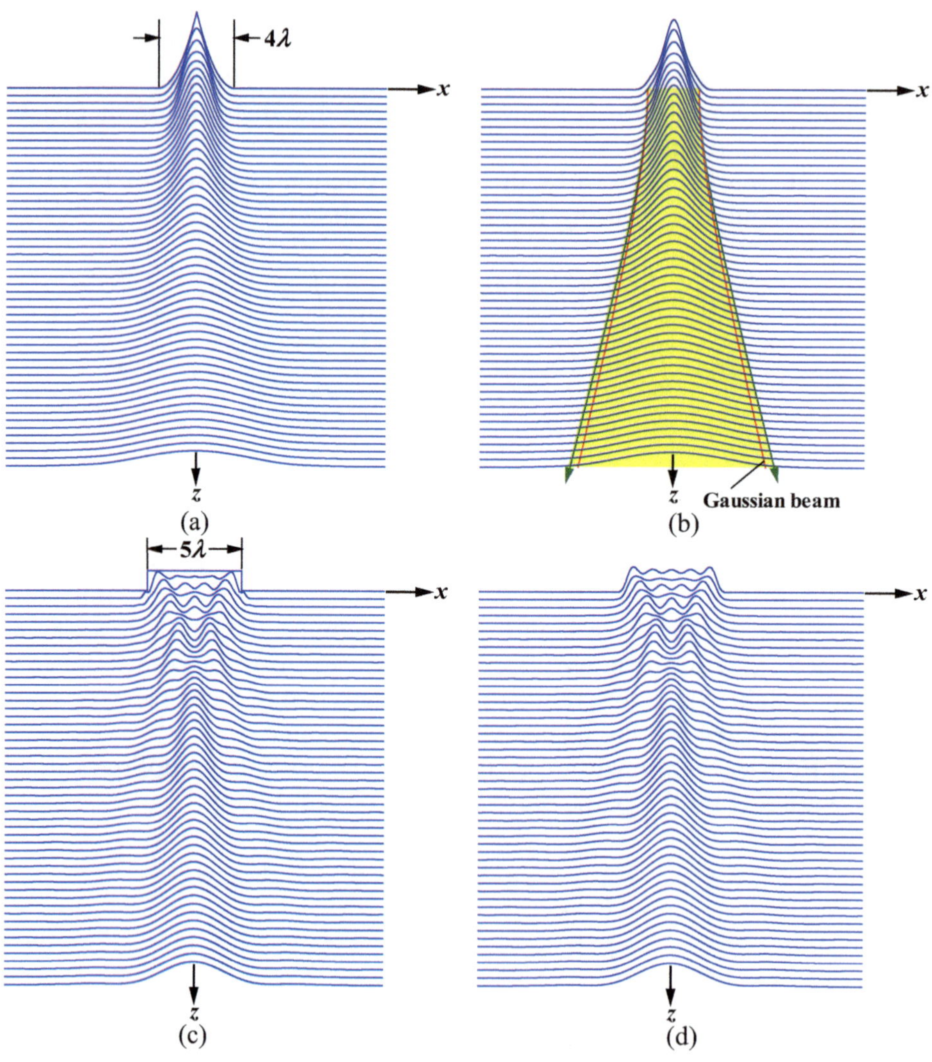

Figure 10.3 z dependence of light intensity in a "tri-sinc^2" beam and "rect-sinc" beam: (a) "tri-sinc^2" beam (including evanescent component), (b) "tri-sinc^2" beam (excluding evanescent component), (c) "rect-sinc" beam (including evanescent component), and (d) "rect-sinc" beam (excluding evanescent component).

Figures 10.3(a) and (c) are calculated including the evanescent component, and Figs. 10.3(b) and (d) are calculated excluding the evanescent component. Since the evanescent component attenuates rapidly, Figs. 10.3(a) and (b) and Figs. 10.3(c) and (d) have almost the same distributions, except for the vicinity of $z = 0$.

In Fig. 10.3(b), the beam diameter defined in the previous chapter is plotted. It can be seen that in this "tri-sinc^2" beam as well, the beam diameter exhibits the same z dependence as in the Gaussian beam, as shown in Eq. (9.21) in the previous chapter. For comparison, the beam diameter of the Gaussian beam is plotted. When comparing beams with the same beam waist, the ratio of beam diameters at a sufficiently far distance corresponds to the M^2 factor.

Side Topic: Is the beam quality of a focused beam spot with a rectangular aperture the worst?

As shown in Table 10.1, in the "rect-sinc" beam, $\sigma_{kx} = \infty$, so according to Eq. (10.4), the M^2 factor has an infinite value, which means that the beam quality is extremely poor. However, since it includes an evanescent component, the far-field pattern does not strictly become a "sinc" function when calculating the propagating component in Eq. (10.8).

Let's consider the opposite case. In other words, consider the case where $f(x)$ at the focal plane is expressed by a "sinc" function as shown below:

$$f(x) = \mathrm{sinc}\left(\frac{x}{a}\right). \tag{10.S1}$$

The Fourier transform of $f(x)$ is given by

$$g(k_x) = \frac{a}{\sqrt{2\pi}}\mathrm{rect}\left(\frac{ak_x}{2\pi}\right). \tag{10.S2}$$

From this equation, if $\pi/a > k$, then all are propagating components. This condition is represented as $a > \lambda/2$. It can be seen from this that a spot diameter on the order of wavelength can be obtained. Nevertheless, when the beam diameter defined from the second-order moment is calculated using Eq. (10.S1), its value becomes infinity.

The situation in which the truncated beam of light emitted from a semiconductor laser is focused with a lens to form a very small spot is also used in optical disk devices, so a beam that forms a "sinc" function at the focal plane is actually used. However, the beam quality of this beam as propagating light is extremely poor according to the definition of Eq. (10.4). Even if the beam diameter defined from the second-order moment is infinity, it can actually be used as an extremely small spot.

The beam diameter defined from the standard deviation is said to be a reasonable definition because the beam waist can be uniquely defined and the beam waist position coincides with the position where the wavefront curvature becomes 0.[1,2] However, as mentioned above, it may be necessary to consider other circumstances depending on the case.

10.4 Conclusion

The M^2 factor, which is a beam quality index, has been discussed for light emitted from an optical waveguide and for a general beam whose near-field and far-field patterns are in a Fourier transform relationship.

References

1. G. Hatakoshi, "Analysis of beam quality factor for semiconductor lasers," *Opt. Review* **10**(4), 307–314 (2003).
2. G. Hatakoshi, "Standards for optical properties and measurements of semiconductor lasers," *O plus E* **30**(9), 953–958 (2008) [in Japanese].

Chapter 11
Wigner Distribution Function

11.1 Introduction

The M^2 factors[1–3] introduced in Chapters 9 and 10 on beam quality, are defined by the beam spatial distribution and the beam divergence angle determined from the distribution of the spatial frequency. The Wigner distribution function expresses both spatial distribution and spatial frequency distribution in phase space.

The Wigner function, introduced by Eugene Paul Wigner, is a real function that connects the wave function in quantum mechanics with the probability distribution of the position coordinates and the momentum in classical phase space.[4] The physical quantity $\hat{A}$, used in quantum mechanics, expressed in phase space without using operators, is the Wigner representation (Wigner function) and is sometimes called the Wigner distribution function, especially when $\hat{A}$ is a density matrix. In his discussion on quantum computers, R. P. Feynman mentioned the density matrix and the Wigner function in connection with the possibility of probabilistically simulating quantum systems with classical computers.[5]

The Wigner distribution function is used in Fourier optics as a function that represents the local spatial frequency distribution in spatial coordinates.[6] This can also be used to define the M^2 factor of the beam quality index. In this chapter, we will discuss the relationship between the Wigner distribution function and the M^2 factor.

11.2 Wigner Distribution Function

For simplicity, the one-dimensional case will be explained. The Wigner distribution function $W(x, p_x)$ for the complex function $\psi(x)$ is given by the following equation, where x is the position coordinate and p_x is the momentum:[6,7]

$$W(x,p_x) = \frac{1}{2\pi}\int_{-\infty}^{\infty} \psi\left(x+\frac{\hbar}{2}\xi_1\right)\psi^*\left(x-\frac{\hbar}{2}\xi_1\right)\exp(-ip_x\xi_1)d\xi_1. \tag{11.1}$$

In the case where $\psi(x)$ is the wave function of a light wave, set the following variables:

$$p_x = \hbar k_x, \tag{11.2}$$

$$\hbar W(x, p_x) \equiv f(x, k_x), \tag{11.3}$$

where $\hbar$ [$= h/(2\pi)$] is the reduced Planck constant (also known as the Dirac constant). The function $f(x, k_x)$ in Eq. (11.3) is represented by

$$f(x, k_x) = \frac{1}{2\pi}\int_{-\infty}^{\infty} \psi\left(x + \frac{\xi}{2}\right)\psi^*\left(x - \frac{\xi}{2}\right)\exp(-ik_x\xi)d\xi. \tag{11.4}$$

Here, $f(x, k_x)$ is a real function and has the following properties:

$$\int_{-\infty}^{\infty} f(x, k_x)dk_x = |\psi(x)|^2, \tag{11.5}$$

$$\int_{-\infty}^{\infty} f(x, k_x)dx = |\Psi(k_x)|^2, \tag{11.6}$$

where $\Psi(k_x)$ is the Fourier transform of $\psi(x)$ and is represented by the following equation:

$$\Psi(k_x) = \frac{1}{\sqrt{2\pi}}\int_{-\infty}^{\infty} \psi(x)\exp(ik_x x)dx. \tag{11.7}$$

11.3 Hermite–Gaussian Beams and the Wigner Distribution Function

Let's take a look at what the Wigner distribution function looks like for a Hermite–Gaussian beam. Substituting Eq. (8.4) in Chapter 8 as $\psi(x)$ into Eq. (11.4),

$$\begin{aligned} f_m(x, k_x) &= \frac{1}{2\pi}\frac{w_0}{w(z)}\exp\left(-\frac{2x^2}{w^2(z)}\right) \\ &\times \int_{-\infty}^{\infty} H_m^+ H_m^- \exp\left(-\frac{\xi^2}{2w^2(z)}\right)\exp\left[-i\xi k_1(x,z)\right]d\xi, \end{aligned} \tag{11.8}$$

where

$$H_m^{\pm} = H_m\left(\frac{\sqrt{2}}{w(z)}\left[x \pm \frac{\xi}{2}\right]\right), \tag{11.9}$$

$$k_1(x, z) = k_x - \frac{kx}{R(z)} = k_x - \frac{z}{z_R}\frac{2x}{w^2(z)}. \tag{11.10}$$

In the following equations, $w(z)$ will be abbreviated as w, and $k_1(x, z)$ will be abbreviated as k_1. Specifically finding $f_m(x, k_x)$ in the case of $m = 0$, 1, and 2 results in the following formulas:

$$f_0(x, k_x) = \frac{w_0}{\sqrt{2\pi}} \exp\left(-\frac{2x^2}{w^2}\right) \exp\left(-\frac{w^2}{2} {k_1}^2\right), \tag{11.11}$$

$$\begin{aligned} f_1(x, k_x) = & \frac{w_0}{\sqrt{2\pi}} \exp\left(-\frac{2x^2}{w^2}\right) \exp\left(-\frac{w^2}{2} {k_1}^2\right) \\ & \times \left(\frac{8}{w^2} x^2 + 2w^2 {k_1}^2 - 2\right), \end{aligned} \tag{11.12}$$

$$\begin{aligned} f_2(x, k_x) = & \frac{w_0}{\sqrt{2\pi}} \exp\left(-\frac{2x^2}{w^2}\right) \exp\left(-\frac{w^2}{2} {k_1}^2\right) \\ & \times \left(\frac{64}{w^4} x^4 - \frac{64}{w^2} x^2 + 32x^2 {k_1}^2 + 8w^2 {k_1}^2 - 4w^4 {k_1}^4 + 8\right). \end{aligned} \tag{11.13}$$

Integrating Eqs. (11.11) to (11.13) with respect to k_x, it is confirmed that they all satisfy Eq. (11.5).

Equations (11.11) to (11.13) are functions of z as well as position x and wave number k_x. That is, the distribution changes depending on the propagation direction z. As mentioned at the beginning of this chapter, the Wigner distribution function is a function that simultaneously expresses the spatial distribution as a function of spatial coordinates and the local spatial frequency distribution at those coordinates. A schematic representation of this local spatial frequency distribution for a Gaussian beam with $m = 0$ is shown in Fig. 11.1.[8] The Wigner distribution function $f_0(x, k_x)$ in the z position of (a), (b), and (c) of this figure is as shown in Fig. 11.2.[8]

As shown in Fig. 11.2, the Wigner distribution function in this case is a function that changes depending on z. This z dependence can be seen in animation in the Excel animations introduced so far [see Appendix 11A]. Figure 11.3 shows an example of the Wigner distribution function displayed using an Excel animation for a Hermite–Gaussian beam with $m = 0$, 1, and 2.

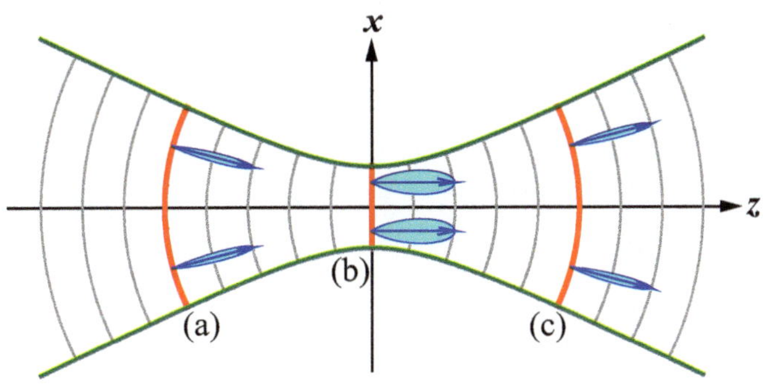

Figure 11.1 Distribution of k_x in a Gaussian beam.

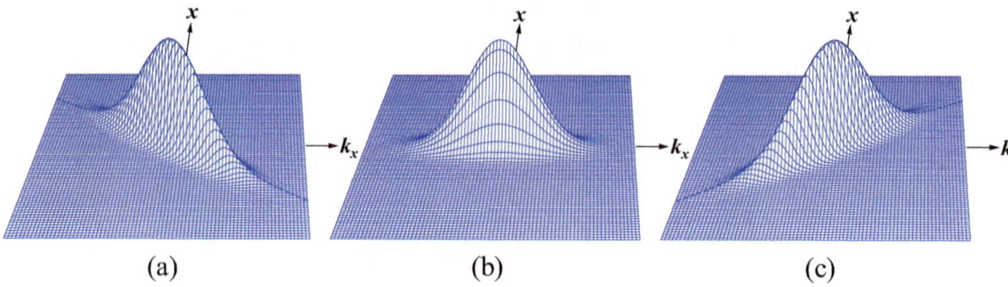

Figure 11.2 Wigner distribution function in the (a), (b), and (c) positions of the Gaussian beam in Fig. 11.1. (a) $z < 0$, (b) $z = 0$, and (c) $z > 0$.

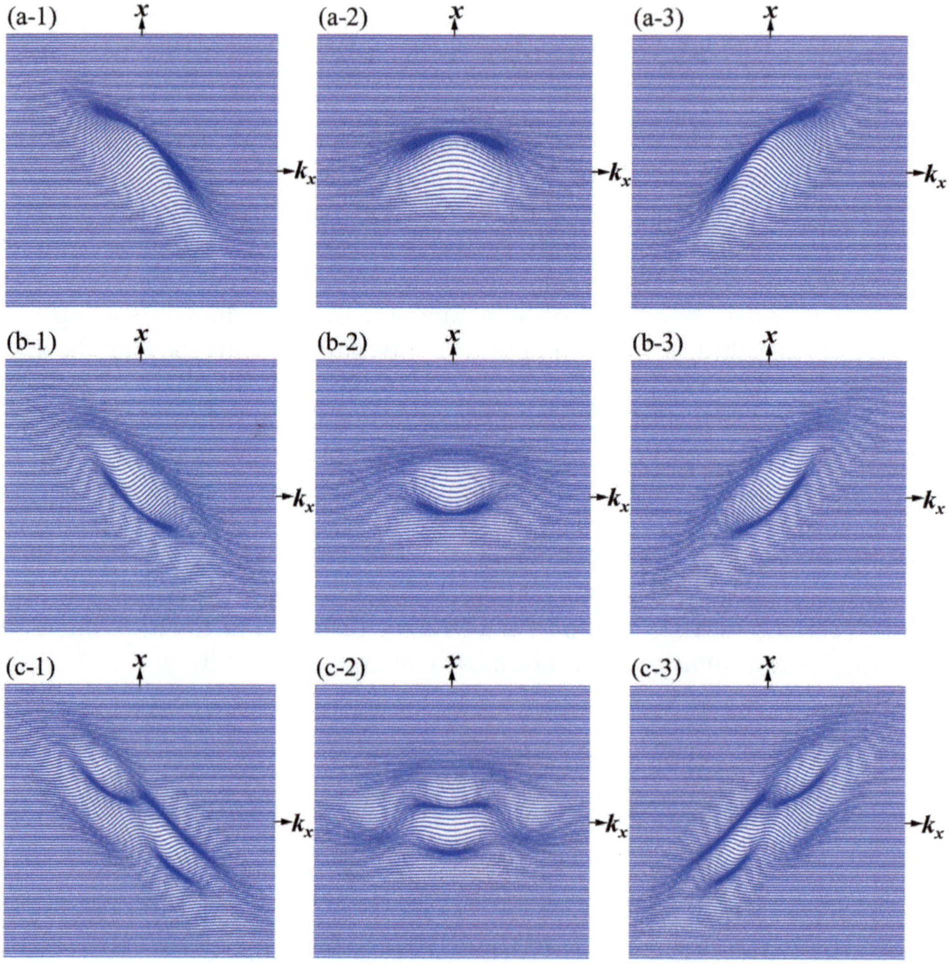

Figure 11.3 Wigner distribution function for Hermite–Gaussian beams. (a) $m = 0$, (b) $m = 1$, and (c) $m = 2$. For each case, $z < 0$ (left), $z = 0$ (center), and $z > 0$ (right).

11.4 Beam Quality and the Wigner Distribution Function

The average position $\bar{x}$ (first-order moment) and the square of the standard deviation σ_x (second-order moment)[9] for calculating the M^2 factor described in Chapter 9 can also be defined using the Wigner distribution function. This is because the Wigner distribution function has the properties given by Eqs. (11.5) and (11.6). For example, ${\sigma_x}^2$ is given by the following equation using the Wigner distribution function $f(x, k_x)$:

$$ {\sigma_x}^2 = \frac{\int (x - \langle x \rangle)^2 f(x, k_x) dx dk_x}{\int f(x, k_x) dx dk_x}, \tag{11.14} $$

where $<x>$ is the average value of x $(= \bar{x})$. In this chapter from here on, second-order moments will similarly be represented by

$$ \langle x^2 \rangle \equiv {\sigma_x}^2, \langle {k_x}^2 \rangle \equiv {\sigma_{kx}}^2. \tag{11.15} $$

The ISO document on the measurement methods of optical characteristics for lasers defines $<x^2>$, $<{\theta_x}^2>$, etc., as the second-order moments of the Wigner distribution function.[10] In the ISO documents, the term "beam propagation ratio" is used for the M^2 factor. As mentioned in Chapter 9, the M^2 factor is defined based on the beam width and beam divergence angle, so the angular wave number k_x is converted into the angle θ_x using the following relationship:

$$ k_x = k \sin \theta_x \cong k\theta_x. \tag{11.16} $$

Note that the angular wave number k and Rayleigh length z_R have the following relationship, as shown in Eq. (7.8) in Chapter 7:

$$ k = 2z_R / {w_0}^2. \tag{11.17} $$

The Rayleigh length will be discussed again in the side topic.

Using the two-dimensional Wigner distribution function $h(x, y, \theta_x, \theta_y)$ expressed as a function of x, y, θ_x, and θ_y, the second-order moment $<x^2>$, for example, is given by the following equation:

$$ \langle x^2 \rangle = \frac{\iiiint (x - \langle x \rangle)^2 \, h(x, y, \theta_x, \theta_y) dx dy d\theta_x d\theta_y}{\iiiint h(x, y, \theta_x, \theta_y) dx dy d\theta_x d\theta_y}. \tag{11.18} $$

If the optical power density in the propagation direction z is $u(x, y, z)$, then from the relationship in Eq. (11.5), Eq. (11.18) is equivalent to the following equation:

$$\langle x^2 \rangle = \frac{\int_{-\infty}^{\infty}\int_{-\infty}^{\infty}(x - \langle x \rangle)^2 u(x, y, z)dxdy}{\int_{-\infty}^{\infty}\int_{-\infty}^{\infty} u(x, y, z)dxdy}. \tag{11.19}$$

Note that in the actual measurement in the ISO document, $u(x, y, z)$ is measured and $<x^2>$ is determined from Eq. (11.19), so $h(x, y, \theta_x, \theta_y)$ itself is not measured. Since the second-order moments such as $<x\theta_x>$ that appear below are also determined from the z dependence of $u(x, y, z)$, what is ultimately measured is the optical power density distribution.

The second-order moments $<x^2>$, $<k_x^2>$, and $<xk_x>$ of the Gaussian beam, whose Wigner distribution function is expressed by Eq. (11.11), are calculated as follows:

$$\langle x^2 \rangle = \frac{w^2(z)}{4}, \tag{11.20}$$

$$\langle k_x^2 \rangle = \frac{1}{w_0^2}, \tag{11.21a}$$

$$\langle xk_x \rangle = \frac{1}{2}\frac{z}{z_R}. \tag{11.22a}$$

Using the relationships of Eqs. (11.16) and (11.17), Eqs. (11.21a) and (11.22a) become the following equations:

$$\langle \theta_x^2 \rangle = \frac{w_0^2}{4z_R^2}, \tag{11.21b}$$

$$\langle x\theta_x \rangle = \frac{1}{4}\frac{w_0^2 z}{z_R^2}. \tag{11.22b}$$

Of course, neither $<k_x^2>$ nor $<\theta_x^2>$ depends on the position z in the propagation direction.

11.5 Beam Matrix

Define the following beam matrix using the second-order moments $<x^2>$, $<\theta_x^2>$, and $<x\theta_x>$:

$$\boldsymbol{P}_x = \begin{pmatrix} \langle x^2 \rangle & \langle x\theta_x \rangle \\ \langle x\theta_x \rangle & \langle \theta_x^2 \rangle \end{pmatrix}. \tag{11.23}$$

Using this, the M^2 factor can be calculated as follows:

$$M_x^2 = \frac{4\pi}{\lambda}\sqrt{\det(\boldsymbol{P}_x)}, \tag{11.24}$$

where $\det(\boldsymbol{P}_x)$ is the determinant of $\boldsymbol{P}_x$, given by

$$\det(\boldsymbol{P}_x) = \langle x^2\rangle\langle \theta_x{}^2\rangle - \langle x\theta_x\rangle^2. \tag{11.25}$$

As can be seen from Eqs. (11.20) and (11.22), $<x^2>$ and $<x\theta_x>$ in Eq. (11.23) are functions of z, but $\det(\boldsymbol{P}_x)$ does not depend on z. As mentioned in Chapter 9, the M^2 factor represents beam characteristics independent of z. By substituting Eqs. (11.20) to (11.22) into Eq. (11.25) and calculating the M^2 factor of the Gaussian beam, it is confirmed that $M_x{}^2 = 1$.

The ray matrix,† which was introduced in Chapter 8, is a matrix representing the transformation of rays in geometric optics and is also called the ray transfer matrix or system matrix. As an example, the ray transfer matrices (denoted here by the symbol $\boldsymbol{S}$) for a propagation of distance d and a thin lens with focal length f are shown again as follows:

$$\boldsymbol{S}_d = \begin{pmatrix} 1 & d \\ 0 & 1 \end{pmatrix}, \tag{11.26}$$

$$\boldsymbol{S}_f = \begin{pmatrix} 1 & 0 \\ -\dfrac{1}{f} & 1 \end{pmatrix}. \tag{11.27}$$

When the beam passes through the optical system with a ray matrix $\boldsymbol{S}$ given by Eqs. (11.26), (11.27), etc., the beam matrix $\boldsymbol{P}_x$ expressed by Eq. (11.23) undergoes the following transformation:[10,11]

$$\boldsymbol{P}_x^{(\mathrm{out})} = \boldsymbol{S}\boldsymbol{P}_x\boldsymbol{S}^{\mathrm{T}}, \tag{11.28}$$

where $\boldsymbol{S}^{\mathrm{T}}$ is the transposed matrix of $\boldsymbol{S}$.

The above explanation is in one dimension in the x direction, but these can be expanded to two dimensions in xy. The aforementioned ISO standard document[10] defines the 10 second-order moments of the Wigner distribution function for general beams and specifies the following beam matrix:

---

†Note: The "ray matrix" is not the "beam matrix" mentioned in this chapter. A "ray" and a "beam" are different.

$$\boldsymbol{P} = \begin{pmatrix} \langle x^2\rangle & \langle xy\rangle & \langle x\theta_x\rangle & \langle x\theta_y\rangle \\ \langle xy\rangle & \langle y^2\rangle & \langle y\theta_x\rangle & \langle y\theta_y\rangle \\ \langle x\theta_x\rangle & \langle y\theta_x\rangle & \langle \theta_x{}^2\rangle & \langle \theta_x\theta_y\rangle \\ \langle x\theta_y\rangle & \langle y\theta_y\rangle & \langle \theta_x\theta_y\rangle & \langle \theta_y{}^2\rangle \end{pmatrix}. \tag{11.29}$$

The effective beam propagation ratio in this case is expressed by the following equation:

$$M_{\text{eff}}{}^2 = \frac{4\pi}{\lambda}[\det(\boldsymbol{P})]^{1/4}. \tag{11.30}$$

From Eqs. (11.24), (11.25), (11.29), and (11.30), if there is no cross term between x and y, that is, if $<xy> = <x\theta_y> = <y\theta_x> = <\theta_x\theta_y> = 0$, then the following equation holds:

$$M_{\text{eff}}{}^2 = (M_x{}^2 M_y{}^2)^{1/2}. \tag{11.31}$$

Furthermore, if $M_x{}^2 = M_y{}^2$, Eq. (11.31) matches Eq. (11.24).

The ray transfer matrix when Eqs. (11.26) and (11.27) are extended to two dimensions is as follows:[11]

$$\boldsymbol{S}_d = \begin{pmatrix} 1 & 0 & d & 0 \\ 0 & 1 & 0 & d \\ 0 & 0 & 1 & 0 \\ 0 & 0 & 0 & 1 \end{pmatrix}, \tag{11.32}$$

$$\boldsymbol{S}_f = \begin{pmatrix} 1 & 0 & 0 & 0 \\ 0 & 1 & 0 & 0 \\ -1/f & 0 & 1 & 0 \\ 0 & -1/f & 0 & 1 \end{pmatrix}. \tag{11.33}$$

Also, in the case of a toric lens with different focal lengths in the x and y directions,

$$\boldsymbol{S}_{fxfy} = \begin{pmatrix} 1 & 0 & 0 & 0 \\ 0 & 1 & 0 & 0 \\ -1/f_x & 0 & 1 & 0 \\ 0 & -1/f_y & 0 & 1 \end{pmatrix}. \tag{11.34}$$

In the case of a cylindrical lens, $f_x = \infty$ ($1/f_x = 0$) or $f_y = \infty$ ($1/f_y = 0$) in Eq. (11.34).

The ray transfer matrix of optical system $\boldsymbol{S}_\alpha$, which represents the rotating $\boldsymbol{S}$ by an angle α in the xy plane, is given as follows:

$$\boldsymbol{S}_\alpha = \boldsymbol{R}_\alpha{}^{\mathrm{T}} \boldsymbol{S} \boldsymbol{R}_\alpha, \tag{11.35}$$

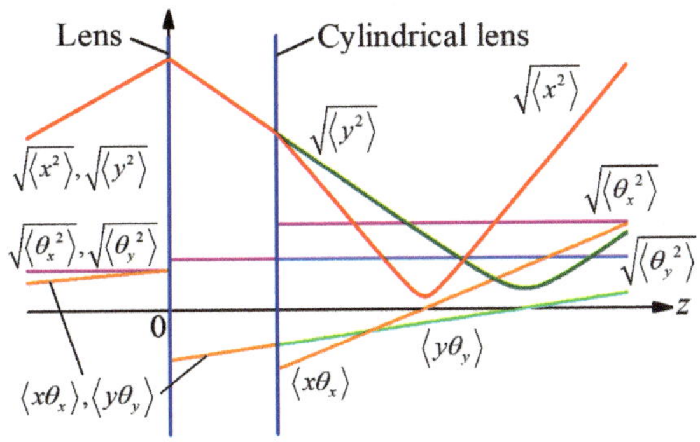

Figure 11.4 *z* dependence of beam matrix elements.

where

$$\boldsymbol{R}_\alpha = \begin{pmatrix} \cos\alpha & \sin\alpha & 0 & 0 \\ -\sin\alpha & \cos\alpha & 0 & 0 \\ 0 & 0 & \cos\alpha & \sin\alpha \\ 0 & 0 & -\sin\alpha & \cos\alpha \end{pmatrix}. \tag{11.36}$$

By applying the beam matrix $\boldsymbol{P}$ in Eq. (11.29) and the ray transfer matrix $\boldsymbol{S}$ in Eqs. (11.32) to (11.35) to Eq. (11.28), the change in the second-order moments, which are the elements of beam matrix $\boldsymbol{P}$, can be calculated.

Figure 11.4 schematically shows how the beam matrix elements of a Gaussian beam change when passing through a lens and passing through a cylindrical lens.

11.6 Relationship with Uncertainty

In Chapter 9, it was stated that the value of the M^2 factor is always greater than or equal to unity. Expressing this using the standard deviation of Eqs. (11.14) and (11.15), it becomes as follows:

$$\sigma_x \sigma_{kx} \geq \frac{1}{2}. \tag{11.37}$$

Using the relationship in Eq. (11.2) and defining $\sigma_x \equiv \Delta x$ and $\sigma_{px} \equiv \Delta p_x$, Eq. (11.37) becomes

$$\Delta x \Delta p_x \geq \frac{1}{2}\hbar. \tag{11.38}$$

This is the expression of the uncertainty principle that is often seen. However, this is not a difficult problem involving measurement but simply an inequality derived from the Fourier transform relationship. It is also derived from the Robertson inequality, Cauchy–Schwartz inequality, etc. In quantum mechanics, the commutation relation between x and p_x is equivalent to Eq. (11.38).

Side Topic: Rayleigh length

The Rayleigh length z_R is defined as follows:

$z_R \equiv$ Distance between the beam waist and the position S where the beam cross-sectional area is twice the value S_0 at the beam waist. (11.S1)

Therefore, the range where $S \leq 2S_0$ is twice the Rayleigh length is also called the depth of focus or confocal parameter. The formula for z_R was given by Eq. (7.11) in Chapter 7 or Eq. (8.16) in Chapter 8, and it is shown again as follows:

$$z_R = \frac{k{w_0}^2}{2} = \frac{\pi {w_0}^2}{\lambda}. \tag{11.S2}$$

On the other hand, using Eqs. (7.37) and (7.38) from Chapter 7,

$$z_R = \frac{d_0}{\Theta}. \tag{11.S3}$$

That is, it can be expressed as the ratio between the beam waist diameter d_0 and the beam divergence angle Θ. The parameters d_0 and Θ are defined by the second-order moment and can be applied to any beam as described in Chapter 9 regarding the virtual Gaussian beam.

Now, the Hermite–Gaussian beam explained in Chapter 7 is expressed as a Gaussian beam with $m = 0$ multiplied by a Hermite polynomial. For this reason, the same formula (11.S2) was used as z_R for higher-order Hermite–Gaussian beams. However, this z_R does not match Eq. (11.S3) for high-order Hermite–Gaussian beams. This is because z_R and w_0 in Eq. (11.S2) are the parameters for the zeroth-order Gaussian beam, and this w_0 is different from the beam width defined by the second-order moment for higher-order Hermite–Gaussian beams.

Therefore, by using the beam waist diameter $d_0 = 2w_0$ defined by the second-order moment as a criterion, and from the definition of the M^2 factor given by Eq. (9.25) in Chapter 9, the beam divergence angle Θ is represented as

$$\Theta = \frac{8M^2}{kd_0} = \frac{4M^2}{kw_0}. \tag{11.S4}$$

Substituting this into equation (11.S3),

$$z_R = \frac{kw_0^2}{2M^2} = \frac{\pi w_0^2}{\lambda M^2}. \tag{11.S5}$$

Comparing Eq. (11.S2) and Eq. (11.S5), it is clear that they differ by a factor of $1/M^2$. Equation (11.S5) shows that for beams with $M^2 > 1$, the Rayleigh length becomes $1/M^2$ times the case for $M^2 = 1$. This is more reasonable since the depth of focus will be reduced by a factor of $1/M^2$ for a beam with poor beam quality. This equation (11.S5) is also consistent with the definition of the Rayleigh length in Eq. (11.S1).

In the literature published so far, Eq. (11.S2) is used as the Rayleigh length equation. Of course, in a Gaussian beam with $M^2 = 1$, all Eqs. (11.S2) to (11.S5) are correct, and there is no problem. The problem is for a virtual Gaussian beam with $M^2 > 1$, in which case Eq. (11.S2) cannot be used for the Rayleigh length equation, and thus Eq. (11.S5) or its equivalent Eq. (11.S3) should be used. It seems that this point has not always been clear.

11.7 Conclusion

The Wigner distribution function, which is a function representing the local spatial frequency distribution in spatial coordinates, was introduced, and its relationship with the M^2 factor was discussed. Also, the relationship between the second-order moment of the Wigner distribution function and the beam matrix was introduced.

References

1. A. E. Siegman, "New developments in laser resonators," *Proc. SPIE* **1224**, 2–14 (1990) [doi: 10.1117/12.18425].
2. A. E. Siegman, "Defining the effective radius of curvature for a nonideal optical beam," *IEEE J. Quantum Electron.* **27**(5), 1146–1148 (1991).
3. G. Hatakoshi, "Standards for optical properties and measurements of semiconductor lasers," *O plus E* **30**(9), 953–958 (2008) [in Japanese].
4. E. Wigner, "On the quantum correction for thermodynamic equilibrium," *Phys. Rev.* **40**(5), 749–759 (1932).
5. R. P. Feynman, "Simulating physics with computers," *International J. Theoretical Physics* **21**(6), 467–488 (1982).
6. M. Takeda, "Wigner distribution function and fractional Fourier optics," *Optical Technology Handbook*, J. Tsujiuchi, et al., eds., Asakura Publishing, Tokyo (2002) [in Japanese].
7. T. L. Curtright, D. B. Fairlie, and C. K. Zachos, *A Concise Treatise on Quantum Mechanics in Phase Space*, World Scientific Publishing (2014).

8. G. Hatakoshi, "Laser beam characteristics and international standards," *Optical and Electro-optical Engineering Contact* **55** (8), 1–3 (2017) [in Japanese].
9. ISO 11146-1: 2021 "Lasers and laser-related equipment—Test methods for laser beam widths, divergence angles and beam propagation ratios—Part 1: Stigmatic and simple astigmatic beams," ISO (2021).
10. ISO 11146-2: 2021 "Lasers and laser-related equipment—Test methods for laser beam widths, divergence angles and beam propagation ratios—Part 2: General astigmatic beams," ISO (2021).
11. ISO/TR 11146-3: 2004 "Lasers and laser-related equipment—Test methods for laser beam widths, divergence angles and beam propagation ratios—Part 3: Intrinsic and geometrical laser beam classification, propagation and details of test methods," ISO (2004).

Appendix 11A Excel File

Appendix I-A contains the following Excel file, which corresponds to Fig. 11.3 in the main text. Refer to the operation instruction for running the animation.

11_Wigner.xlsm

Instruction	
Wigner_HG	Fig. 11.3

The input parameter m is the order of the Hermite polynomials. Animation can be executed for each case of $m = 0$, 1, and 2 in Fig. 11.3.

Chapter 12
Optical Multilayer Film Part 1

12.1 Introduction

Laminated structures of optical thin films, typified by dielectric multilayer films, are used in a variety of optical components such as reflectors, bandpass filters, and multiplexers/demultiplexers for the wavelength division multiplexing (WDM) system. In this chapter, the principles of the optical multilayer film and the analysis methods[1–4] using characteristic matrices will be explained, and calculations of reflectance and transmittance using Excel will be introduced.

12.2 Characteristic Matrices for Optical Multilayer Films

12.2.1 Electric and magnetic fields in multilayer films

Consider a multilayer film with N layers as shown in Fig. 12.1. The refractive indices on the incident side and the output side are n_{in} and n_{out}, respectively, and the refractive index and thickness of the j-th layer are n_j and d_j, respectively. In addition, the electric fields of the light traveling in the $+z$ direction and the light traveling in the $-z$ direction in each layer are written as

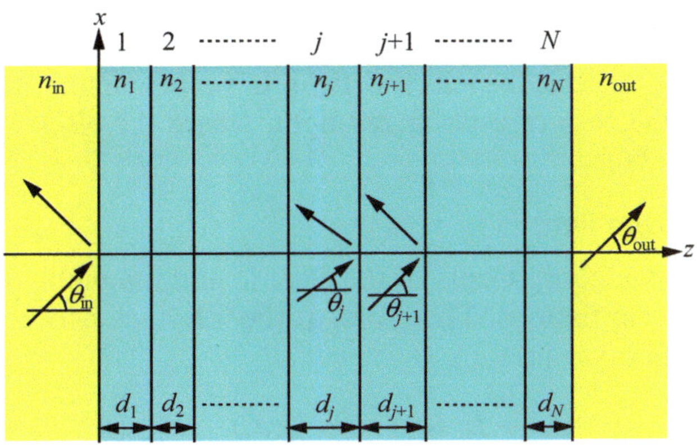

Figure 12.1 Parameters of optical multilayer film.

$\boldsymbol{E}_j^+$ and $\boldsymbol{E}_j^-$, respectively, and the azimuthal angle (the angle with the optical axis) of the light traveling in the $+z$ direction is θ. From Snell's law,

$$n_{\text{in}} \sin\theta_{\text{in}} = n_j \sin\theta_j = n_{\text{out}} \sin\theta_{\text{out}}. \tag{12.1}$$

$\boldsymbol{E}_j^+$ and $\boldsymbol{E}_j^-$ at the j-th layer can be written as follows:

$$\boldsymbol{E}_j^+(x,z) = \boldsymbol{E}_{0j}^+ \exp(-i\text{k}_0 n_j[x\sin\theta_j + z\cos\theta_j]), \tag{12.2}$$

$$\boldsymbol{E}_j^-(x,z) = \boldsymbol{E}_{0j}^- \exp(-i\text{k}_0 n_j[x\sin\theta_j - z\cos\theta_j]). \tag{12.3}$$

Let us denote the electric field components parallel to the boundary surface at $x=0$ as $E_j^+(z)$ and $E_j^-(z)$, respectively. Assuming that $E_j^+(z)$ and $E_j^-(z)$ at the left end of the j-th layer at $x=0$ are $E_j^+(0)$ and $E_j^-(0)$, respectively, and the electric field components at the right end (also at $x=0$) are $E_j^+(d_j)$ and $E_j^-(d_j)$, respectively. Then the following relationship holds for them:

$$E_j^+(0) = e^{i\phi_j} E_j^+(d_j), \tag{12.4}$$

$$E_j^-(0) = e^{-i\phi_j} E_j^-(d_j), \tag{12.5}$$

where

$$\phi_j = k_0 n_j d_j \cos\theta_j. \tag{12.6}$$

Now, if the multilayer film is considered as a single optical element, the light with the electric field of E_{in}^- is reflected, and the light with the electric field of E_{out}^+ is transmitted, when the light with the electric field of E_{in}^+ is incident on this element. Therefore, in order to calculate the reflectance and transmittance of this multilayer film, it is sufficient to know the relationship between E_{in}^+, E_{in}^-, and E_{out}^+ under the conditions of $E_{\text{out}}^- = 0$. For this purpose, the relationship between E_j^+, E_j^-, E_{j+1}^+, and E_{j+1}^- is required. In the following two subsections, this relationship will be determined separately for the case of s-polarized light, where the electric field is perpendicular to the plane of incidence, and the case of p-polarized light, where the electric field is within the plane of incidence.

12.2.1.1 s-polarized light

At an interface, the components of the electric and magnetic fields parallel to the interface are continuous. Therefore, at the interface between the j-th layer and the $(j+1)$-th layer $(z=d_j)$,

$$E_j^+(d_j) + E_j^-(d_j) = E_{j+1}^+(d_j) + E_{j+1}^-(d_j), \tag{12.7}$$

$$q_j^{\mathrm{s}}\{E_j^+(d_j) - E_j^-(d_j)\} = q_{j+1}^{\mathrm{s}}\{E_{j+1}^+(d_j) - E_{j+1}^-(d_j)\}, \tag{12.8}$$

$$q_j^{\mathrm{s}} = \frac{n_j}{\mu_{\mathrm{r},j}} \cos\theta_j, \tag{12.9a}$$

where $\mu_{\mathrm{r},j}$ is the relative magnetic permeability of the j-th layer.

Now, define A_j and B_j as follows:

$$A_j = E_j^+(0) + E_j^-(0), \tag{12.10}$$

$$B_j = q_j^{\mathrm{s}}\{E_j^+(0) - E_j^-(0)\}. \tag{12.11}$$

The parameters A_j and B_j correspond to the tangential component of the electric field and the tangential component of the magnetic field (divided by $\sqrt{\varepsilon_0/\mu_0}$), respectively, at the left end of the j-th layer. From Eqs. (12.4), (12.5), (12.10), and (12.11),

$$A_j = e^{i\phi_j} E_j^+(d_j) + e^{-i\phi_j} E_j^-(d_j), \tag{12.12}$$

$$B_j = q_j^{\mathrm{s}}\{e^{i\phi_j} E_j^+(d_j) - e^{-i\phi_j} E_j^-(d_j)\}. \tag{12.13}$$

From Eqs. (12.7) and (12.8), express $E_j^+(d_j)$ and $E_j^-(d_j)$ as functions of $E_{j+1}^+(d_j)$ and $E_{j+1}^-(d_j)$, and substitute them into Eqs. (12.12) and (12.13). Then the following equations are obtained:

$$A_j = (\cos\phi_j)A_{j+1} + i\frac{\sin\phi_j}{q_j^{\mathrm{s}}} B_{j+1}, \tag{12.14}$$

$$B_j = i(\sin\phi_j) q_j^{\mathrm{s}} A_{j+1} + \cos\phi_j B_{j+1}. \tag{12.15}$$

12.2.1.2 p-polarized light

For p-polarized light, use the following equations instead of Eqs. (12.10) and (12.11):†

$$A_j = \{E_j^+(0) + E_j^-(0)\}\cos\theta_j, \tag{12.16}$$

$$B_j = \{E_j^+(0) - E_j^-(0)\}\frac{n_j}{\mu_{\mathrm{r},j}}. \tag{12.17}$$

Similarly to the case of s-polarized light, the following equations can be obtained:

---

†Note: As mentioned in Chapter 4, the sign of E_j^- changes depending on the direction of the electric field of the reflected wave. Here, when $\theta = 0$, the direction of the electric field of the reflected wave was chosen so that the signs of the amplitude reflection coefficients were the same for s-polarized light and p-polarized light [corresponding to Fig. 4.4(b) in Chapter 4].

$$A_j = (\cos\phi_j)A_{j+1} + i\frac{\sin\phi_j}{q_j^{\mathrm{p}}}B_{j+1}, \tag{12.18}$$

$$B_j = i(\sin\phi_j)q_j^{\mathrm{p}}A_{j+1} + \cos\phi_j B_{j+1}, \tag{12.19}$$

$$q_j^{\mathrm{p}} = \frac{n_j}{\mu_{\mathrm{r},j}}\frac{1}{\cos\theta_j}. \tag{12.20a}$$

Since the relative permeability of most materials is unity, Eqs. (12.9a) and (12.20a) in that case become as follows:

$$q_j^{\mathrm{s}} = n_j \cos\theta_j, \tag{12.9b}$$

$$q_j^{\mathrm{p}} = \frac{n_j}{\cos\theta_j}. \tag{12.20b}$$

12.2.2 Characteristic matrix

As seen by comparing Eqs. (12.14) and (12.15) with Eqs. (12.18) and (12.19), they have the same exact form. The only differences are the parameters defined in Eqs. (12.9) and (12.20). Therefore, if these two parameters are collectively written as q_j, Eqs. (12.14), (12.15), (12.18), and (12.19) can be represented as follows:

$$\begin{pmatrix} A_j \\ B_j \end{pmatrix} = \boldsymbol{M}_j \begin{pmatrix} A_{j+1} \\ B_{j+1} \end{pmatrix}, \tag{12.21}$$

$$\boldsymbol{M}_j = \begin{pmatrix} \cos\phi_j & i\dfrac{\sin\phi_j}{q_j} \\ iq_j \sin\phi_j & \cos\phi_j \end{pmatrix}. \tag{12.22}$$

The matrix in Eq. (12.22) is called a characteristic matrix or a transfer matrix and expresses the relationship between the electric field and magnetic field of the j-th layer and those of the $(j+1)$-th layer.[†] From this, the relationship between A_{in} and B_{in} on the input side and A_{out} and B_{out} on the output side is represented by

$$\begin{pmatrix} A_{\mathrm{in}} \\ B_{\mathrm{in}} \end{pmatrix} = \boldsymbol{M} \begin{pmatrix} A_{\mathrm{out}} \\ B_{\mathrm{out}} \end{pmatrix}, \tag{12.23}$$

[†]Note: The sign of the off-diagonal term in the characteristic matrix changes depending on whether the time dependence of the electric field is set to $\exp(+i\omega t)$ or $\exp(-i\omega t)$. In this book, it is set as $\exp(+i\omega t)$.

$$M = \prod_{j=1}^{N} M_j. \tag{12.24}$$

Since the characteristic matrix M_j of each layer can be calculated if n_j, d_j, and θ_j are given, the characteristic matrix of the entire multilayer film can be obtained using Eq. (12.24).

12.2.3 Reflectance and transmittance of multilayer films

Equation (12.23) gives the relationship between A_{in} and B_{in} on the incident side and A_{out} and B_{out} on the output side, so the reflectance and transmittance of the multilayer film can be determined from this. In the following two subsections, the reflectance and transmittance for s-polarized light and p-polarized light are derived.

12.2.3.1 s-polarized light

The amplitude reflection coefficient r_s and amplitude transmission coefficient t_s of the multilayer film for s-polarized light are defined below:

$$r_s = \frac{E_{in}^-}{E_{in}^+}, \tag{12.25}$$

$$t_s = \frac{E_{out}^+}{E_{in}^+}. \tag{12.26}$$

Consider the case where s-polarized light with an electric field amplitude of unity is incident on a multilayer film. From the definitions of Eqs. (12.10) and (12.11), and Eqs. (12.25) and (12.26), the following relations are obtained:

$$A_{in} = 1 + r_s, \tag{12.27a}$$

$$B_{in} = q_{in}^s (1 - r_s), \tag{12.27b}$$

$$A_{out} = t_s, \tag{12.27c}$$

$$B_{out} = q_{out}^s t_s. \tag{12.27d}$$

Then Eq. (12.23) becomes

$$\begin{pmatrix} 1 + r_s \\ q_{in}^s (1 - r_s) \end{pmatrix} = \begin{pmatrix} M_{11} & M_{12} \\ M_{21} & M_{22} \end{pmatrix} \begin{pmatrix} t_s \\ q_{out}^s t_s \end{pmatrix}. \tag{12.28}$$

From this, r_s and t_s can be found as follows:

$$r_{\rm s} = \frac{q_{\rm in}^{\rm s}(M_{11} + q_{\rm out}^{\rm s}M_{12}) - (M_{21} + q_{\rm out}^{\rm s}M_{22})}{q_{\rm in}^{\rm s}(M_{11} + q_{\rm out}^{\rm s}M_{12}) + (M_{21} + q_{\rm out}^{\rm s}M_{22})}, \tag{12.29}$$

$$t_{\rm s} = \frac{2q_{\rm in}^{\rm s}}{q_{\rm in}^{\rm s}(M_{11} + q_{\rm out}^{\rm s}M_{12}) + (M_{21} + q_{\rm out}^{\rm s}M_{22})}. \tag{12.30}$$

From this, the reflectance $R_{\rm s}$ and transmittance $T_{\rm s}$ can be calculated using the following equations:

$$R_{\rm s} = |r_{\rm s}|^2, \tag{12.31}$$

$$T_{\rm s} = \frac{q_{\rm out}^{\rm s}}{q_{\rm in}^{\rm s}}|t_{\rm s}|^2 = \frac{n_{\rm out}\cos\theta_{\rm out}}{n_{\rm in}\cos\theta_{\rm in}}|t_{\rm s}|^2. \tag{12.32}$$

12.2.3.2 p-polarized light

In the case of p-polarized light, similarly to Eqs. (12.25) and (12.26), the amplitude reflection coefficient $r_{\rm p}$ and amplitude transmission coefficient $t_{\rm p}$ are given by

$$r_{\rm p} = \frac{E_{\rm in}^-}{E_{\rm in}^+}, \tag{12.33}$$

$$t_{\rm p} = \frac{E_{\rm out}^+}{E_{\rm in}^+}. \tag{12.34}$$

Here, define $t'_{\rm p}$ as follows:

$$t'_{\rm p} = \frac{\cos\theta_{\rm out}}{\cos\theta_{\rm in}}t_{\rm p}. \tag{12.35}$$

If the electric field component of the incident p-polarized light is $E_{\rm in}^+\cos\theta_{\rm in} = 1$, then from Eqs. (12.16), (12.17), (12.33), and (12.34), equations in exactly the same form as Eqs. (12.27a) to (12.27d) are obtained for $r_{\rm p}$ and $t'_{\rm p}$. Therefore, the equation corresponding to Eq. (12.28) is

$$\begin{pmatrix} 1 + r_{\rm p} \\ q_{\rm in}^{\rm p}(1 - r_{\rm p}) \end{pmatrix} = \begin{pmatrix} M_{11} & M_{12} \\ M_{21} & M_{22} \end{pmatrix} \begin{pmatrix} t'_{\rm p} \\ q_{\rm out}^{\rm p}t'_{\rm p} \end{pmatrix}. \tag{12.36}$$

The equations corresponding to Eqs. (12.29) and (12.30) will have the same form. That is,

$$r_{\rm p} = \frac{q_{\rm in}^{\rm p}(M_{11} + q_{\rm out}^{\rm p}M_{12}) - (M_{21} + q_{\rm out}^{\rm p}M_{22})}{q_{\rm in}^{\rm p}(M_{11} + q_{\rm out}^{\rm p}M_{12}) + (M_{21} + q_{\rm out}^{\rm p}M_{22})}, \tag{12.37}$$

$$t'_{\rm p} = \frac{2q_{\rm in}^{\rm p}}{q_{\rm in}^{\rm p}(M_{11} + q_{\rm out}^{\rm p}M_{12}) + (M_{21} + q_{\rm out}^{\rm p}M_{22})}. \tag{12.38}$$

The reflectance $R_{\rm p}$ and the transmittance $T_{\rm p}$ can be calculated using the following equations in the same way as Eqs. (12.31) and (12.32):

$$R_{\rm p} = |r_{\rm p}|^2, \tag{12.39}$$

$$T_{\rm p} = \frac{q_{\rm out}^{\rm p}}{q_{\rm in}^{\rm p}}|t'_{\rm p}|^2. \tag{12.40}$$

Using the original $t_{\rm p}$, Eq. (12.40) becomes

$$T_{\rm p} = \frac{n_{\rm out}\cos\theta_{\rm out}}{n_{\rm in}\cos\theta_{\rm in}}|t_{\rm p}|^2. \tag{12.41}$$

It has the same form as Eq. (12.32).

In actual calculations, by using $r_{\rm p}$ and $t'_{\rm p}$, it is possible to simply switch the parameter q_j to Eq. (12.9) or Eq. (12.20) and perform the calculation using the same equations.

12.2.4 The case for $N = 0$

When the number of layers N is 0, as can be seen from Fig. 12.1 (on page 165), reflection and refraction occur at the interface between two media with refractive indices $n_{\rm in}$ and $n_{\rm out}$. Let's confirm this.

When $N = 0$, the characteristic matrix is an identity matrix. In this case, for s-polarized light, Eqs. (12.29) and (12.30) are represented as follows:

$$r_{\rm s} = \frac{q_{\rm in}^{\rm s} - q_{\rm out}^{\rm s}}{q_{\rm in}^{\rm s} + q_{\rm out}^{\rm s}} = \frac{(n_{\rm in}/\mu_{\rm r,in})\cos\theta_{\rm in} - (n_{\rm out}/\mu_{\rm r,out})\cos\theta_{\rm out}}{(n_{\rm in}/\mu_{\rm r,in})\cos\theta_{\rm in} + (n_{\rm out}/\mu_{\rm r,out})\cos\theta_{\rm out}}, \tag{12.42}$$

$$t_{\rm s} = \frac{2q_{\rm in}^{\rm s}}{q_{\rm in}^{\rm s} + q_{\rm out}^{\rm s}} = \frac{2(n_{\rm in}/\mu_{\rm r,in})\cos\theta_{\rm in}}{(n_{\rm in}/\mu_{\rm r,in})\cos\theta_{\rm in} + (n_{\rm out}/\mu_{\rm r,out})\cos\theta_{\rm out}}. \tag{12.43}$$

Naturally, this matches the equations for reflection and refraction at the interface between two media [Eqs. (4.9) and (4.11) in Chapter 4].

In the case of p-polarized light, from Eqs. (12.37), (12.38), and (12.35),

$$r_{\rm p} = \frac{q_{\rm in}^{\rm p} - q_{\rm out}^{\rm p}}{q_{\rm in}^{\rm p} + q_{\rm out}^{\rm p}} = \frac{(n_{\rm in}/\mu_{\rm r,in})\cos\theta_{\rm out} - (n_{\rm out}/\mu_{\rm r,out})\cos\theta_{\rm in}}{(n_{\rm in}/\mu_{\rm r,in})\cos\theta_{\rm out} + (n_{\rm out}/\mu_{\rm r,out})\cos\theta_{\rm in}}, \tag{12.44}$$

$$t_{\rm p} = \frac{\cos\theta_{\rm in}}{\cos\theta_{\rm out}}\frac{2q_{\rm in}^{\rm p}}{q_{\rm in}^{\rm p} + q_{\rm out}^{\rm p}} = \frac{(n_{\rm in}/\mu_{\rm r,in})\cos\theta_{\rm in}}{(n_{\rm in}/\mu_{\rm r,in})\cos\theta_{\rm out} + (n_{\rm out}/\mu_{\rm r,out})\cos\theta_{\rm in}}. \tag{12.45}$$

This also matches the equation for reflection and refraction for p-polarized light. However, the direction of the electric field of the reflected wave was chosen here so that $r_s = r_p$ when $\theta = 0$, so the sign of Eq. (12.44) is opposite to that of Eq. (4.8) in Chapter 4. Equation (12.45) is the same as Eq. (4.10).

12.2.5 Calculation program using characteristic matrix

As mentioned in Sections 12.2.2 and 12.2.3, the reflectance and transmittance of a multilayer film can be calculated using the following procedure.

i) Given n_j, d_j, and θ_j for each layer, use Eq. (12.4) and find M_j in Eq. (12.22).
ii) Find the characteristic matrix of the entire multilayer film using Eq. (12.24).
iii) Find r_s and t_s (for s-polarized light) or r_p and t'_p (for p-polarized light) using Eqs. (12.29) and (12.30) or Eqs. (12.37) and (12.38).
iv) Calculate the reflectance and transmittance using Eqs. (12.31) and (12.32) or Eqs. (12.39) and (12.40).

This procedure can be easily performed using Excel's VBA, for example. Some of the specifications of the created Excel program are shown below:

(a) Specify the type of expression for film thickness d_j of each layer in one of the following ways:
[1] $n_j d_j/\lambda$, [2] $n_j d_j$, [3] d_j.
For structures based on $\lambda/4$ layers such as distributed Bragg reflectors (DBRs), it is easier to input $n_j d_j/\lambda$ than to input d_j itself.
(b) The reflectance and transmittance can be plotted as a function of the following variables:
[1] wavelength λ, [2] photon energy E_p, [3] incident angle θ.
(c) The Excel program is applicable to absorbent materials as well.
(d) The real and imaginary parts of the refractive index of each layer can be given as a function of wavelength or photon energy.

Appendix 12A shows how to use the Excel program.

12.3 Calculation Example

Calculation examples of various multilayer films using the above Excel program will be introduced in the next chapter. Here, just one simple example of a DBR will be shown.

DBR is a multilayer film in which a high refractive index layer (H) and a low refractive index layer (L) are periodically laminated, each having an optical thickness of $\lambda/4$. If the input side material is denoted by A and the output side material (or substrate) is denoted by B, then the multilayer film structure is expressed as follows:

$$A||HLHL\text{-----}HLH||B, \tag{12.46a}$$

$$A||LHLH\text{-----}LHL||B. \tag{12.47a}$$

Alternatively, this can be written as follows:

$$A||(HL)^N H||B, \tag{12.46b}$$

$$A||(LH)^N L||B. \tag{12.47b}$$

Equations (12.46b) and (12.47b) indicate that this multilayer film has a structure in which $(N+0.5)$ pairs of (HL) or (LH) combinations are laminated. Note that the last layer (+0.5) is to improve symmetry and is not necessarily required.

Figure 12.2 shows an example of the reflection characteristics of a DBR using TiO_2 ($n_r = 2.3$) as the high refractive index layer and SiO_2 ($n_r = 1.46$) as

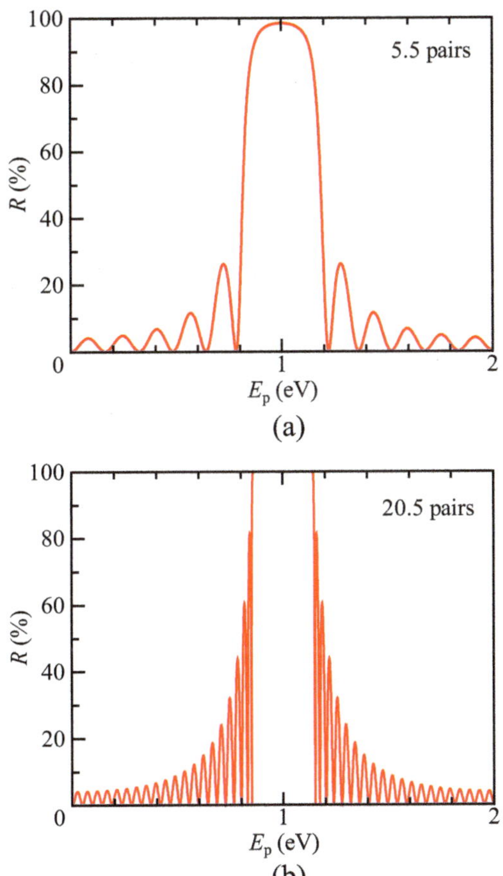

Figure 12.2 Examples of the reflectance characteristics of a DBR as calculated in Excel.

the low refractive index layer. The photon energy at the center wavelength was set to $E_{p0} = 1$ eV (corresponding to $\lambda_0 = 1240$ nm), and both the input and output sides were made of glass (BK7, $n_r = 1.5$). Using these parameters, the following structure was adopted:

$$\mathrm{BK7}||(\mathrm{HL})^N\mathrm{H}||BK7. \tag{12.48}$$

The choice of BK7 on both sides assumes that the DBR manufactured on a glass substrate will be used by bonding it to another glass component (prism, rod lens, etc.). The case of Fig. 12.2(a) is a DBR of 5.5 pairs, and the case of Fig. 12.2(b) is a DBR of 20.5 pairs. As can be seen from the figure, as the pair numbers increase, the reflectance at the center wavelength increases and the boundary of the reflection band becomes steeper. Furthermore, the period of ripples appearing on both sides becomes smaller. In this example, by setting the horizontal axis as the photon energy E_p, the reflectance characteristics become symmetrical with respect to E_{p0}.

Side Topic: Complex number programming

As can be seen from the characteristic matrix represented by Eq. (12.22), the matrix elements are complex numbers, so calculations are performed as complex number operations. In programming languages such as C++ and FORTRAN, complex numbers can be used as variables, and operations on complex numbers can be performed as they are. Therefore, the program can be constructed by simply writing the formulas that appear in the text almost exactly as they are.

On the other hand, Excel also has functions for complex number operations such as IMPRODUCT and IMSQRT, but they are difficult to handle because they require complex numbers to be input as strings such as "5 + 3i". Also, Excel's VBA has a library for complex number operations as an add-in, but the calculation speed seems to be slow.

After all, the program needs to be constructed by yourself, and it is quite a complicated process. For example, when calculating the square root, the argument can be found using Excel's ATAN2, but VBA does not have ATAN2 and only has the Atn function, which is equivalent to ATAN. The results for ATAN and ATAN2 are not necessarily the same, so it is necessary to differentiate between cases. Due to these circumstances, programs often become complicated and are not as clean as FORTRAN.

In the examples shown here, the refractive index of each layer is assumed to be constant regardless of the wavelength, but as mentioned in Section 12.2.5, calculations that take into account the wavelength dispersion of the refractive index are also possible. Regarding wavelength dispersion, Sellmeier's equation was introduced in Chapter 5. It should be noted that the refractive index of the material used for the multilayer film depends on the manufacturing process and is not necessarily the same as the value in bulk.

12.4 Conclusion

The analysis method for optical multilayer films using characteristic matrices was outlined, and a program for calculating reflectance and transmittance using Excel was introduced. In the next chapter, bandpass filters and various multilayer filters calculated using this program will be introduced.

References

1. H. A. Macleod, *Thin-Film Optical Filters, 4th Ed.*, CRC Press (2010).
2. K. Kuroda, *Physical Optics—Propagation of light waves in a medium*, Asakura Publishing (2011) [in Japanese].
3. Y. Honguh, *Landscape of Wave Optics, Multilayer*, Advanced Communication Media, Tokyo (2013) [in Japanese].
4. K. Iga and G. Hatakoshi, Chapter 25, *Treasure Microbox of Optoelectronics II*, Advanced Communication Media Tokyo (2018) [in Japanese].

Appendix 12A Calculation of Reflectance and Transmittance of Multilayer Films Using Excel

Appendix I-A contains an Excel file for calculating the reflectance and transmittance of multilayer films. The operation procedure for this Excel program is shown below.

Table 12.AP1 on page 177 shows an example of inputting an Excel table. This is the DBR structure shown in Fig. 12.2(a) of the main text. In Fig. 12.2, the reflectance was plotted against the photon energy E_p, but here the mode is set to plot the reflectance against the wavelength λ.

Enter the structure of the multilayer film in the three columns A to C. The input parameters are as follows (where Md is mode):

N: The number of layers in the multilayer film. This is automatically calculated and entered in cell E3, so the input of 1 is enough.

M: Number of materials used.

θ_0: Angle of incidence (design value) [deg]

θ: Angle of incidence when calculating [deg]
In the case where $\theta \neq \theta_0$, calculate the reflectance and transmittance for incident angles that differ from the design values.

λ_0: Center wavelength (design value) [μm]

λ: Wavelength when the refractive index is given as a function of this wavelength [μm].

※ When Md – x (described later) is 1, give the values of photon energy [unit: eV] E_{p0} and E_p instead of λ_0, and λ.

λ_1: Lower limit of wavelength range to calculate [μm]

λ_2: Upper limit of wavelength range to calculate [μm]

※ When Md – x is 1, give the values of E_{p1} and E_{p2}.

※ When Md – θ is 1, give the values of θ_1 and θ_2.

n_{div}: Number of divisions when calculating (dividing λ_1 to λ_2 into n_{div})

Md – p: Mode of polarization
0: s-polarization, 1: p-polarization, 2: random polarization

※ The results are the same if the angle of incidence is 0.

Md – θ: If Md – $\theta = 0$, set the wavelength or the photon energy as a variable.
If Md – $\theta = 1$, set the angle of incidence as a variable.

Md – x: If Md – $x = 0$, set the wavelength as a variable.
If Md – $x = 1$, set the photon energy as a variable.

Md – n_r: If Md – $n_r = 0$, the refractive index and the extinction coefficient are fixed values.
If Md – $n_r = 1$, use the refractive index and the extinction coefficient as functions of the wavelength or the photon energy.

Md – d: Type of expression for film thickness
0: $n_r d/\lambda$, 1: $n_r d$, 2: d

※ The unit is in μm for the case of Md – $d = 1$ or 2.

R/T: Vertical axis to plot (reflectance or transmittance)
[Refractive index and extinction coefficient of materials]
Give parameters in the same number of lines as the number of materials M.
Column A is for a comment and is not related to the calculation.

n_r: Real part of the refractive index

κ: Extinction coefficient
Parameters n_r and κ can be given as functions of wavelength or photon energy. In this case, κ is given as a function of the value of cell B7 using, for example, Sellmeier's formula as follows (example for SiO_2):
= SQRT(1 + 0.6961663/(1 – (0.0684043/B7)^2) + 0.4079426/(1 – (0.1162414/B7)^2) + 0.8974794/(1 – (9.896161/B7)^2)).

Table 12.AP1 Part of an Excel spreadsheet for calculating reflectance and transmittance of multi-layer films.

A	B	C	D	E	F	G
DBR	(Comment)		Expand	DBR		
N	M	Md-p		N	M	Md-p
1	3	0	Clear1	11	3	0
θ_0	θ	Md-θ		θ_0	Θ	Md-θ
0	0	0		0	0	0
λ_0	Λ	Md-x		λ_0	Λ	Md-x
1.24	1.24	0		1.24	1.24	0
λ_1	λ_2	n_{div}		λ_1	λ_2	n_{div}
0.6	2.6	2000		0.6	2.6	2000
Md-n_r	Md-d	R/T		Md-n_r	Md-d	R/T
0	0	R		0	0	R
Blank						
Mater.	n_r	κ		Mater.	n_r	κ
BK7	1.5	0		BK7	1.5	0
TiO_2	2.3	0		TiO_2	2.3	0
SiO_2	1.46	0		SiO_2	1.46	0
Blank						
Layer	nd/λ	M		Layer	nd/λ	M
In	0	1		0	0	1
Repeat	5	-1		1	0.25	2
H	0.25	2		2	0.25	3
L	0.25	3		3	0.25	2
End-R		-1		4	0.25	3
H	0.25	2		5	0.25	2
Out	0	3		6	0.25	3
Blank				7	0.25	2
				8	0.25	3
				9	0.25	2
				10	0.25	3
				11	0.25	2
				12	0	1

When Md $-$ $n_r = 1$, this function calculates the reflectance/transmittance in consideration of wavelength dispersion. In addition to the example above, it can be defined by functions in VBA.

[Layer structure]

Column A is for a comment and is not related to the calculation.

Film thickness: Give $n_r d/\lambda$ or $n_r d$ or d, depending on the value of Md $-$ d. m: Material number ($1 \leq m \leq M$)

In the case of $m \leq 0$, the rows until the next $m \leq 0$ appearance are repeated a number of times, which is given by the value in column B of that row.

After entering the structure in the three columns A to C, click the [Expand] button to expand the structure to the three columns E to G (see Table 12.AP1). You can also enter the structure directly into columns E to G at the start. The [Expand] button also plots the film thickness and refractive index distribution (Fig. 12.AP1), allowing you to confirm the input structure.

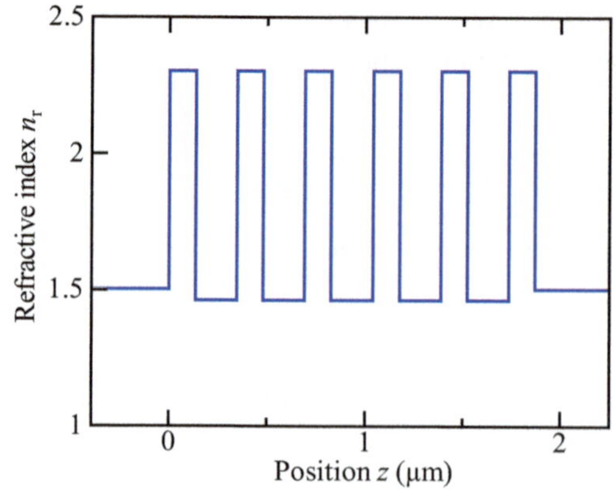

Figure 12.AP1 Film thickness and refractive index distribution of input structure.

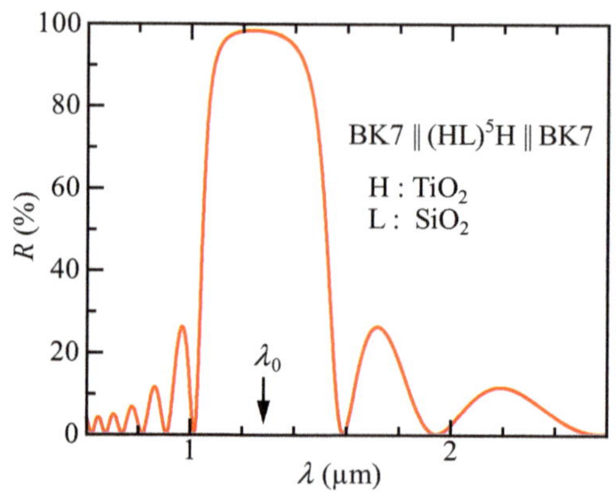

Figure 12.AP2 Plot of reflectance calculation results.

After the structure is expanded in columns E to G, click the [MLF] button to calculate the reflectance and transmittance, and plot the reflectance or transmittance specified by R/T. Figure 12.AP2 shows the results. This is the same as Fig. 12.2(a) in the main text, but since the horizontal axis is the wavelength, it is not symmetrical about the center wavelength.

The contents of the Excel file are as follows.

12_MLF1.xlsm

Instruction	
DBR1	Fig. 12.2
DBR2	Table 12.AP1, Fig. 12.AP1, Fig. 12.AP2

Chapter 13
Optical Multilayer Film Part 2

13.1 Introduction

In the previous chapter, the principles of optical multilayer films and the analysis methods[1–3] that use characteristic matrices were outlined, and an Excel program for calculating reflectance and transmittance was introduced. This chapter introduces various filters that can be calculated using this Excel program.

13.2 Single-Layer Film

Before calculating the multilayer film, let's take a look at what the characteristic matrix looks like in a simple structure where the number of layers N is 1, that is, in the case of a single-layer film.

Here, for simplicity, the case of normal incidence is considered. The characteristic matrix of a single-layer film with refractive index n_1 and thickness d_1 is given by the following equation from Eq. (12.20) in the previous chapter:

$$\boldsymbol{M} = \begin{pmatrix} \cos\phi_1 & i\dfrac{\sin\phi_1}{n_1} \\ in_1\sin\phi_1 & \cos\phi_1 \end{pmatrix}. \tag{13.1}$$

$$\phi_1 = \frac{2\pi}{\lambda}n_1 d_1. \tag{13.2}$$

Also, from Eqs. (12.29) to (12.32) in the previous chapter, the reflectance and transmittance are given as follows:

$$R = \frac{{n_1}^2(n_{\text{in}} - n_{\text{out}})^2\cos^2\phi_1 + ({n_1}^2 - n_{\text{in}}n_{\text{out}})^2\sin^2\phi_1}{{n_1}^2(n_{\text{in}} + n_{\text{out}})^2\cos^2\phi_1 + ({n_1}^2 + n_{\text{in}}n_{\text{out}})^2\sin^2\phi_1}, \tag{13.3}$$

$$T = \frac{4n_{\text{in}}n_{\text{out}}{n_1}^2}{{n_1}^2(n_{\text{in}} + n_{\text{out}})^2\cos^2\phi_1 + ({n_1}^2 + n_{\text{in}}n_{\text{out}})^2\sin^2\phi_1}. \tag{13.4}$$

In the following two subsections, the case where the optical thickness $n_1 d_1$ is $\lambda/4$ and the case where the $n_1 d_1$ is $\lambda/2$ are considered as special cases of a single-layer film.

13.2.1 λ/4 film

When the optical thickness of the film is λ/4, the parameter given by Eq. (12.4) in the previous chapter is $\phi_1 = \pi/2$, so the characteristic matrix of Eq. (13.1) becomes as follows:

$$\boldsymbol{M} = \begin{pmatrix} 0 & \dfrac{i}{n_1} \\ in_1 & 0 \end{pmatrix}. \tag{13.5}$$

In this case, Eqs. (13.3) and (13.4) are

$$R = \frac{(n_1{}^2 - n_{\text{in}}n_{\text{out}})^2}{(n_1{}^2 + n_{\text{in}}n_{\text{out}})^2}, \tag{13.6}$$

$$T = \frac{4n_{\text{in}}n_{\text{out}}n_1{}^2}{(n_1{}^2 + n_{\text{in}}n_{\text{out}})^2}. \tag{13.7}$$

As can be seen from the two above equations, the situation where $R = 0$ and $T = 1$ is realized if the following conditions are satisfied:

$$n_1{}^2 = n_{\text{in}}n_{\text{out}}. \tag{13.8}$$

This is the principle of anti-reflection (AR) film.

In reality, it is difficult to obtain a material that strictly satisfies Eq. (13.8). Also, non-reflection occurs only at the design wavelength; at other wavelengths, the optical thickness of the film is no longer λ/4, so it is not possible to achieve non-reflection over a wide wavelength range with a single-layer film.

As an example, Fig. 13.1 shows the reflectance when a λ/4 layer of MgF_2 ($n_r = 1.3785$ at $\lambda = 550$ nm) is formed on an SF6 glass substrate ($n_r = 1.8119$ at $\lambda = 550$ nm). In this example, $R = 0.057\%$ at the center wavelength of 550 nm, and a low reflectance of less than 3% is obtained in the range of 400 nm to 930 nm. Note that this calculation was plotted using the Excel multilayer film calculation introduced in the previous chapter, and the wavelength dispersions of the refractive indices for SF6 glass and MgF_2 were taken into account using Sellmeyer's equation (see Chapter 5).

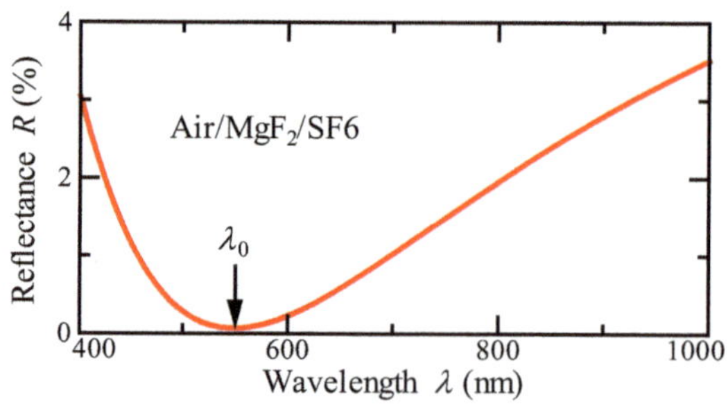

Figure 13.1 Reflection characteristics of MgF_2 λ/4 film on SF6 glass substrate.

13.2.2 λ/2 film

When the optical thickness of the film is λ/2, $\phi_1 = \pi$, so from Eqs. (13.3) and (13.4),

$$R = \frac{(n_{in} - n_{out})^2}{(n_{in} + n_{out})^2}, \tag{13.9}$$

$$T = \frac{4n_{in}n_{out}}{(n_{in} + n_{out})^2}. \tag{13.10}$$

In this case, the reflectance will become zero if the following conditions are satisfied:

$$n_{in} = n_{out}. \tag{13.11}$$

This is a case where the refractive indices on the incident side and the exit side are the same.

The situation in which the reflectance becomes zero is not limited to λ/2 films and is valid if the optical thickness n_1L of the film is $m(\lambda/2)$, where m is an integer. This condition is given by

$$\lambda = 2n_1L/m. \tag{13.12}$$

That is, the reflectance becomes zero for a wavelength λ that satisfies the above equation (see Fig. 13.2). This is the principle of an etalon.

The transmission characteristics of an etalon vary depending on the end face reflectance R (not the reflectance of the etalon but the reflectance at the two-layer interface). Figure 13.2 shows the cases where the etalon material is SiO_2 ($R = 3.5\%$) and TiO_2 ($R = 19.4\%$). As can be seen, the larger R is, the greater the contrast is. The actual etalon used has multilayer films formed on both sides to achieve a reflectance of nearly 100%. This will be discussed in Section 13.4.

The wavelength interval Δλ that satisfies Eq. (13.12) is called the free spectral range (FSR). When determining Δλ, it is necessary to consider the wavelength dispersion of the refractive index n_r as shown below:

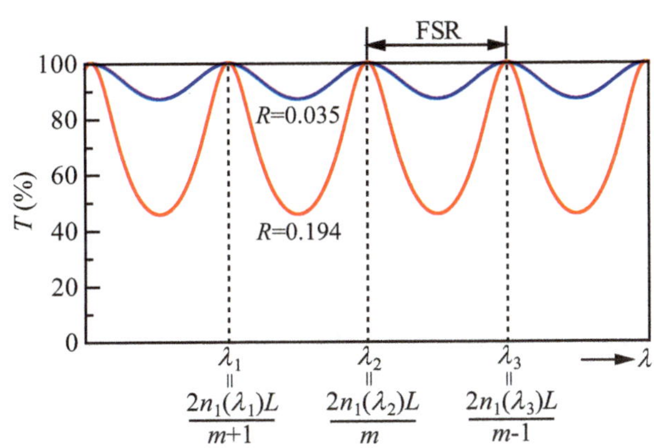

Figure 13.2 Transmission characteristics of $m(\lambda/2)$ film.

$$\Delta\lambda = \frac{\lambda^2}{2n_g L}, \tag{13.13}$$

$$n_g \equiv n_r - \lambda \frac{dn_r}{d\lambda}. \tag{13.14}$$

The parameter n_g defined by Eq. (13.14) is called the group refractive index or group index.

13.3 Distributed Bragg Reflector

13.3.1 Dielectric multilayer film DBR

In the previous chapter, a calculation example for a dielectric multilayer distributed Bragg reflector (DBR) with the following structure was introduced:

$$\mathrm{BK7} \parallel (\mathrm{HL})^N\mathrm{H} \parallel \mathrm{BK7}. \tag{13.15}$$

The reflectance of such a structure can be calculated relatively easily for the center wavelength. From Eq. (13.5), the characteristic matrices of H and L are represented, respectively, as follows:

$$\boldsymbol{M}_\mathrm{H} = \begin{pmatrix} 0 & \dfrac{i}{n_\mathrm{H}} \\ in_\mathrm{H} & 0 \end{pmatrix}, \tag{13.16}$$

$$\boldsymbol{M}_\mathrm{L} = \begin{pmatrix} 0 & \dfrac{i}{n_\mathrm{L}} \\ in_\mathrm{L} & 0 \end{pmatrix}. \tag{13.17}$$

Therefore, the characteristic matrix $\boldsymbol{M}_\mathrm{HL}$ for the structure with HL is

$$\boldsymbol{M}_\mathrm{HL} = \boldsymbol{M}_\mathrm{H}\boldsymbol{M}_\mathrm{L} = \begin{pmatrix} -\dfrac{n_\mathrm{L}}{n_\mathrm{H}} & 0 \\ 0 & -\dfrac{n_\mathrm{H}}{n_\mathrm{L}} \end{pmatrix}. \tag{13.18}$$

From this, the characteristic matrix $\boldsymbol{M}_N$ for the structure with $(\mathrm{HL})^N$ is given by

$$\boldsymbol{M}_N = (\boldsymbol{M}_\mathrm{HL})^N = \begin{pmatrix} (-1)^N\left(\dfrac{n_\mathrm{L}}{n_\mathrm{H}}\right)^N & 0 \\ 0 & (-1)^N\left(\dfrac{n_\mathrm{H}}{n_\mathrm{L}}\right)^N \end{pmatrix}. \tag{13.19}$$

The characteristic matrix $\boldsymbol{M}_{N+0.5}$ for the structure with $(\mathrm{HL})^N\mathrm{H}$ is represented by

$$\boldsymbol{M}_{N+0.5} = (\boldsymbol{M}_{\mathrm{HL}})^N \boldsymbol{M}_{\mathrm{H}} = \begin{pmatrix} (-1)^N \left(\frac{n_{\mathrm{L}}}{n_{\mathrm{H}}}\right)^N & 0 \\ 0 & (-1)^N \left(\frac{n_{\mathrm{H}}}{n_{\mathrm{L}}}\right)^N \end{pmatrix} \begin{pmatrix} 0 & \frac{i}{n_{\mathrm{H}}} \\ i n_{\mathrm{H}} & 0 \end{pmatrix}$$

$$= \begin{pmatrix} 0 & i(-1)^N \frac{{n_{\mathrm{L}}}^N}{{n_{\mathrm{H}}}^{N+1}} \\ i(-1)^N \frac{{n_{\mathrm{H}}}^{N+1}}{{n_{\mathrm{L}}}^N} & 0 \end{pmatrix}. \tag{13.20}$$

In the case where $n_{\mathrm{in}} = n_{\mathrm{out}} = n_0$, the reflectance R is given by the following formula:

$$R = \left\{ \frac{n_0(M_{11} + n_0 M_{12}) - (M_{21} + n_0 M_{22})}{n_0(M_{11} + n_0 M_{12}) + (M_{21} + n_0 M_{22})} \right\}^2$$

$$= \left\{ \frac{(n_0/n_{\mathrm{H}})^2 - (n_{\mathrm{H}}/n_{\mathrm{L}})^{2N}}{(n_0/n_{\mathrm{H}})^2 + (n_{\mathrm{H}}/n_{\mathrm{L}})^{2N}} \right\}^2. \tag{13.21}$$

Figure 13.3 shows the dependence of reflectance R on N for this DBR. In the figure, the reflectance for the $(N + 0.5)$ pair structure BK7 || $(\mathrm{HL})^N$ H || BK7 given by Eq. (13.21) is indicated by ○, and the reflectance for the N pair structure BK7 || $(\mathrm{HL})^N$ || BK7 is indicated by □. As can be seen from the figure, as N increases, the reflectance increases, and for $N > 10$ (not shown in this figure), a reflectance close to unity is obtained. Figure 12.2 in the previous chapter shows the reflectance for BK7 || $(\mathrm{HL})^N$ H || BK7 structures with 5.5 pairs and 20.5 pairs.

Figure 13.4 shows the dependence of the reflectance on the angle of incidence for the above DBR with 20.5 pairs.

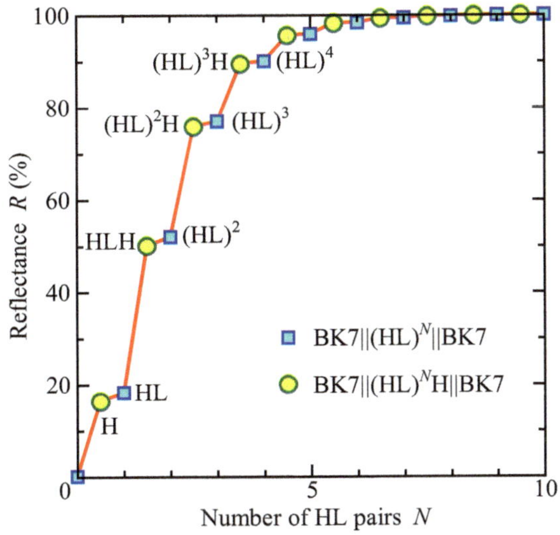

Figure 13.3 Reflectance at the center wavelength of DBR.

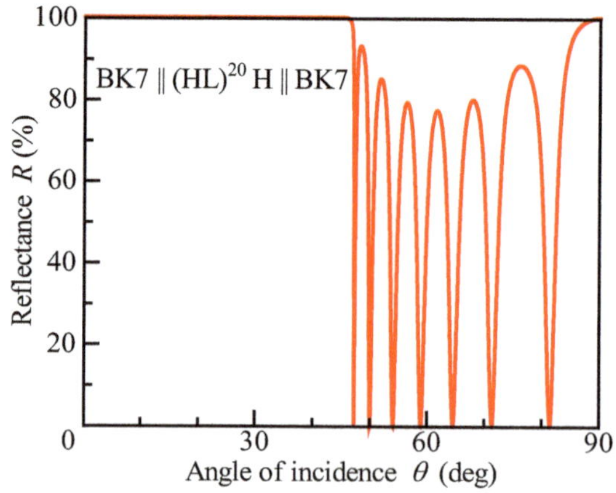

Figure 13.4 Incident angle dependence of DBR reflectance.

13.3.2 Semiconductor multilayer film DBR

DBRs are also used in vertical-cavity surface-emitting lasers (VCSELs). As an example of a DBR using a semiconductor multilayer film, Fig. 13.5 was calculated using the reflection characteristics of a GaAs/AlAs DBR for a 980-nm semiconductor laser. In the calculation, the $Ga_{0.7}Al_{0.3}As$ layer was set as the input side, and the GaAs substrate was set as the output side. Figure 13.5(a) is calculated by fixing the refractive index of the semiconductor to the value at the center wavelength. As shown in this figure, when the reflectance is plotted as a function of the photon energy E_p, it becomes symmetrical with respect to the center wavelength E_{p0}, which is similar to the diagram shown in Fig. 12.2 in the previous chapter. On the other hand, if the wavelength dispersion of the refractive index is considered, the result will be as shown in Fig. 13.5(b). The reason the reflectance is smaller than that in Fig. 13.5(a) in the region where E_p is large is due to the appearance of absorption in the GaAs layer. This figure is based on the Excel multilayer film calculation program introduced in the previous chapter, and the refractive index model[4,5] described in Chapter 6 was used for the refractive index of GaAlAs.[†]

13.4 Etalon

An etalon made of a single-layer $m(\lambda/2)$ film was explained in Section 13.2.2. An actual etalon uses multilayer films to increase the reflectance on both sides to nearly 100%, increasing the steepness of wavelength selection.

[†]Note: The parameters for GaAs were based on Ref. 4, and the parameters for $Ga_{1-x}Al_xAs$ were based on Ref. 5. For parameters whose dependence on composition x is unknown, the values for GaAs were used as they were. Therefore, some areas do not necessarily match the measured data.

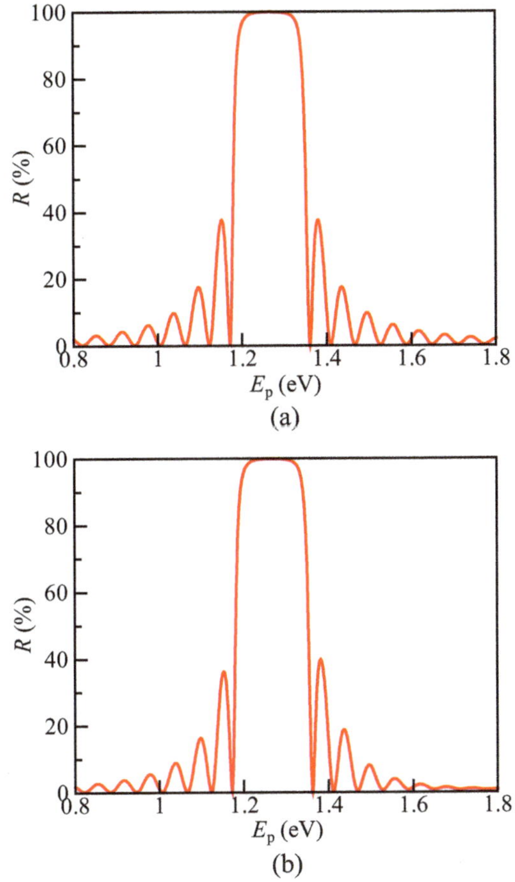

Figure 13.5 Reflection characteristics of a GaAs/AlAs DBR (a) with a fixed refractive index and (b) considering wavelength dispersion of the refractive index.

Figure 13.6(a) shows an example of the structure of an etalon with DBRs provided on both sides of a SiO_2 substrate. In this example, the DBR was set to 3.5 pairs to make it easier to see. Even with such a small number of DBR layers, relatively steep wavelength selectivity can be obtained, as shown in Fig. 13.6(b). Even steeper characteristics can be obtained by increasing the number of DBR layers, in which case, the wavelength division must be made quite fine when calculating using an Excel program.

Another example of an etalon is a structure with an air gap between the DBR substrates, as shown in Fig. 13.7. In this case, the air gap plays the role of an $m(\lambda/2)$ layer. Since the temperature dependence of the refractive index of air is smaller than that of glass, this etalon has the advantage that the temperature dependence of the transmission peak wavelength is small. The transmittance characteristics of the etalon in Fig. 13.7 are almost the same as those in Fig. 13.6(b).

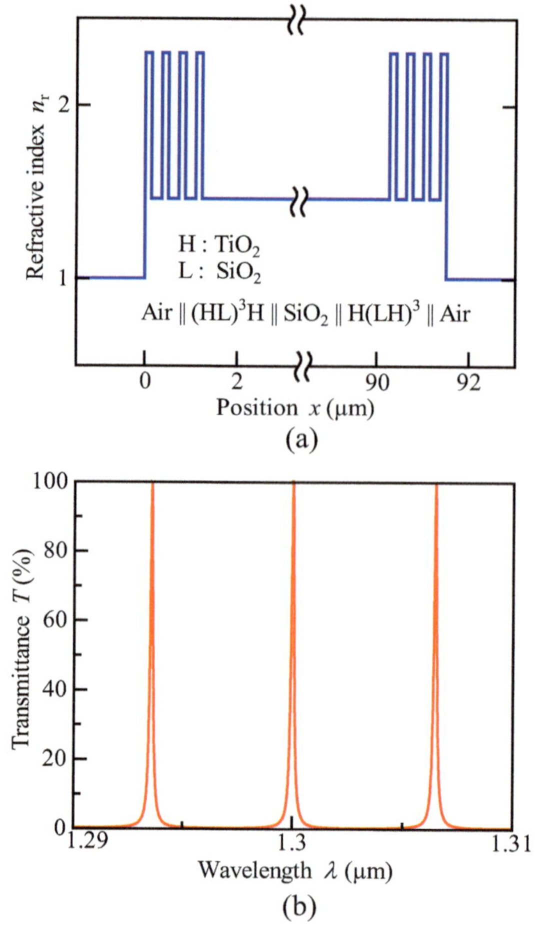

Figure 13.6 Example of structure and characteristics of etalon with DBR on both sides. (a) Layer structure and (b) transmittance characteristics.

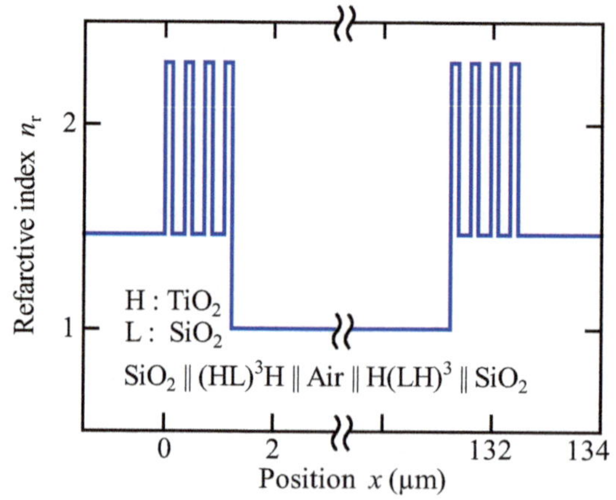

Figure 13.7 Structure example of air gap etalon.

13.5 Wavelength Range Selection Filters

13.5.1 Bandpass filters

As mentioned above, an etalon is a type of bandpass filter (BPF). However, a BPF like the one shown in Fig. 13.6 has a very narrow passband width. The bandwidth can be widened by reducing the number of DBR layers and the reflectance, but in this case, as shown in Fig. 13.2, the reflectance in the reflection band does not become very large, so the performance as a BPF is not high.

It is possible to increase the reflectance in the reflection band and widen the passband width by providing multiple cavities. In the following discussion, as in the case of DBR in Section 13.3.1, the high refractive index layer (TiO_2) with an optical thickness of $\lambda/4$ is represented by H, and the low refractive index layer (SiO_2) is represented by L. First, let A be a structure in which a cavity (2L) with an optical thickness of $\lambda/2$ is sandwiched between 1.5 pairs of DBRs (HLH). That is,

$$A = \text{HLH(2L)HLH}. \tag{13.22}$$

Consider the following structure[6,7] in which multiple structures of A are provided:

$$\text{BK7} \parallel \text{ALA} \parallel \text{BK7}, \tag{13.23a}$$

$$\text{BK7} \parallel \text{ALALA} \parallel \text{BK7}, \tag{13.23b}$$

$$\text{BK7} \parallel \text{ALALALA} \parallel \text{BK7}, \tag{13.23c}$$

$$\text{BK7} \parallel \text{ALALALALA} \parallel \text{BK7}. \tag{13.23d}$$

Figure 13.8 shows the calculated transmission characteristics for these structures. As can be seen from the figure, increasing the number of cavities makes the boundary between the passband and reflection band steeper, but the ripple in the passband also increases. The three-cavity structure shown in Fig. 13.8(b) shows relatively good characteristics as a BPF.[6,7]

As mentioned above, the passband width and steepness of BPF depend on the number of $\lambda/2$ cavities and the number of DBR layers. In the BPF design, it is necessary to consider not only the transmission characteristics but also the tolerance to film thickness changes[6] and the tolerance to refractive index changes[8] during fabrication.

13.5.2 Long-wavelength pass filters and short-wavelength pass filters

In addition to the BPF mentioned above, long-wavelength pass filters (LWPF) and short-wavelength pass filters (SWPF) are used for applications such as optical multiplexers/demultiplexers.[6,7,9] The long-wavelength

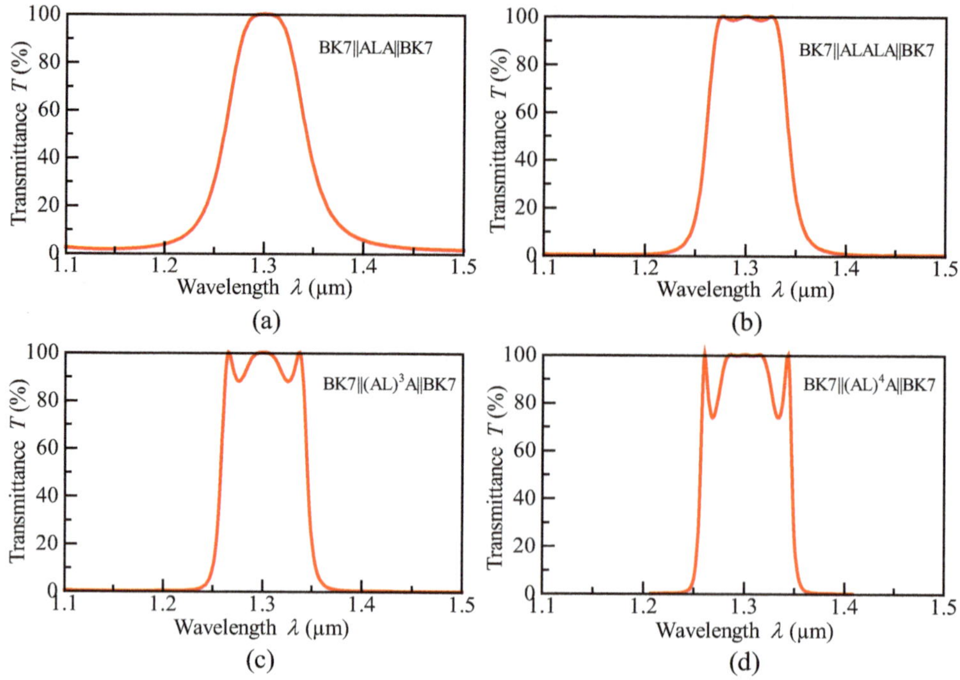

Figure 13.8 Transmission characteristics of the BPF expressed by Eqs. (13.23a) to (13.23d).

side and short-wavelength side of the reflection band in the DBR described in Section 13.3.1 function as a LWPF and SWPF, respectively, but as shown in Fig. 12.2 in the previous chapter, there is a large ripple on either side. Therefore, such a DBR cannot be used as a practical LWPF or SWPF in its current state.

A structure similar to the DBR structure, $((\mathrm{H}/2)\mathrm{L}(\mathrm{H}/2))^m$ or $((\mathrm{L}/2)\mathrm{H}(\mathrm{L}/2))^m$, where m is an integer, can be used as the basic structure of LWPF and SWPF.[6] These are structures where the thickness of the outermost layers of $(\mathrm{HL})^N\mathrm{H}$ or $(\mathrm{LH})^N\mathrm{L}$ are halved to H/2 or L/2. However, even with this structure, ripples still remain. There have been reports of cases in which ripples were further reduced by introducing an AR (anti-reflection) structure to the outermost layer.[6] There is also a method of optimizing the film thickness of the outer few layers as a parameter to eliminate ripples. To perform the optimization, the simplex method, simulated annealing, etc., can be used.

An example of a 27-layer structure in Table 13.1 shows how ripples are reduced. Column (a) in the table is a simple DBR structure of $(\mathrm{HL})^{13}\mathrm{H}$. Column (b) shows the structure where the thickness of the outermost layer is H/2 and is the basic structure of the LWPF described above. Column (c) shows a structure optimized to reduce ripple by changing the thickness of each of the six outer layers. The transmission characteristics for these three types of

Table 13.1 27-layer LWPF structure.

Layer No.	(a)	(b)	(c)
	BK7	BK7	BK7
1	H	0.5 H	0.388 H
2	L	L	0.964 L
3	H	H	0.916 H
4	L	L	0.884 L
5	H	H	0.968 H
6	L	L	0.980 L
7 ⋮ 21	$(HL)^7H$	$(HL)^7H$	$(HL)^7H$
22	L	L	0.980 L
23	H	H	0.968 H
24	L	L	0.884 L
25	H	H	0.916 H
26	L	L	0.964 L
27	H	0.5 H	0.388 H
	BK7	BK7	BK7

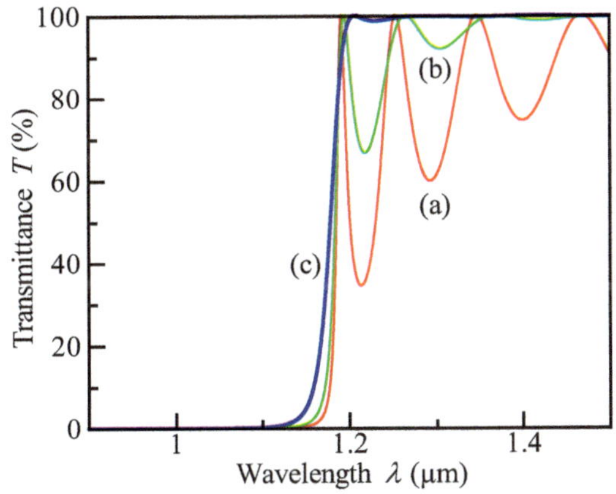

Figure 13.9 Transmission characteristics of the LWPF in Table 13.1.

structures are shown in Fig. 13.9. In Fig. 13.9(b), the ripple around 1.2 μm has been reduced by about half compared to Fig. 13.9(a). Figure 13.9(c), which has been further optimized, shows that the ripple has been significantly reduced.

13.6 Various Filters

Depending on the application, such as display or colorimetry, filters with special spectral characteristics may be used. In a multilayer film consisting of N layers, there are N parameters (degrees of freedom) in the case where the refractive index of each layer is fixed and the thickness is varied. Using this flexibility, it is possible to design filters with various characteristics. Several methods have been reported to solve the "inverse problem" of determining the parameters of a multilayer film from desired spectral characteristics, such as the Fourier transform method[10,11] and the needle method.[12,13]

Sullivan et al.[13,14] have reported the design and fabrication of filters with various characteristics using Nb_2O_5/SiO_2 multilayer films. As an example, a color matching function filter in CIE *XYZ* color space was reported. The color matching functions $\bar{x}(\lambda)$, $\bar{y}(\lambda)$, and $\bar{z}(\lambda)$ representing the tristimulus values in the *XYZ* color space are shown in Fig. 13.10(a). An example has been reported in which 31 layers of a Nb_2O_5/SiO_2 multilayer film were designed for $\bar{x}(\lambda)$.[13] Although not the same as this, an example of the transmission characteristics of the $\bar{x}(\lambda)$ filter designed as a 31-layer structure

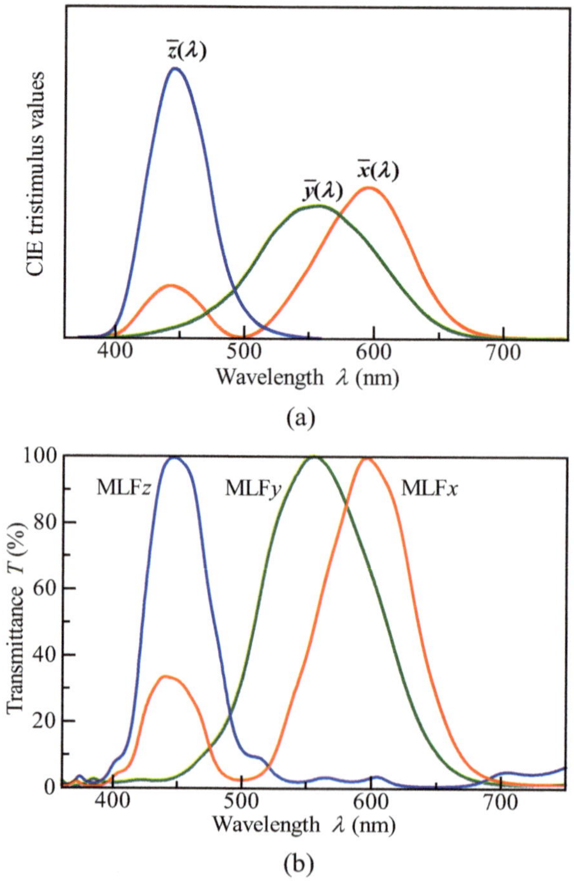

Figure 13.10 (a) Color matching functions in CIE *XYZ* color space and (b) transmission characteristics of $\bar{x}(\lambda)$, $\bar{y}(\lambda)$, and $\bar{z}(\lambda)$ filters designed with a Nb_2O_5/SiO_2 multilayer film.

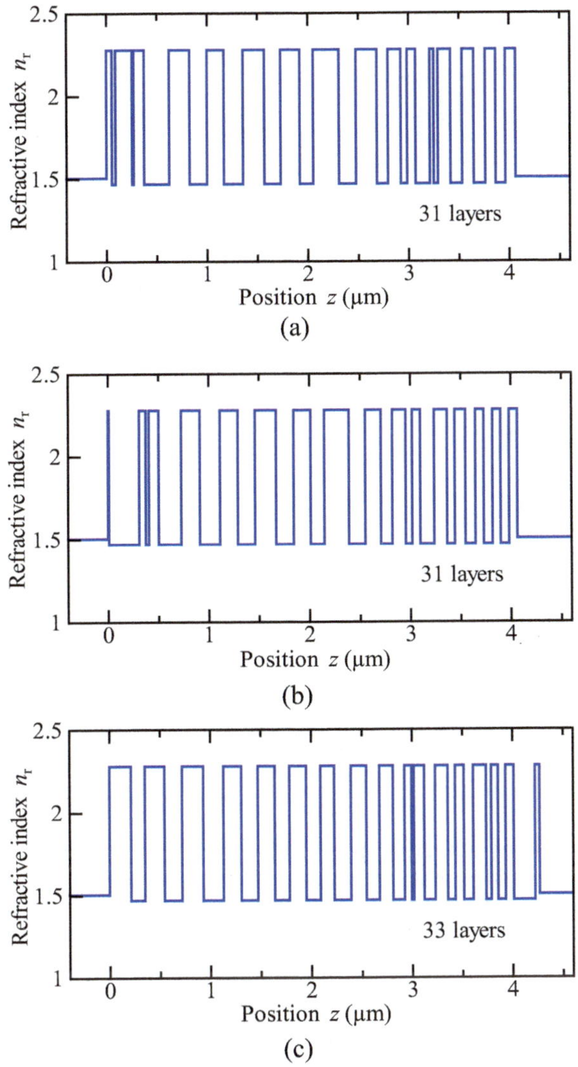

Figure 13.11 Layer structures for (a) the $\bar{x}(\lambda)$ filter, (b) $\bar{y}(\lambda)$ filter, and (c) $\bar{z}(\lambda)$ filter.

of Nb_2O_5/SiO_2 is shown in Fig. 13.10(b). The same figure also shows examples of $\bar{y}(\lambda)$ filters (31 layers) and $\bar{z}(\lambda)$ filters (33 layers). In these calculations, the wavelength dispersion formula for the refractive index of Nb_2O_5 and SiO_2 reported in Ref. 13 was used. The designed layer structures for these filters are shown in Fig. 13.11.

As another example, Sullivan et al. presented a demonstration filter whose reflection characteristics were designed to have a silhouette shape of the Taj Mahal. It is also a 61-layer dielectric multilayer filter made of Nb_2O_5 and SiO_2, and it was designed using the Fourier transform method. It was presented at the International Thin Film Conference held in India.[14]

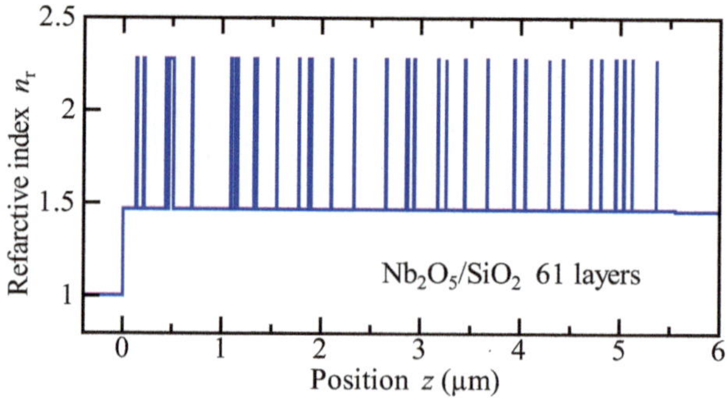

(a)

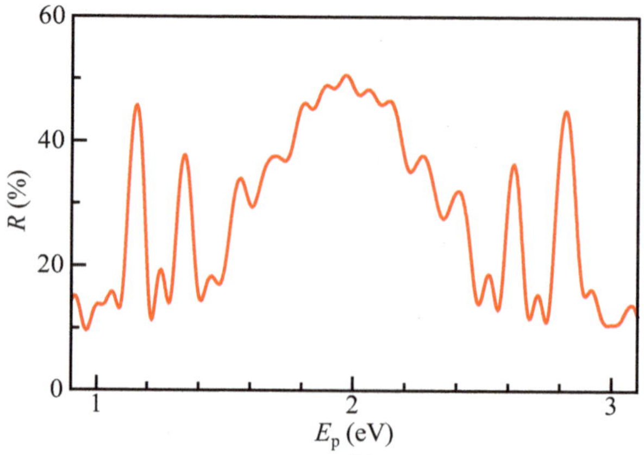

(b)

Figure 13.12 (a) Layer structure of a Taj Mahal filter designed by Sullivan et al.[13] (b) Reflection characteristics calculated with an Excel program. Plotted with permission based on data from B. T. Sullivan and J. A. Dobrowolski[13] © Optica Publishing Group.

An example of designing the same Taj Mahal filter using the needle method has also been reported.[13] The layer structure is reported in Ref. 13 and is reproduced in Fig. 13.12. As in the case of Fig. 13.11, the wavelength dispersion of the refractive index of Nb_2O_5 and SiO_2 is taken into account. Note that this layered structure has two layers with a thickness of 0, so it is essentially a 57-layer structure. In the Excel program, an error will not occur even if the thickness is 0.

Side Topic: Filter design as an inverse problem

Using the characteristic matrix method introduced in the previous chapter, given the layer structure of a multilayer film, the reflection and transmission characteristics of that structure can be easily calculated. On the other hand, it is not so easy to find a multilayer film structure for given

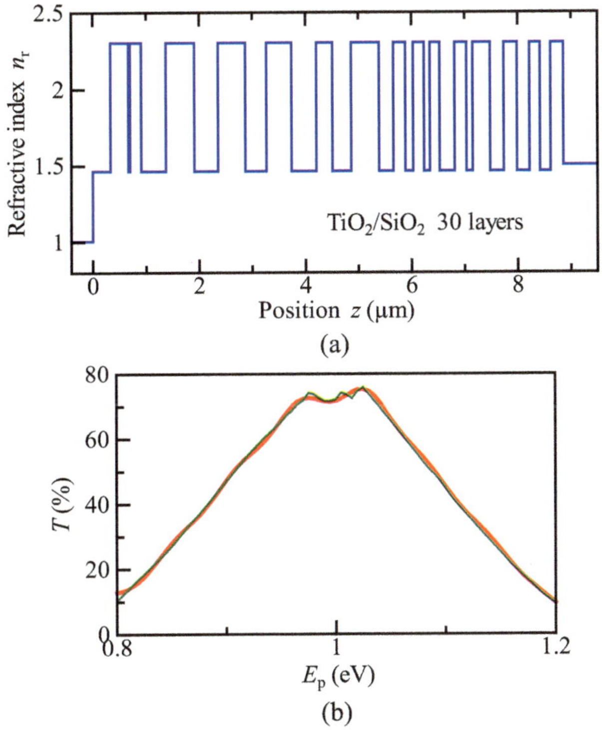

Figure 13.S1 (a) Layer structure of a Mt. Fuji filter and (b) the target function (thin line) used for fitting and reflection characteristics (thick line) for the layer structure in (a).

reflection or transmission characteristics. As mentioned in the main text, this involves solving an inverse problem.

The Taj Mahal shape shown in Fig. 13.12 has many fine undulations, so it uses a structure with a large number of layers: 61 layers. A shape with simple transmission characteristics can reduce the number of layers, but fitting does not necessarily become easier. Figure 13.S1 shows an example of a Mt. Fuji-shaped filter composed of 30 TiO_2/SiO_2 multilayer films. Figure 13.S1(b) shows the target function (thin line) used for fitting, but with 30 layers, it is not possible to fit even the finer undulations of the mountain shape.

As a fitting method, when the number of layers N is sufficiently large, the number of parameters is N, corresponding to the film thicknesses of each layer, so it is possible to select N points as the value of reflectance or transmittance and solve simultaneous N-parameter equations. For example, a multivariable Newton method can be used for this. However, convergence is not necessarily guaranteed. It depends on how close the initial value can be set. There are several ways to provide that initial value.

Although the Fourier transform method is an approximation, it uses the Fourier transform of transmission characteristics to determine the refractive index distribution in the layer thickness direction and converts it into a binary refractive index and the thickness of each layer. Using this as an initial value, the film thickness of each layer is further optimized.

On the other hand, a method called the needle method increases the number of layers to improve the degree of fitting. This involves inserting a very thin layer (needle layer) and determining its insertion position so that the fitting evaluation function is the smallest. Normally, as soon as one layer is inserted, the transmission characteristics of the entire structure change significantly. For very thin layers, there is no significant change in transmission characteristics. The method is to increase the number of layers and improve the degree of fitting by repeating the process of inserting a needle layer, optimizing the overall layer thickness, and inserting more needle layers.

13.7 Conclusion

The characteristics of various filters can be analyzed using an Excel program based on the optical multilayer film analysis method using characteristic matrices. This chapter introduced examples of DBR, etalon, BPF, LWPF, etc., designed by using the Excel program.

References

1. H. A. Macleod, *Thin-Film Optical Filters, 4th Ed.*, CRC Press (2010).
2. K. Kuroda, *Physical Optics—Propagation of light waves in a medium*, Asakura Publishing (2011) [in Japanese].
3. K. Iga and G. Hatakoshi, Chapter 25, *Treasure Microbox of Optoelectronics II*, Advanced Communication Media (2018) [in Japanese].
4. S. Adachi, *Properties of Group-IV, III-V, and II-VI Semiconductors*, Wiley Series in Materials for Electronic & Optoelectronic Applications, John Wiley & Sons (2005).
5. S. Adachi, "GaAs, AlAs, and $Al_xGa_{1-x}As$: Material parameters for use in research and device applications," *J. Appl. Phys.* **58**(3), R1–R29 (1985).
6. J. Minowa, Y. Fujii, and N. Miyamoto, "Dielectric multilayer filters for WDM transmission systems," *Tech. Report IEICE*, OQE82-14, 65–72 (1982) [in Japanese].
7. J. Minowa and Y. Fujii, "Dielectric multilayer thin-film filters for WDM transmission systems," *J. Lightwave Tech.* **LT-1**(1), 116–121 (1983).

8. G. Hatakoshi, K. Ide, S. Ohshima, A. Yoshinaga, and T. Ito, "Analysis of center wavelength shift in dielectric multilayer filters," *Tech. Report IEICE*, OQE84-106, 75–80 (1984) [in Japanese].
9. H. Mineo, S. Ashikawa, K. Yamamura, and G. Hatakoshi, "Low crosstalk optical multiplexer/demultiplexer for bidirectional WDM transmission," *National Conference Record, the Institute of Electronics and Communication, Optical and radio wave electronics* **359**, 2–103 (1984) [in Japanese].
10. J. A. Dobrowolski and D. Lowe, "Optical thin film synthesis program based on the use of Fourier transforms," *Appl. Opt.* **17**(19), 3039–3050 (1978).
11. P. G. Verly, "Fourier transform technique with refinement in the frequency domain for the synthesis of optical thin films," *Appl. Opt.* **35**(25), 5148–5154 (1996).
12. A. V. Tikhonravov and J. A. Dobrowolski, "Quasi-optimal synthesis for antireflection coatings: a new method," *Appl. Opt.* **32**(22), 4265–4275 (1993).
13. B. T. Sullivan and J. A. Dobrowolski, "Implementation of a numerical needle method for thin-film design," *Appl. Opt.* **35**(28), 5484–5492 (1996).
14. B. T. Sullivan, J. A. Dobrowolski, G. Clarke, T. Akiyama, N. Osborne, M. Ranger, L. Howe, A. Matsumoto, Y. Song, and K. Kikuchi, "Manufacture of complex optical multilayer filters using an automated deposition system," *Vacuum* **51**(4), 647–654 (1998).

Appendix 13A Excel File

Appendix I-A contains the following Excel file. Each tab corresponds to a figure and/or table in the main text.

13_MLF2.xlsm

Instructions	
AR	Fig. 13.1
mλ2	Fig. 13.2
DBR-θ	Fig. 13.4
DBR-GaAlAs	Fig. 13.5
Etalon1	Fig. 13.6
Etalon2	Fig. 13.7
BPF	Fig. 13.8
LWPF1	Table 13.1(a), Fig. 13.9(a)
LWPF2	Table 13.1(b), Fig. 13.9(b)

LWPF3	Table 13.1(c), Fig. 13.9(c)
CIE-x	Fig. 13.10(b), Fig. 13.11(a)
CIE-y	Fig. 13.10(b), Fig. 13.11(b)
CIE-z	Fig. 13.10(b), Fig. 13.11(c)
TajMahal	Fig. 13.12
Mt. Fuji	Fig. 13.S1

Chapter 14
Optical Waveguide Part 1

14.1 Introduction

In the optical multilayer film described in Chapters 12 and 13, the reflectance and transmittance of light traveling in the direction perpendicular to the layer interface were analyzed using a characteristic matrix. On the other hand, planar optical waveguides[1–4] have structures in which light travels in a direction parallel to the layer interface. In fact, the boundary conditions for optical multilayer films and planar optical waveguides are exactly the same, and the formula for the characteristic matrix can be applied almost directly to multilayer structure planar optical waveguides. Utilizing this, it is possible to determine the waveguide mode of an arbitrary multilayer planar optical waveguide. In this chapter, the method for calculating modes in a three-layer optical waveguide, which is the basis of a planar optical waveguide, will be discussed.

14.2 Modes in Optical Waveguides

14.2.1 Guided plane waves in planar optical waveguides

Optical waveguides were briefly introduced in Chapter 2, where the solutions for cylindrical waves were described. In this chapter, the more basic guided plane waves will be discussed. As shown in Fig. 14.1, the propagating direction of the guided plane wave is set as z, and the stacking direction of the optical waveguide is set as x. Note that the coordinate system is different from the one used in Chapter 2.

Guided plane waves include the TE (transverse-electric) mode and TM (transverse-magnetic) mode. This "transverse" refers to the plane perpendicular to the direction of propagation; therefore, the xy plane in Fig. 14.1 corresponds to the transverse plane. The mode in which the electric field $\boldsymbol{E}$ is in the transverse plane is called the TE mode, and the mode in which the magnetic field $\boldsymbol{H}$ is in the transverse plane is called the TM mode. Although the derivation is omitted here, it is derived that in the TE mode, the electric field has only the E_y component, and in the TM mode, the magnetic field has only the H_y component.

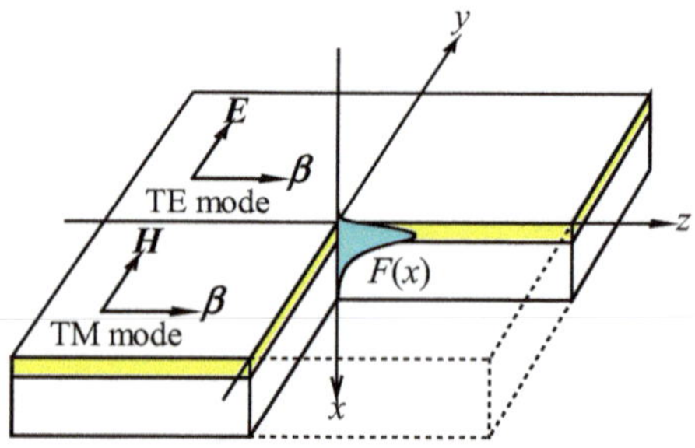

Figure 14.1 Optical waveguide and coordinate system.

Now, if the time dependence of the guided plane wave is $\exp(i\omega t)$ and the effective propagation constant in the z direction is β, then E_y (for the TE mode) or H_y (for the TM mode) can be written as follows:

$$E_y, H_y = F(x)\exp(-i\beta z + i\omega t). \tag{14.1}$$

The function $F(x)$ in the above equation represents the waveguide mode and is the same as the $F(z)$ that appeared in Chapter 2. By substituting Eq. (14.1) into the wave equation derived from Maxwell's equations, the equation satisfied by $F(x)$ can be obtained. In a planar optical waveguide consisting of a layered structure of uniform media, $F(x)$ satisfies the following equation:

$$\frac{d^2}{dx^2}F(x) + \{k_0{}^2 n^2(x) - \beta^2\}F(x) = 0. \tag{14.2}$$

Using this $F(x)$, the electric field $\boldsymbol{E}$ and magnetic field $\boldsymbol{H}$ can be expressed by the following equations:

(TE mode)

$$\boldsymbol{E} = \begin{pmatrix} 0 \\ 1 \\ 0 \end{pmatrix} F(x)\exp(-i\beta z + i\omega t), \tag{14.3}$$

$$\boldsymbol{H} = \frac{1}{\omega\mu_0}\begin{pmatrix} -\beta \\ 0 \\ i(d/dx) \end{pmatrix} F(x)\exp(-i\beta z + i\omega t). \tag{14.4}$$

(TM mode)

$$\boldsymbol{H} = \begin{pmatrix} 0 \\ 1 \\ 0 \end{pmatrix} F(x)\exp(-i\beta z + i\omega t), \tag{14.5}$$

$$\boldsymbol{E} = \frac{1}{\omega\varepsilon_0 n^2(x)} \begin{pmatrix} \beta \\ 0 \\ -i(d/dx) \end{pmatrix} F(x)\exp(-i\beta z + i\omega t). \tag{14.6}$$

From Eqs. (14.3) and (14.4), it can be seen that in the TE mode, the electric field has only a y component, but the magnetic field has a z component (longitudinal component) in addition to the x component (transverse component).† Similarly, from Eqs. (14.5) and (14.6), the electric field has a z component in the TM mode.

Here, consider the time-averaged Poynting vector of the following equation:

$$\boldsymbol{S} = \frac{1}{4}(\boldsymbol{E} \times \boldsymbol{H}^* + \boldsymbol{E}^* \times \boldsymbol{H}). \tag{14.7}$$

Equation (14.7) has only a z component, which is given by

$$S_z^{\mathrm{TE}} = \frac{1}{2\omega\mu_0}\mathrm{Re}(\beta)|F(x)|^2, \tag{14.8}$$

$$S_z^{\mathrm{TM}} = \frac{1}{2\omega\varepsilon_0}\mathrm{Re}\left(\frac{\beta}{n^2(x)}\right)|F(x)|^2. \tag{14.9}$$

As can be seen from Eq. (14.9), the Poynting vector in the TM mode is discontinuous due to the refractive index step at the interface.††

14.2.2 Three-layer optical waveguides

Let us consider an optical waveguide with a three-layer structure consisting of cladding/core/cladding as the simplest planar optical waveguide. The term three layers here refers to the number of layers including the cladding on both sides.††† In the example in Fig. 14.1, the air layer corresponds to the cladding on one side.

Let the refractive index of each layer be n_1, n_2, and n_3. From Eq. (14.2), $F(x)$ is represented as $\cos(\kappa x - \phi)$ in the core, and as $\exp(\gamma_1 x)$ and $\exp(-\gamma_3 x)$ in

†Note: See the side topic in Chapter 2 for the terms "transverse" and "longitudinal."

††Note: Since the relative permeability $\mu_\mathrm{r} = 1$ is assumed here, there is no discontinuity in the Poynting vector of the TE mode at the interface. However, if μ_r changes from layer to layer, a discontinuity due to a step in the permeability will appear in the TE mode.

†††Note: In the optical multilayer film described in the previous chapter, the input side medium and the output side medium were not counted in the number of layers, so the number of layers in a planar optical waveguide is two more than in the case of an optical multilayer film. Therefore, it will be referred to here as a three-layer optical waveguide rather than a single-layer optical waveguide. It is assumed that the thickness of the cladding layers on both sides of the three layers is infinite. Although this is not possible in reality, it corresponds to an approximation when the cladding layer thickness is sufficiently larger than $1/\gamma_i$.

cladding layers 1 and 3, respectively. Then the parameters κ, γ_1, and γ_3 are given by the following equations:

$$\kappa = \sqrt{k_0{}^2 n_2{}^2 - \beta^2}, \tag{14.10a}$$

$$\gamma_1 = \sqrt{\beta^2 - k_0{}^2 n_1{}^2}, \tag{14.11a}$$

$$\gamma_3 = \sqrt{\beta^2 - k_0{}^2 n_3{}^2}. \tag{14.12a}$$

Let the effective propagation constant β be expressed as

$$\beta = k_0 n_{\mathrm{eq}}. \tag{14.13}$$

Then Eqs. (14.10a) to (14.12a) become

$$\kappa = k_0 \sqrt{n_2{}^2 - n_{\mathrm{eq}}{}^2}, \tag{14.10b}$$

$$\gamma_1 = k_0 \sqrt{n_{\mathrm{eq}}{}^2 - n_1{}^2}, \tag{14.11b}$$

$$\gamma_3 = k_0 \sqrt{n_{\mathrm{eq}}{}^2 - n_3{}^2}. \tag{14.12b}$$

The index n_{eq} defined by Eq. (14.13) is the equivalent refractive index† of the optical waveguide.

Using the above parameters κ, γ_1, and γ_3, the mode function $F(x)$ can be expressed as follows:

$$F(x) = \begin{cases} \cos\phi_1 \exp(\gamma_1 x) & (x \le 0) \\ \cos(\kappa x - \phi_1) & (0 \le x \le h), \\ \cos\phi_3 \exp(-\gamma_3 [x - h]) & (x \ge h) \end{cases} \tag{14.14}$$

where h is the thickness of the core. The phases ϕ_1 and ϕ_3 are related by the following equation:

$$\phi_3 = \kappa h - \phi_1. \tag{14.15}$$

From the derivative of $F(x)$ or the boundary condition with respect to H_z (for the TE mode) or E_z (for the TM mode), ϕ_1 and ϕ_3 are represented by

†Note: Some literature calls the equivalent refractive index the effective refractive index. The terms "equivalent" and "effective" are often used interchangeably. In Chapters 1 and 13, the group refractive index was discussed. This is sometimes called the effective refractive index. There is no doubt that both the equivalent refractive index and the group refractive index are a type of "effective refractive index," but the names sometimes cause confusion. In this book, n_{eq} is called the equivalent refractive index, and n_{g} is called the group refractive index.

$$\tan\phi_1 = \frac{\alpha_1}{\alpha_2}\frac{\gamma_1}{\kappa}, \tag{14.16}$$

$$\tan\phi_3 = \frac{\alpha_3}{\alpha_2}\frac{\gamma_3}{\kappa}, \tag{14.17}$$

where α_i ($i = 1, 2, 3$) is given by

$$\alpha_i = \begin{cases} 1 & \text{(TE mode)} \\ 1/n_i^2 & \text{(TM mode)} \end{cases}. \tag{14.18}$$

From Eq. (14.16),

$$\phi_1 = \tan^{-1}\left(\frac{\alpha_1}{\alpha_2}\frac{\gamma_1}{\kappa}\right). \tag{14.19}$$

The phase ϕ_3 can be represented by a similar equation. When compared to ϕ_1, ϕ_3 has degrees of freedom with integer multiples of π. Therefore, ϕ_3 is expressed by

$$\phi_3 = \tan^{-1}\left(\frac{\alpha_3}{\alpha_2}\frac{\gamma_3}{\kappa}\right) + m\pi. \tag{14.20}$$

From Eqs. (14.15), (14.19), and (14.20),

$$\kappa h = \tan^{-1}\left(\frac{\alpha_1}{\alpha_2}\frac{\gamma_1}{\kappa}\right) + \tan^{-1}\left(\frac{\alpha_3}{\alpha_2}\frac{\gamma_3}{\kappa}\right) + m\pi. \tag{14.21}$$

As shown in Eqs. (14.10) to (14.12), κ, γ_1, and γ_3 in the above equation are all functions of β. Therefore, Eq. (14.21) is the eigenvalue equation that determines the effective propagation constant β of the guided mode. The number m corresponds to the mode order.

Equation (14.21) cannot be solved analytically for β, but it can be easily solved numerically using Newton's method introduced in Chapter 2. Let $f(\beta)$ be a function that should be 0:

$$f(\beta) \equiv -\kappa h + \tan^{-1}\left(\frac{\alpha_1}{\alpha_2}\frac{\gamma_1}{\kappa}\right) + \tan^{-1}\left(\frac{\alpha_3}{\alpha_2}\frac{\gamma_3}{\kappa}\right) + m\pi. \tag{14.22}$$

The derivative $f' = df/d\beta$ used in Newton's method is a simple formula as shown below:

$$\frac{df}{d\beta} = \frac{\beta}{\kappa}\left(h + \frac{q_1}{\gamma_1} + \frac{q_3}{\gamma_3}\right), \tag{14.23}$$

where q_i ($i = 1, 3$) is given by

$$q_i = \begin{cases} 1 & \text{(TE mode)} \\ \dfrac{k_0{}^2}{\gamma_i{}^2/n_i{}^2 + \beta^2/n_2{}^2} & \text{(TM mode)} \end{cases}. \tag{14.24}$$

Using f and f' in Eqs. (14.22) and (14.23), β can be found by Newton's method using Excel as introduced in Chapter 2. The mode function $F(x)$ can be found by substituting β into Eq. (14.14). Note that when finding n_{eq} instead of β, the derivative f' is given by $df/dn_{\text{eq}} = k_0 df/d\beta$.

14.2.3 Symmetrical three-layer optical waveguides

In the case of a symmetrical three-layer optical waveguide in which the cladding on both sides is made of the same material,

$$\gamma_1 = \gamma_3 \equiv \gamma, \tag{14.25}$$

$$\frac{\alpha_1}{\alpha_2} = \frac{\alpha_3}{\alpha_2} \equiv p. \tag{14.26}$$

Then Eq. (14.21) becomes

$$\kappa h = 2\tan^{-1}\left(p\frac{\gamma}{\kappa}\right) + m\pi. \tag{14.27}$$

The mode function $F(x)$ can be found by substituting the solution of this eigenvalue equation into Eq. (14.14).

Figure 14.2 shows an example of the mode functions in a symmetric three-layer optical waveguide. In order to show the difference between the TE mode and TM mode, a case with a large refractive index difference between the core and cladding is shown. The parameters used are $n_1 = n_3 = 1.45$ (SiO_2), $n_2 = 3.6$ (Si), $h = 0.8$ μm, and $\lambda = 1.3$ μm. As can be seen from the figure, for the TM mode, the differential of $F(x)$, which is proportional to H_y, becomes discontinuous at the interface. Furthermore, as shown in Eq. (14.9), the magnitude of the Poynting vector $\boldsymbol{S}$, which is proportional to the light intensity, also becomes discontinuous at the interface. This situation is shown in Fig. 14.3.

Now, let us rewrite the eigenvalue equation of Eq. (14.27) using dimensionless quantities. Set the core thickness h as

$$h = 2a. \tag{14.28}$$

Using a in Eq. (14.28), introduce the following parameters:

$$V \equiv k_0 a\sqrt{n_2{}^2 - n_1{}^2} = k_0 a n_2\sqrt{2\Delta}, \tag{14.29}$$

$$\Delta \equiv \frac{n_2{}^2 - n_1{}^2}{2n_2{}^2} \cong \frac{n_2 - n_1}{n_2}, \tag{14.30}$$

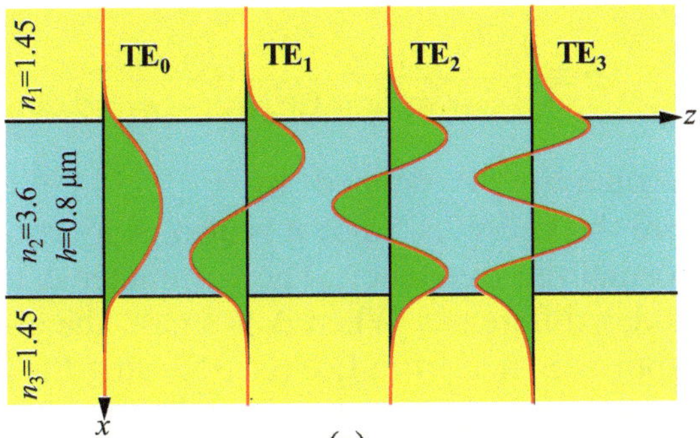

(a)

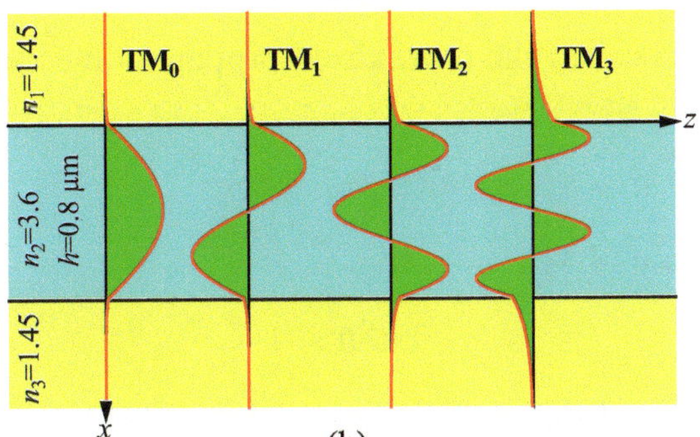

(b)

Figure 14.2 Mode functions in a symmetric three-layer optical waveguide: (a) TE mode and (b) TM mode.

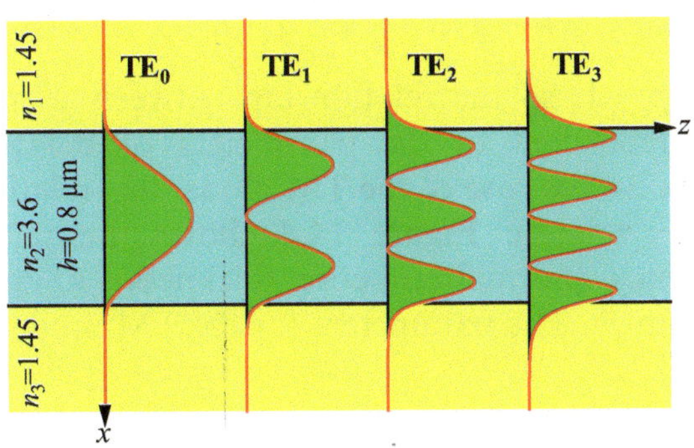

(a)

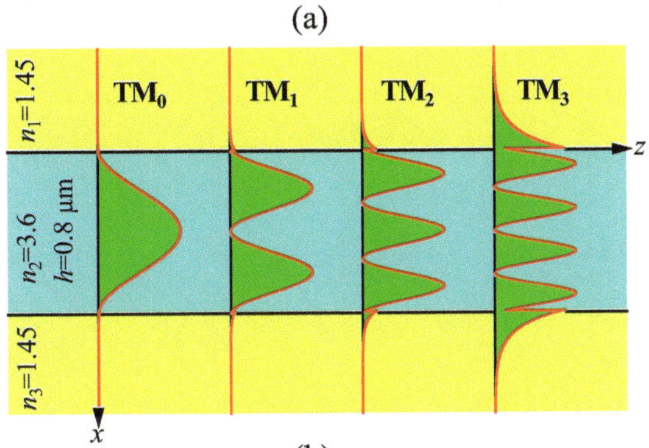

(b)

Figure 14.3 S_z in a symmetric three-layer optical waveguide: (a) TE mode and (b) TM mode.

$$b \equiv \frac{n_{\text{eq}}{}^2 - n_1{}^2}{n_2{}^2 - n_1{}^2} = \frac{\beta^2 - k_0{}^2 n_1{}^2}{k_0{}^2(n_2{}^2 - n_1{}^2)}. \tag{14.31}$$

The dimensionless quantity V defined by Eq. (14.29) is the normalized parameter of the core thickness $2a$. Since it is also proportional to $k_0 = 2\pi/\lambda$, it is also called the normalized frequency. The parameter Δ in Eq. (14.30) is the relative refractive index difference. When Δ is small, the effective propagation constant β can be approximated from Eq. (14.31) using the following equation:

$$\beta \cong k_0\{n_1 + b(n_2 - n_1)\}. \tag{14.32}$$

For this reason, b is called the normalized propagation constant. The range of equivalent refractive index n_{eq} is

$$n_1 < n_{\text{eq}} < n_2. \tag{14.33}$$

Therefore, from Eq. (14.31),

$$0 < b < 1. \tag{14.34}$$

Now, using V and b defined in Eqs. (14.29) and (14.31), Eq. (14.27) can be expressed as follows:[3,4]

$$V = \frac{1}{\sqrt{1-b}}\left\{\tan^{-1}\left(p\sqrt{\frac{b}{1-b}}\right) + \frac{m\pi}{2}\right\}. \tag{14.35}$$

In the TE mode, $p = 1$, so the relationship between V and b expressed by Eq. (14.35) is independent of the parameters of the film thickness and the refractive index. In the case of the TM mode, this relationship depends on the value of p $(= n_1{}^2/n_2{}^2)$. Figure 14.4 shows the relationship between normalized frequency (normalized core thickness) V and the normalized propagation constant b. The parameters for the TM mode are the same as in Figs. 14.2 and 14.3.

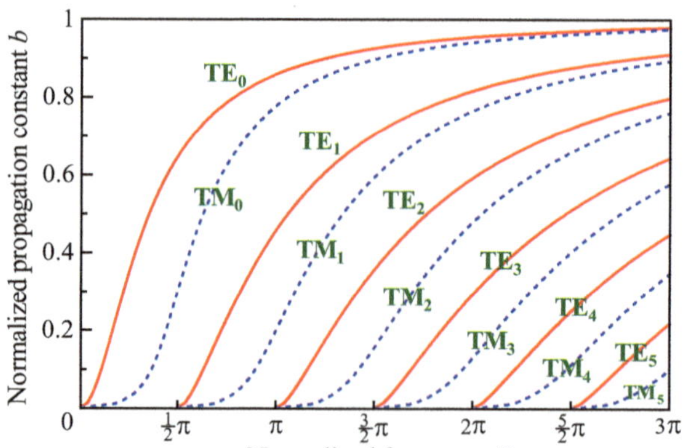

Figure 14.4 *b–V* curves of a symmetrical three-layer optical waveguide ($p = 6.16$).

Since V is defined as a function of b in Eq. (14.35), Fig. 14.4 can be easily plotted by giving b and finding V. Actually, it is necessary to find the effective propagation constant β by giving the optical waveguide parameters. Therefore, it is necessary to find the solution for b by giving V in Eq. (14.35). However, similar to Eq. (14.21), Eq. (14.35) cannot be solved analytically for b, so it must be solved numerically using Newton's method.

The eigenvalue equations expressed by Eqs. (14.21) and (14.35) are suitable for solving using Newton's method because the solution is uniquely determined when the mode order m is given.

14.2.4 Asymmetric three-layer optical waveguides

For a three-layer optical waveguide with an asymmetric structure, the formula in Section 14.2.2 can be used as is. An example of a mode function is shown in Fig. 14.5, where a glass optical waveguide on a SiO_2 substrate is shown. The parameters are $n_1 = 1.0$ (air), $n_2 = 1.53$ (7059 glass), $n_3 = 1.46$ (SiO_2), $h = 3.0$ μm, and $\lambda = 0.65$ μm. Although only the TE mode is shown, the difference between the TE mode and the TM mode is not very large because the difference in refractive index between the core and the cladding is small.

Dimensionless quantities can also be defined in the case of an asymmetric three-layer optical waveguide. The range of equivalent refractive index n_{eq} is given by

$$n'_3 \equiv \max(n_1, n_3) < n_{eq} < n_2. \tag{14.36}$$

Using the above n'_3, define b as follows:

$$b \equiv \frac{{n_{eq}}^2 - {n_3}'^2}{{n_2}^2 - {n_3}'^2}. \tag{14.37}$$

In addition, a parameter b_a representing the asymmetry is defined below:

$$b_a \equiv \frac{{n_3}'^2 - {n_1}'^2}{{n_2}^2 - {n_3}'^2}, \tag{14.38}$$

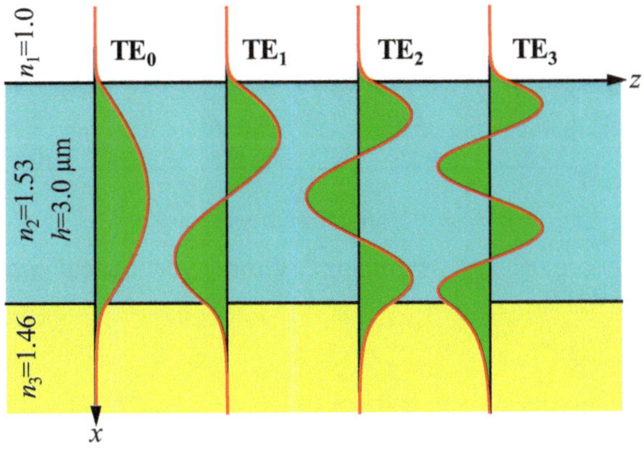

Figure 14.5 Mode functions for the TE mode in an asymmetric three-layer optical waveguide.

where

$$n_1' \equiv \min(n_1, n_3). \tag{14.39}$$

Using these parameters, Eq. (14.21) can be rewritten as follows:[3]

$$V = \frac{1}{2\sqrt{1-b}}\left\{\tan^{-1}\left(p_1\sqrt{\frac{b+b_a}{1-b}}\right) + \tan^{-1}\left(p_3\sqrt{\frac{b}{1-b}}\right) + \frac{m\pi}{2}\right\}, \tag{14.40}$$

where

$$p_i = \begin{cases} 1 & \text{(TE mode)} \\ {n_2}^2/n_i'^2 & \text{(TM mode)} \end{cases}. \tag{14.41}$$

Figure 14.6 shows the b–V curve of the asymmetric three-layer optical waveguide expressed by Eq. (14.40). The refractive index parameters are the same as in Fig. 14.5. A characteristic feature of the asymmetric three-layer optical waveguide is that there is a cutoff thickness or cutoff frequency for the fundamental modes (TE_0 mode and TM_0 mode). In the symmetrical three-layer optical waveguide shown in Fig. 14.4, $V \to 0$ in both the TE_0 mode and TM_0 mode when $b \to 0$, but in the case of the asymmetric three-layer optical waveguide, it does not become 0. By setting $m = 0$ and $b = 0$ in Eq. (14.40), the cutoff film thickness is given by the following equation:†

$$V_{\text{cutoff}} = \frac{1}{2}\tan^{-1}\left(p_1\sqrt{b_a}\right). \tag{14.42}$$

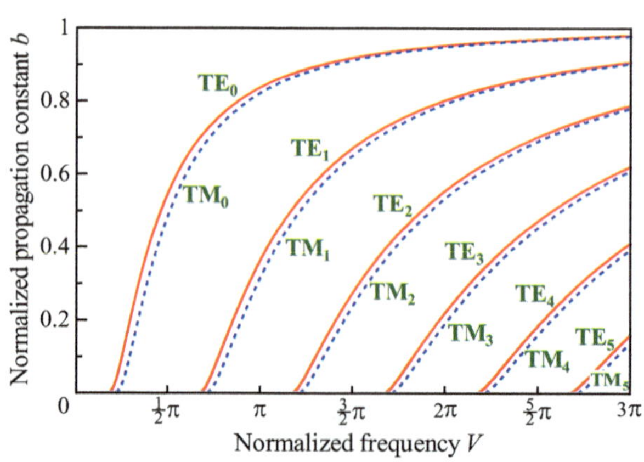

Figure 14.6 b–V curve of an asymmetric three-layer optical waveguide ($p_1 = 2.34$, $p_3 = 1.10$, and $b_a = 5.41$).

†Note: In the case of an asymmetric optical waveguide, in Eqs. (14.29) and (14.30), which express the relationship between V and the core thickness $2a$, n_3' in Eq. (14.36) is used instead of n_1.

Side Topic: Ray image of the waveguide mode

Diagrams that explain the waveguide mode as a ray of light that travels in a zigzag pattern while repeating total reflection in the core layer are often seen (see Fig. 14.S1). Although it has been pointed out that such a diagram is misleading since waveguide modes are not light rays in the first place, this diagram may also be somewhat useful in explaining waveguide modes in optical waveguides. Figure 14.S1 shows the case of the TE mode. It can be seen from this diagram that the electric field must be in the y direction (perpendicular to the plane of the paper) because the light wave where the electric field vector is only in the transverse plane is s-polarized. That is, the electric field component in the TE mode is only E_y. Also, in this case, as explained in the main text, it can be seen that the magnetic field vector has a longitudinal component (z component). As mentioned in Chapter 2, the Poynting vector of this ray does not coincide with the z direction, but since there are waves traveling in the $+x$ direction and waves traveling in the $-x$ direction, the time-averaged Poynting vector is in the z direction, which is the propagation direction of the guided mode.

Considering total internal reflection when a wave packet is incident on an interface, the position of the reflected wave packet shifts by $2d_s$ as shown in the figure. This is the Goos–Hänchen shift.[4] By setting $h_s = d_s \tan\theta$, it appears as if the light ray was reflected from a surface shifted by h_s. This h_s corresponds to the penetration of the waveguide mode into the cladding layer, and in the case of the TE mode, $h_s = 1/\gamma$. Note that the values of h_s in layers 1 and 3 are $1/\gamma_1$ and $1/\gamma_3$, respectively, in the case of an asymmetric optical waveguide. The value h_{eff}, which is the sum of the core layer thickness h and the amount of penetration distance, is called the effective film thickness[1] of the three-layer optical waveguide.

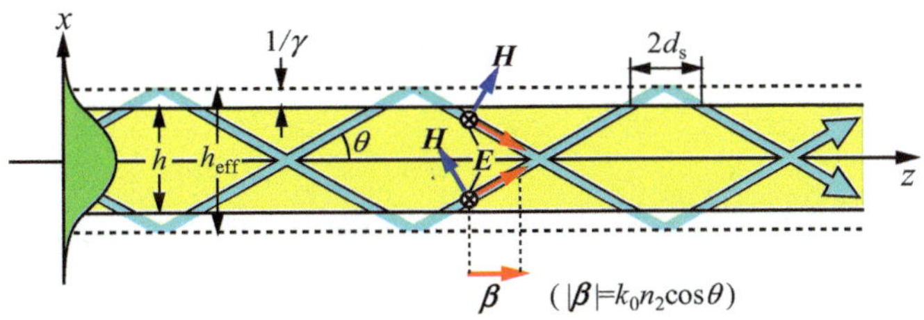

Figure 14.S1 Zigzag ray image of the TE guided mode.

14.3 Propagation of Guided Lightwave

An Excel animation that displays the propagation of a Gaussian beam was introduced in Chapter 8. The propagation of a guided lightwave can be displayed in the same way. Since the mode function $F(x)$ does not depend on the coordinate z in the propagation direction, the state of propagation cannot be seen even if $F(x)$ is plotted. Therefore, the amplitude including the vibration term, i.e., $F(x)\exp(-i\beta z + i\omega t)$, is needed for visualization of guided lightwave propagation. The result is shown in Fig. 14.7. The parameters are $n_1 = n_3 = 1.46$, $n_2 = 1.53$, $h = 3.0$ μm, and $\lambda = 0.65$ μm. In the Excel animation, you can see how the lightwave propagates as it oscillates.

Now, in a multimode optical waveguide, multiple waveguide modes may propagate simultaneously. In this case, the effective propagation constant differs for each waveguide mode. For example, when two modes propagate, the summed mode amplitude $F_{12}(x, z, t)$ is as follows:

$$F_{12}(x,z,t) = F_1(x)\exp(-i\beta_1 z + i\omega t) + F_2(x)\exp(-i\beta_2 z + i\omega t). \quad (14.43)$$

The magnitude of the Poynting vector in this case is

$$S_z(x,z) \propto |F_{12}(x,z,t)|^2 = |F_1(x)|^2 + |F_2(x)|^2 + 2F_1(x)F_2(x)\cos(\Delta\beta z), \quad (14.44)$$

$$\Delta\beta \equiv \beta_1 - \beta_2. \quad (14.45)$$

As shown in Eq. (14.11) an oscillation term $\cos(\Delta\beta z)$ appears. In other words, intermode interference occurs.

Figure 14.8 shows how the TE_0 mode and TE_1 mode propagate simultaneously in a two-mode optical waveguide with $h = 1.0$ μm. At $z = 0$,

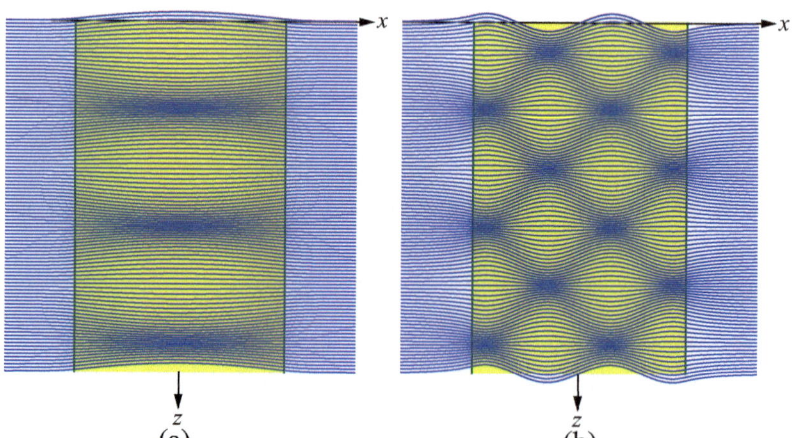

Figure 14.7 Propagation of the guided lightwave: (a) TE_0 mode and (b) TE_3 mode.

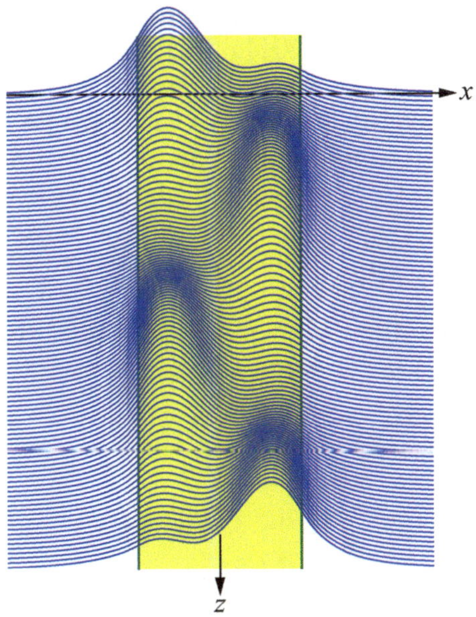

Figure 14.8 Intermode interference due to simultaneous propagation of the TE_0 mode and TE_1 mode.

there is a region of high light intensity on the left side of the optical waveguide, but as it propagates, it moves to the right and then returns to the left again, which is repeated. This is actually the same principle as that under which the directional coupler operates.[5] The coupling length in this case is $\pi/\Delta\beta$. Note that it is difficult to display both the vibration state of each mode and the interference between modes because $\Delta\beta$ is smaller than β, but this is possible if the division in the z direction is made quite fine.

14.4 Conclusion

This chapter introduced a mode calculation method and an example of a calculation using Excel for a three-layer optical waveguide, which is the basis of a planar optical waveguide. All of the expressions shown in this chapter can be extended to complex numbers. In the next chapter, mode calculations for semiconductor optical waveguides with gain/loss will be introduced.

References

1. G. Hatakoshi, "Optical waveguide," Chapter III-4, *Optical Engineering Handbook*, T. Ose, H. Saito, S. Tanaka, J. Tsujiuchi, T. Namioka, eds., Asakura Publishing, Tokyo (1986) [in Japanese].
2. Y. Kokubun and G. Hatakoshi, "Waveguide optics," Chapter II-3, *Microoptics Handbook*, Asakura Publishing, Tokyo (1995) [in Japanese].

3. K. Kuroda, *Physical Optics—Propagation of light waves in a medium*, Asakura Publishing, Tokyo (2011) [in Japanese].
4. K. Iga and G. Hatakoshi, Chapter 7, *Treasure Microbox of Optoelectronics I*, Advanced Communication Media, Tokyo (2018) [in Japanese].
5. K. Iga and G. Hatakoshi, Chapter 36, *Treasure Microbox of Optoelectronics III*, Advanced Communication Media, Tokyo (2019) [in Japanese].

Appendix 14A Excel Files

Appendix I-A contains the following Excel files. Each tab in the files corresponds to a figure/s in the main text.

14_Waveguide1.xlsx	
Instruction	
WG1	Fig. 14.2, Fig. 14.3
WG2	Fig. 14.5
14_Waveguide2.xlsm	
Instruction	
WG3	Fig. 14.7
WG4	Fig. 14.8

Chapter 15
Optical Waveguide Part 2

15.1 Introduction

In the previous chapter, the mode analysis method in a three-layer optical waveguide, which is the basis of a planar optical waveguide, was described, and calculations using Newton's method in Excel were introduced. This calculation method can be extended to cases where the parameters are complex numbers. Therefore, in this chapter, the modes of a three-layer optical waveguide with a complex refractive index distribution will be discussed.

15.2 Waveguide Modes in Semiconductor Lasers

A semiconductor laser can be regarded as a complex refractive index optical waveguide with gain and absorption. Although general semiconductor laser structures are not necessarily three layered, let's first find the waveguide mode when this structure is approximated as a three-layer optical waveguide.

15.2.1 Gain-guided structure as a p–n junction

A p–n junction structure is a structure in which p-type and n-type semiconductors made of the same material are joined. Since the materials are the same, the real part of the refractive index is the same over the entire region. Therefore, the eigenvalue of the waveguide mode cannot be obtained from the distribution of the real part of the refractive index alone. Due to the gain generated in the junction area, a waveguide mode appears as an eigenvalue due to the distribution of the imaginary part of the refractive index. Generally, this gain distribution is not a stepped distribution, but here it is approximated with a three-layer structure as shown in Fig. 15.1.

Since the structure shown in Fig. 15.1 is the same three-layer planar optical waveguide as in the previous chapter, it is possible to find a solution using Newton's method using the same eigenvalue equation. In the previous chapter, the function f and its derivative f' both as functions of β were used in Newton's method. Although the method is almost the same, f and f' will be expressed as

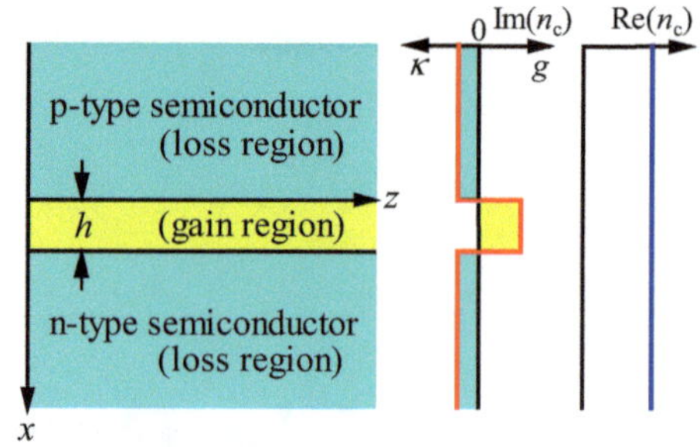

Figure 15.1 p–n junction structure expressed as a three-layer planar optical waveguide.

functions of β^2 and will be used to find the solution for β^2 $(= {k_0}^2 {n_{eq}}^2)$ using Newton's method. In this case, f and its derivative are given below:

$$f(\beta^2) \equiv -\kappa h + \tan^{-1}\left(\frac{\alpha_1}{\alpha_2}\frac{\gamma_1}{\kappa}\right) + \tan^{-1}\left(\frac{\alpha_3}{\alpha_2}\frac{\gamma_3}{\kappa}\right) + m\pi, \tag{15.1}$$

$$\frac{df}{d(\beta^2)} = \frac{1}{2\kappa}\left(h + \frac{q_1}{\gamma_1} + \frac{q_3}{\gamma_3}\right). \tag{15.2}$$

Equations (15.1) and (15.2) correspond to Eqs. (14.22) and (14.23) in the previous chapter. However, since the refractive index of each layer is a complex number here, the parameters β, κ, γ_1, γ_3, α_1, α_2, α_3, q_1, and q_3 in Eqs. (15.1) and (15.2) are all complex numbers. In this chapter, the complex refractive index will be expressed as n_c, and its real and imaginary parts will be expressed as n and g. That is,

$$n_c = n + ig. \tag{15.3}$$

In Eq. (15.3), $g > 0$ corresponds to gain, and $g < 0$ corresponds to absorption. When $g < 0$, it has the following relationship with the extinction coefficient κ:

$$g = -\kappa. \tag{15.4}$$

Although it is not particularly difficult to calculate Eqs. (15.1) and (15.2) in the complex number domain, the real and the imaginary parts of $\tan^{-1}(z)$, which appear in Eq. (15.1), are represented by the following equations:

$$\mathrm{Re}[\tan^{-1}(a + bi)] = \frac{1}{2}\{\mathrm{Atan}\,2(1 + b, a) + \mathrm{Atan}\,2(1 - b, a)\}, \tag{15.5}$$

$$\mathrm{Im}[\tan^{-1}(a + bi)] = \frac{1}{2}\left\{\ln\sqrt{a^2 + (1 + b^2)} - \ln\sqrt{a^2 + (1 - b^2)}\right\}. \tag{15.6}$$

Newton's method using Excel spreadsheets was introduced in Chapter 2. The three-layer optical waveguide in the previous chapter was also calculated using this method. Newton's method can also be executed in a spreadsheet when the parameters are complex numbers, but it becomes a little more

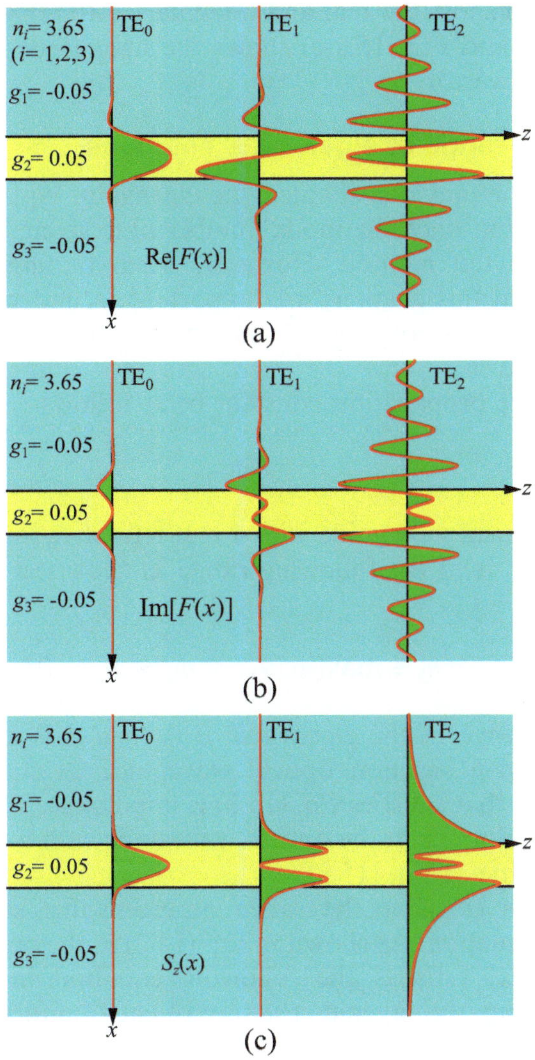

Figure 15.2 Waveguide modes in a three-layer complex-refractive-index optical waveguide with a simplified p–n junction structure: (a) Re[$F(x)$], (b) Im[$F(x)$], and (c) $S_z(x)$.

complicated because Excel cannot handle complex numbers as one variable. Therefore, for Newton's method of complex numbers, Excel VBA was used.†

Figure 15.2 shows the waveguide mode in a GaAs p–n homojunction calculated using Eqs. (15.1) and (15.2). The parameters used are $\lambda = 0.86$ μm, $n_1 = n_2 = n_3 = 3.65$, $g_2 = 0.05$, $g_1 = g_3 = -0.05$, and $h = 1$ μm. Note that these parameters do not necessarily correspond to the actual p–n junction. Figures

†Note: As described in the side topic in Chapter 12, Excel VBA cannot handle complex numbers as a single variable, so the program is more complicated than in FORTRAN, just as in the case of spreadsheet calculations.

15.2(a) and 15.2(b) show the real and imaginary parts of the mode function $F(x)$, respectively, and Fig. 15.2(c) shows the magnitude $S_z(x)$ of the time-averaged Poynting vector.

15.2.2 Existence range of eigenvalue solutions

As mentioned in the previous chapter, the eigenvalue equation of Eq. (15.1) has a unique solution once the mode number m is given, meaning it can be solved by the Newton method. However, when finding solutions on the complex plane as in this chapter, it is important to understand the range in which solutions exist, as convergence may not occur, depending on how the initial values are given.

Let the effective propagation constant be as follows:

$$\beta = k_0 n_{\mathrm{eq}}. \tag{15.7}$$

In a three-layer optical waveguide with a real refractive index, the equivalent refractive index n_{eq} is in the following range as shown in Eq. (14.36) in the previous chapter:

$$n_3' \equiv \max(n_1, n_3) < n_{\mathrm{eq}} < n_2. \tag{15.8}$$

In this case, the range of the eigenvalue β $(= k_0 n_{\mathrm{eq}})$ is clearly determined. However, in the p–n junction optical waveguide as shown in Fig. 15.2, $n_1 = n_2 = n_3$, so n_{eq} that satisfies Eq. (15.8) is impossible. Therefore, it is clear that the range of solutions in the optical waveguide with a complex refractive index is not the same as Eq. (15.8).

Therefore, let's consider the range of eigenvalue solutions from the underlying wave equation. As shown in Eq. (14.2) in the previous chapter, the mode function $F(x)$ satisfies the following equation in a planar optical waveguide consisting of a layered structure of each uniform medium:

$$\frac{d^2}{dx^2} F(x) + \{k_0{}^2 n_c{}^2(x) - \beta^2\} F(x) = 0. \tag{15.9}$$

The reason for using the condition "uniform medium" here is that Eq. (15.9) is for the TE mode, and in the TM mode, the equation is generally as follows:

$$n_c{}^2(x) \frac{d}{dx} \left\{ \frac{1}{n_c{}^2(x)} \frac{dF(x)}{dx} \right\} + \{k_0{}^2 n_c{}^2(x) - \beta^2\} F(x) = 0. \tag{15.10}$$

If $n_c(x)$ is constant in each layer, Eq. (15.10) becomes the same as Eq. (15.9) except for the boundary.

Now, by multiplying Eq. (15.9) by $F^*(x)$ and integrating it, the following equation is obtained:

$$\int_{-\infty}^{\infty} F^*(x)\frac{d^2}{dx^2}F(x)dx + \int_{-\infty}^{\infty}\{k_0{}^2 n_c{}^2(x) - \beta^2\}|F(x)|^2 dx = 0. \qquad (15.11)$$

Applying partial integration to the first term and using the fact that $F(x) \to 0$ at $x = \pm\infty$, Eq. (15.11) becomes as follows for the case of the TE mode:

$$\beta^2 \int_{-\infty}^{\infty} |F|^2 dx = k_0{}^2 \int_{-\infty}^{\infty} n_c{}^2(x)|F|^2 dx - \int_{-\infty}^{\infty}\left|\frac{dF}{dx}\right|^2 dx. \qquad (15.12)$$

From this, the following relational expression regarding the real part $\mathrm{Re}(\beta^2)$ and imaginary part $\mathrm{Im}(\beta^2)$ of β^2 can be obtained:

$$\mathrm{Re}(\beta^2)\int_{-\infty}^{\infty} |F|^2 dx = k_0{}^2 \int_{-\infty}^{\infty} \mathrm{Re}(n_c{}^2)|F|^2 dx - \int_{-\infty}^{\infty}\left|\frac{dF}{dx}\right|^2 dx, \qquad (15.13)$$

$$\mathrm{Im}(\beta^2)\int_{-\infty}^{\infty} |F|^2 dx = k_0{}^2 \int_{-\infty}^{\infty} \mathrm{Im}(n_c{}^2)|F|^2 dx. \qquad (15.14)$$

From Eqs. (15.7), (15.13), and (15.14), it is derived that n_{eq} is in the following range:[1,2]

$$\mathrm{Re}(n_{eq}{}^2) \leq [\mathrm{Re}(n_c{}^2)]_{max}, \qquad (15.15)$$

$$[\mathrm{Im}(n_c{}^2)]_{min} \leq \mathrm{Im}(n_{eq}{}^2) \leq [\mathrm{Im}(n_c{}^2)]_{max}. \qquad (15.16)$$

Equation (15.16) determines the upper and lower limits for the imaginary part of $n_{eq}{}^2$, but for the real part of $n_{eq}{}^2$, only an upper limit is given from Eq. (15.15). In other words, it is not necessary that $[\mathrm{Re}(n_c{}^2)]_{min} \leq \mathrm{Re}(n_{eq}{}^2)$ as in Eq. (15.16).

Figure 15.3 shows the distribution of $n_c{}^2$ and $n_{eq}{}^2$ for the p–n junction structure shown in Fig. 15.2. The filled area in the figure is the range in which the solutions given by Eqs. (15.15) and (15.16) exist, and all eigenvalues including higher-order modes are within this area. Since the constituent material is only GaAs, $\mathrm{Re}(n_c{}^2)$ is equal in all layers with the parameters shown in Fig. 15.2. On the other hand, the waveguide mode is within the eigenvalue existence range shown in the figure, and in this example, there are three solutions. In all three cases, $\mathrm{Re}(n_{eq}{}^2) \leq \mathrm{Re}(n_c{}^2)$, so in a real refractive index optical waveguide, some of the modes classified as radiation modes exist as discrete modes. The mode that actually oscillates is the TE_0 mode (fundamental mode), which has the largest gain among these modes.

Since the derivation of Eqs. (15.15) and (15.16) only assumes Eq. (15.9), the range of existence of the solution shown by this equation is not limited to a three-layer optical waveguide, but it is also applicable to the TE mode for any multilayer structure planar optical waveguide. The double heterostructure,

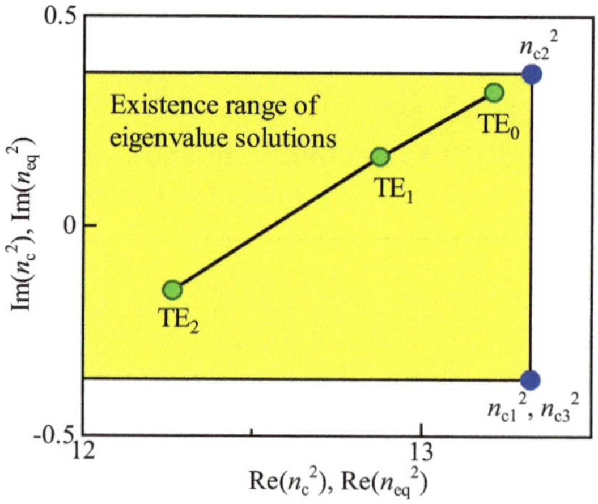

Figure 15.3 Existence range of eigenvalue solutions and distribution of n_c^2 and n_{eq}^2 in the p–n junction structure optical waveguide in Fig. 15.2.

which is a typical structure for semiconductor lasers, is often treated as a three-layer optical waveguide consisting of an active layer and cladding layers on both sides. However, in reality, there are substrates, contact layers, electrodes, etc., on the outside, so strictly speaking, it is necessary to analyze the optical waveguide including these.

Figure 15.4 shows a typical waveguide structure for a GaAlAs or InGaAlP semiconductor laser.[2] The characteristic feature of such a waveguide structure is that the outermost layer has a large absorption, and the real part of the refractive index is the largest in the outermost layer. If the thickness of the cladding layer or contact layer cannot be made sufficiently large due to lattice mismatch or other reasons, it is also necessary to include an electrode outside the cladding layer or contact layer in the optical waveguide structure.

The structure shown in Fig. 15.4 is no longer a three-layer optical waveguide, so it must be analyzed as a general multilayer planar optical

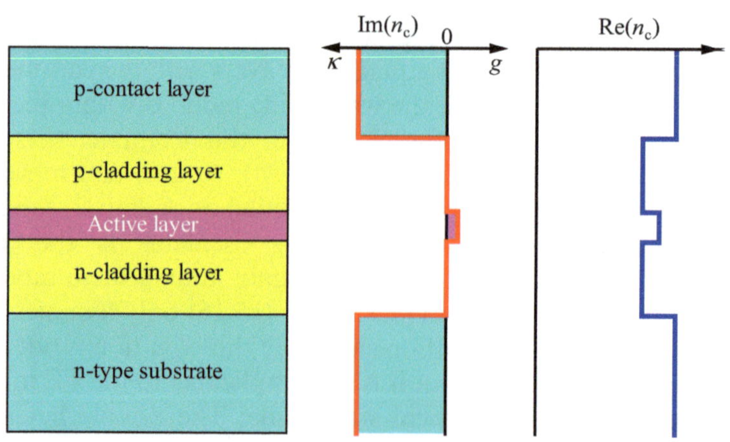

Figure 15.4 Typical waveguide structure for GaAlAs and InGaAlP semiconductor lasers.

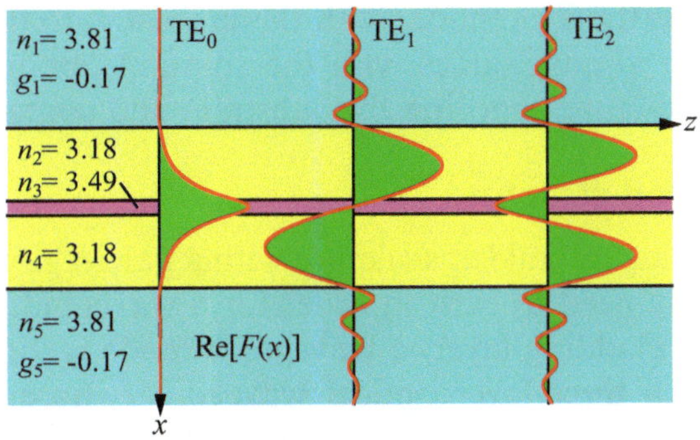

Figure 15.5 Calculation examples of a waveguide mode in an InGaAlP semiconductor laser.

waveguide. This can be done using the formula for the multilayer film characteristic matrix mentioned at the beginning of the previous chapter. Figure 15.5 shows an example of the waveguide mode determined in this way. Assuming an InGaAlP semiconductor laser with $\lambda = 670$ nm, it has a double heterostructure consisting of an InGaP active layer ($h_3 = 0.1$ μm) and an InGaAlP cladding layer ($h_2 = h_4 = 0.5$ μm), and the substrate and contact layer are made of GaAs. There are nine TE modes in this optical waveguide. Figure 15.5 shows the real part of the mode function $F(x)$ in the three low-order modes.

Even in this structure, the range of TE mode solutions is still given by Eqs. (15.15) and (15.16). Figure 15.6 shows an example of the distribution of n_c^2 and n_{eq}^2 for the waveguide structure shown in Fig. 15.5. As can be seen from this figure, all $\mathrm{Re}(n_{eq}^2)$ are in the region smaller than the $\mathrm{Re}(n_c^2)$ of the constituent materials, and all except the fundamental mode are smaller than the $\mathrm{Re}(n_c^2)$ of the cladding layer material. In this case as well, it is the

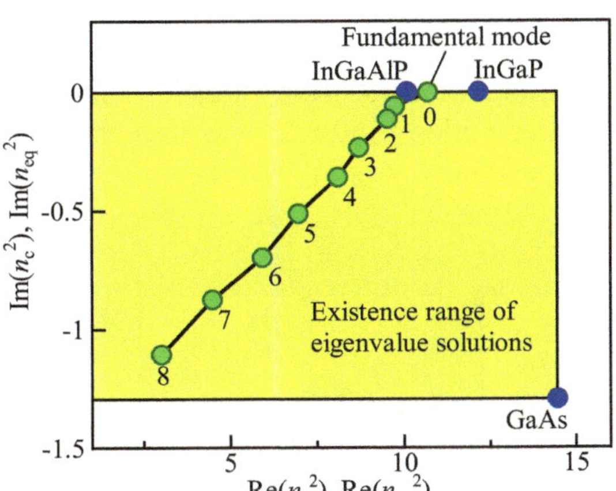

Figure 15.6 Range of eigenvalue solutions and distribution of n_c^2(•) and n_{eq}^2(•) in the InGaAlP semiconductor laser in Fig. 15.5.

fundamental mode (the TE_0 mode) with the highest gain that oscillates. This fundamental mode almost agrees with the solution obtained using the real refractive index structure, ignoring the substrate and contact layer.

15.2.3 Loss-guided structure

Figure 15.1 shows a gain-guided structure with a gain region. The gain region does not necessarily need to exist in order for a waveguide mode to form. If there is loss in the cladding layer, a waveguide mode is formed.

Let's consider a three-layer optical waveguide whose core is AlAs and cladding is GaAs. Since the refractive index of AlAs is lower than that of GaAs, this structure has a core and cladding that are opposite to those of a typical optical waveguide. Therefore, normally no waveguide mode is formed. However, in the visible region, GaAs exhibits absorption loss, resulting in a loss-guided structure. For example, at the wavelength $\lambda = 0.5$ μm, the refractive index of GaAs is $n_c = 4.3 - 0.4i$, and it has a large extinction coefficient. On the other hand, AlAs is transparent with $n_c = 3.3$.

Figure 15.7 shows the mode function of this GaAs/AlAs/GaAs optical waveguide with core thickness $h = 0.8$ μm. Here, only Re[$F(x)$] is shown for TE_0 to TE_2. It can be seen that the confinement is stronger than in the case of Fig. 15.2, and this is due to the large absorption of GaAs.

In the optical waveguide shown in Fig. 15.7, there are 19 TE modes, TE_0 to TE_{18}. The distribution of n_{eq}^2 for these modes is shown in Fig. 15.8. As in Figs. 15.3 and 15.6, all modes are within the range of eigenvalue solutions. As can be seen from Fig. 15.8, $\mathrm{Re}(n_{eq}^2)$ becomes negative in higher-order modes. This is because $\mathrm{Im}(n_{eq})$ takes a large negative value, and $\mathrm{Re}(n_{eq})$ itself is all positive.

15.3 Facet Reflectance and Far-Field Pattern

15.3.1 Facet reflectance

In a semiconductor laser, the oscillation threshold changes depending on the facet reflectance for the guided mode, so it is necessary to calculate the facet

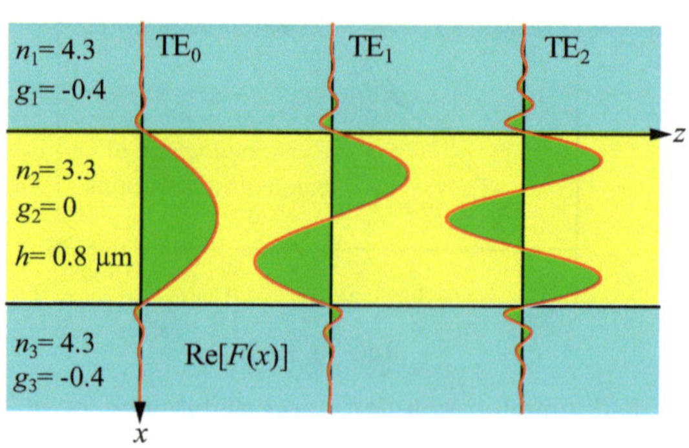

Figure 15.7 Waveguide mode in the GaAs/AlAs/GaAs optical waveguide ($\lambda = 0.5$ μm).

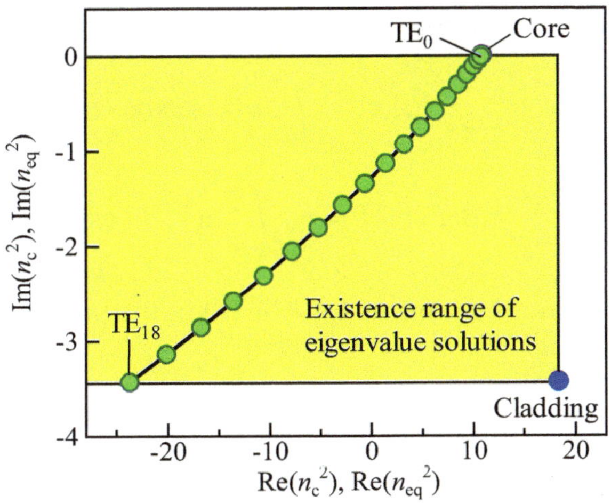

Figure 15.8 Range of eigenvalue solutions and distribution of n_c^2(•) and n_{eq}^2(•) in the GaAs/AlAs/GaAs optical waveguide in Fig. 15.7.

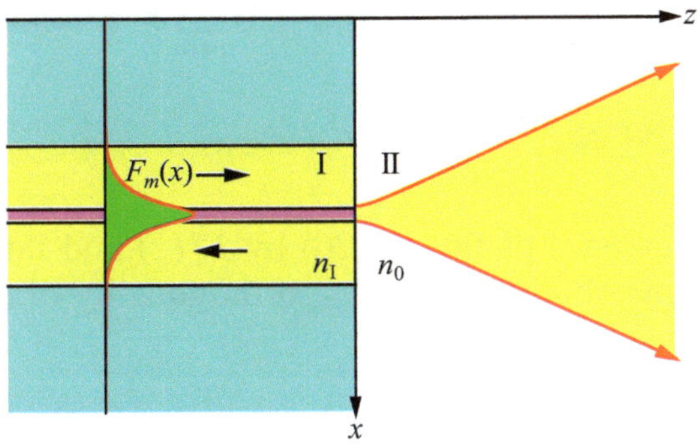

Figure 15.9 Connection of the waveguide mode (area I) and radiation mode (area II) at the optical waveguide facet.

reflectance. This facet reflectance can be determined from the boundary condition at $z = 0$ in the arrangement shown in Fig. 15.9.[1,4,5]

In Fig. 15.9, the y-component of the electric field of mode m in region I and the y-component of the electrical field in region II (in the case of the TE mode) are expressed as follows. (Note that, in the case of the TM mode, this corresponds to the y-component of the magnetic field.)

$$E_y(x) = \begin{cases} F_m(x)e^{-i\beta_m z} + \sum\limits_n r_n F_n(x) e^{i\beta_n z} & (z \le 0) \\ \int_{-\infty}^{\infty} a(k_x)\exp(-ik_x x - ik_z z)dk_x & (z \ge 0) \end{cases}, \tag{15.17}$$

where

$$k_z(k_x) = \sqrt{k_0{}^2 n_0{}^2 - k_x{}^2}. \tag{15.18}$$

From the boundary condition at the facet $z = 0$,

$$F_m(x) + \sum_n r_n F_n(x) = \int_{-\infty}^{\infty} a(k_x) \exp(-ik_x x) dk_x, \tag{15.19}$$

$$\alpha(x)\{\beta_m F_m(x) - \sum_n r_n \beta_n F_n(x)\} = \alpha_0 \int_{-\infty}^{\infty} a(k_x) k_z \exp(-ik_x x) dk_x, \tag{15.20}$$

where $\alpha(x)$ and α_0 are the same as Eq. (14.18) in the previous chapter and are represented by

$$\alpha(x) = \begin{cases} 1 & \text{(TE mode)} \\ 1/n^2(x) & \text{(TM mode)} \end{cases}. \tag{15.21}$$

Equation (15.19) shows that $a(k_x)$ is the Fourier transform of the left side. That is,

$$a(k_x) = (1 + r_m) f_m(k_x) + \sum_{n \neq m} r_n f_n(k_x), \tag{15.22}$$

$$f_n(k_x) = \frac{1}{2\pi} \int_{-\infty}^{\infty} F_n(x) \exp(ik_x x) dx. \tag{15.23}$$

By multiplying both sides of Eq. (15.20) by $F_n^*(x)$ and integrating by x, the following equation is obtained from the orthogonality of the modes and the relationship of Eq. (15.23):

$$\beta_m(\delta_{mn} - r_m) \int_{-\infty}^{\infty} \alpha(x) |F_m(x)|^2 dx = 2\pi\alpha_0 \int_{-\infty}^{\infty} a(k_x) k_z f_n^*(k_x) dk_x. \tag{15.24}$$

Since there are as many equations of Eq. (15.24) as there are modes, r_m can be found by solving them simultaneously. In general, the amplitude of reflected waves other than those in the same mode as the incident wave is small, so only the reflected waves with $n = m$ are considered. Then the following equation can be derived from Eqs. (15.22) and (15.24):

$$\frac{1 - r_m}{1 + r_m} = \frac{2\pi\alpha_0 \int_{-\infty}^{\infty} k_z(k_x) |f_m(k_x)|^2 \, dk_x}{\beta_m \int_{-\infty}^{\infty} \alpha(x) |F_m(x)|^2 \, dx}. \tag{15.25}$$

For comparison, in the case of plane wave reflection described in Chapter 4, the z components of the angular wave vector in media I and II are expressed as follows:

$$k_{z1} = k_0 n_1 \cos\theta_i, \tag{15.26a}$$

$$k_{z2} = k_0 n_2 \cos\theta_t. \tag{15.26b}$$

Then, from Eqs. (4.8) and (4.9) in Chapter 4,

$$\frac{1-r}{1+r} = \frac{\alpha_2 k_{z2}}{\alpha_1 k_{z1}}. \tag{15.27}$$

This is similar to Eq. (15.25). Comparing Eq. (15.25) and Eq. (15.27), it is found that in Eq. (15.25), $k_{z1} = \beta_m$, α_1 is replaced by the integral weighted by $|F_m(x)|^2$, and k_{z2} is replaced by the integral weighted by the square of the absolute value of the Fourier component.

Using r_m given by Eq. (15.25), the reflectance R_m is expressed by the following equation:

$$R_m = |r_m|^2. \tag{15.28}$$

An example of calculating the dependence of reflectance on core thickness h for a semiconductor optical waveguide is shown in Fig. 15.10.[1,6] As can be seen from this figure, the reflectance for the TE mode is generally higher than the reflectance for the TM mode. As the core thickness h approaches 0, the equivalent refractive index n_{eq} asymptotically approaches the cladding refractive index n_1, as can be seen from Fig. 14.4 in the previous chapter. At this time, the guided mode approaches a mode in which the light propagates through the cladding perpendicular to the facet, and R asymptotes the reflectance determined by the cladding and air.[1] Also, at this time, the phase velocity and group velocity[6] of the guided mode both asymptotically approach c/n_1. On the other hand, when the core thickness is infinite, $n_{\text{eq}} \rightarrow n_2$, and R asymptotes the reflectance determined by the core and air. At this time, both the phase velocity and group velocity asymptotically approach c/n_2.

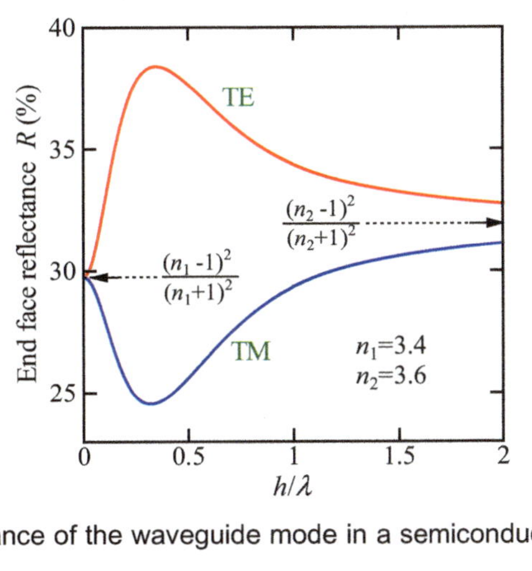

Figure 15.10 Reflectance of the waveguide mode in a semiconductor optical waveguide.

15.3.2 Far-field pattern

The Fourier transform of the mode function given by Eq. (15.23) corresponds to the amplitude of the far-field pattern. Since the mode function $F(x)$ in the planar optical waveguide is expressed by a trigonometric function or an exponential function in each region, this Fourier transform can be obtained analytically. In the case of a three-layer optical waveguide, this can be concretely expressed as follows:

$$\begin{aligned} f(k_x) &= \frac{1}{2\pi}\int_{-\infty}^{\infty} F(x)\exp(ik_x x)dx \\ &= \frac{1}{2\pi}\left(\frac{\cos\phi_1}{\gamma_1 + ik_x} + \frac{\cos\phi_3}{\gamma_3 - ik_x}e^{ik_x h}\right) + \frac{1}{2\pi(\kappa^2 - k_x{}^2)} \\ &\quad \times \left\{\kappa\sin\phi_1 - ik_x\cos\phi_1 + (\kappa\sin\phi_3 + ik_x\cos\phi_3)e^{ik_x h}\right\}. \end{aligned} \tag{15.29}$$

Using this $f(k_x)$, the intensity distribution $I_{\mathrm{FFP}}(\theta)$ of the far-field pattern is expressed by the following equation:

$$I_{\mathrm{FFP}}(\theta) = |g(\theta)f(k_0\sin\theta)|^2, \tag{15.30}$$

where $g(\theta)$ is an angle correction term, and there is no problem in using it as a constant in the paraxial region where θ is small.

Examples of calculated far-field patterns are shown in Figs. 15.11 and 15.12. Figure 15.11 is for the SiO_2/Si/SiO_2 optical waveguide shown in Fig. 14.2 in the previous chapter, and Fig. 15.12 is for the GaAs gain-guided structure shown in Fig. 15.2 in this chapter, both of which show the three modes TE_0 to TE_2.

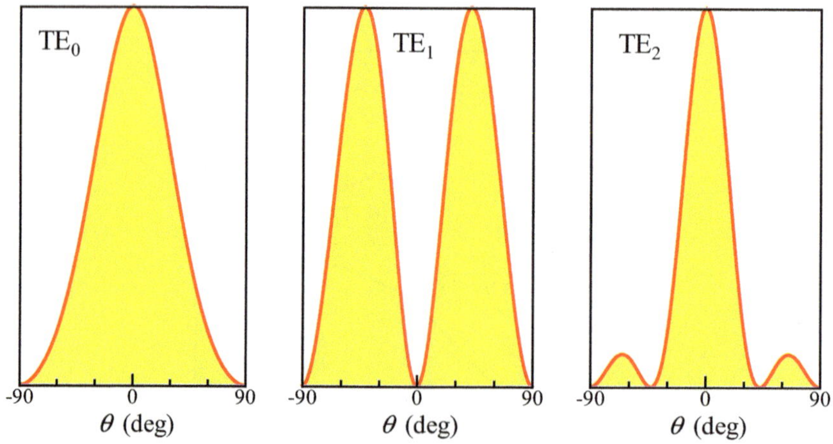

Figure 15.11 Far-field pattern of a SiO_2/Si/SiO_2 optical waveguide (Fig. 14.2 in the previous chapter).

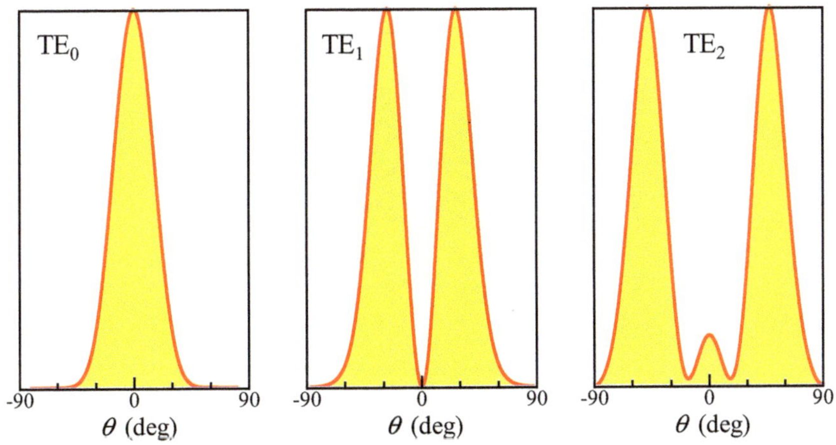

Figure 15.12 Far-field pattern of a GaAs gain-guided structure (Fig. 15.2).

15.4 Conclusion

A method for calculating guided modes in a three-layer optical waveguide with a complex refractive index distribution and examples of calculations using Excel were introduced. Specifically, guided modes in semiconductor optical waveguides that have gain and absorption have been discussed. Another typical example of a material with high absorption is metal. This will be discussed in the next chapter.

Side Topic: Convergence of solutions by Newton's method in the complex plane

Newton's method is a powerful method for finding solutions, but it may not converge depending on the initial values. Newton's method using Eqs. (15.1) and (15.2) finds solutions on the complex plane. Of course, if starting from an initial value close to the solution, it will converge easily, but since the solution is not known from the beginning, setting the initial value is important.

Figure 15.S1 shows the state of convergence when finding the solution for the TE$_0$ mode in Fig. 15.8 of the main text. Figure 15.S1(a) is an example where a solution was obtained after a relatively large number of iterations; it took 45 iterations to converge. The location indicated by the number 0 indicates the initial value. This initial value is within the range of eigenvalues. However, it is not a necessary and sufficient condition that the initial value be within the existence range of the eigenvalue solutions. As can be seen from the figure, it jumps out of the solution range in the

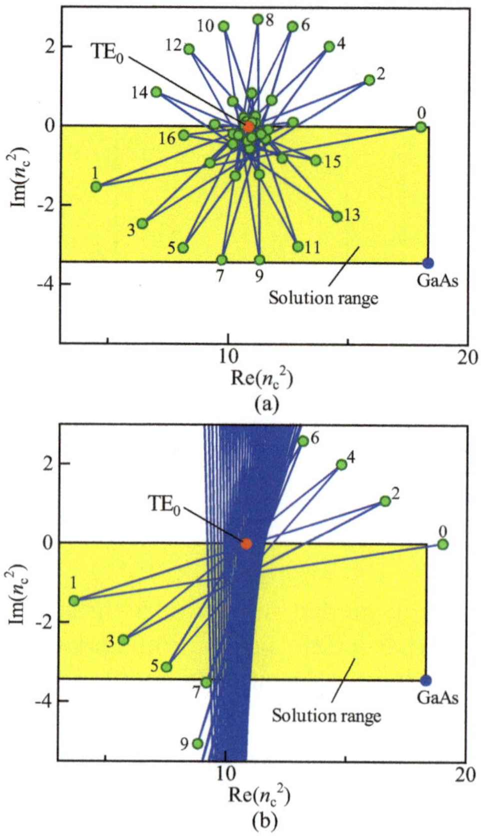

Figure 15.S1 Convergence of the solution on the complex plane using Newton's method: (a) converged 45 times and (b) diverged.

middle of calculation, but it eventually converges. Therefore, for example, although the point indicated by the number 2 is not within the solution range, it should converge even if this point is used as the initial value.

On the other hand, in Fig. 15.S1(b), the initial value is only slightly different from that in Fig. 15.S1(a), but in this case, it diverges without converging and no solution is obtained. In this case, the initial value is outside the solution range, but points 1, 3, and 5 during the calculation are within the solution range. Even if the initial value is 1, 3, or 5, it will still diverge.

From the above, it follows that whether by starting from within or outside the existence range of the solution, there are cases where it converges and cases where it does not. Therefore, even if the solution range is known, is it of use in calculations?

Then, let us consider a method in which, when $\mathrm{Im}(n_c^2)$ exceeds the range of solutions during the iterative calculation, the calculation can proceed by replacing it with the maximum or minimum value of $\mathrm{Im}(n_c^2)$. Figure 15.S2 shows the convergence of the solution calculated using this method. Figure 15.S2(a) is the case with the same initial value as Fig. 15.S1(a), and Fig. 15.S2(b) is the case where point 5 in the middle of Fig. 15.S1(b) is set as the initial value. The latter corresponds to the case in which convergence does not occur in Fig. 15.S1. Points 2, 4, 6, 8, and 10 in Fig. 15.S2(a) and points 1, 3, and 5 in Fig. 15.S2(b) were all replaced by the maximum value of $\mathrm{Im}(n_c^2)$ during the calculation. As can be seen from this figure, the solution converges in all cases, and the number of iterations is small. Therefore, in this method, it is extremely useful to understand the range of solutions. Note that the range of solutions in the TM mode is different from that in the TE mode. This will be introduced in the next chapter.

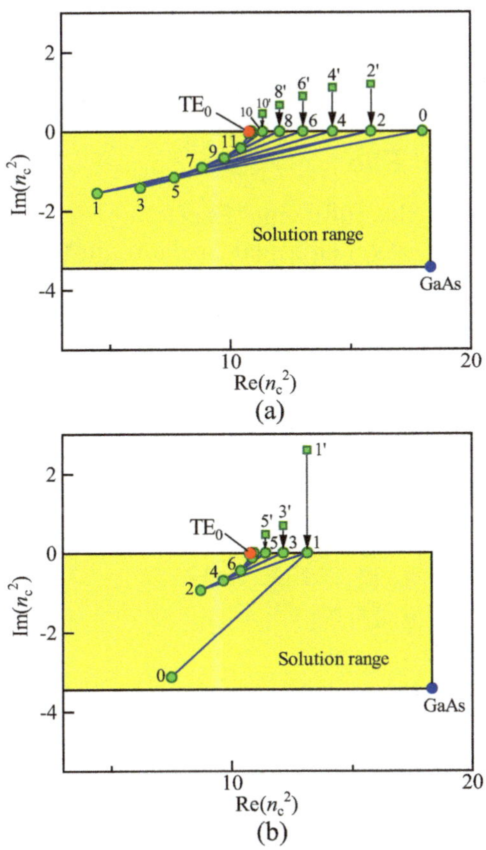

Figure 15.S2 Convergence of the solution on the complex plane using Newton's method: (a) converged 17 times and (b) converged 12 times.

References

1. G. Hatakoshi, "Semiconductor lasers," Chapter IV-1, *Microoptics Handbook*, Asakura Publishing, Tokyo (1995) [in Japanese].
2. G. Hatakoshi, M. Onomura, and M. Ishikawa, "Optical, electrical and thermal analysis for GaN semiconductor lasers," *Int. J. Numer. Model.* **14**(4), 303–323 (2000).
3. K. Iga and G. Hatakoshi, Chapter 16, *Treasure Microbox of Optoelectronics I*, Advanced Communication Media, Tokyo (2018) [in Japanese].
4. F. K. Reinhart, I. Hayashi, and M. B. Panish, "Mode reflectivity and waveguide properties of double-heterostructure injection lasers," *J. Appl. Phys.* **42**(11), 4466–4479 (1971).
5. T. Ikegami, "Reflectivity of mode at facet and oscillation mode in double-heterostructure injection lasers," *IEEE J. Quantum Electron.* **QE-8**(6), 470–476 (1972).
6. K. Iga and G. Hatakoshi, Chapter 7, *Treasure Microbox of Optoelectronics I*, Advanced Communication Media, Tokyo (2018) [in Japanese].

Appendix 15A Excel File

Appendix I-A contains the following Excel file. Each tab corresponds to a figure/s in the main text. As mentioned in the main text, in this Excel file, the convergence calculation of Newton's method is performed using a VBA program rather than a spreadsheet. Since Fig. 15.11 is a real refractive index optical waveguide, it can be calculated using Excel as described in the previous chapter, but it can also be calculated as the complex refractive index optical waveguide as described in this chapter by setting the imaginary part to 0.

15_Waveguide3.xlsm

Instruction	
WG5	Fig. 15.2, Fig. 15.12
WG6	Fig. 15.7
WG7	Fig. 15.11

Chapter 16
Surface Plasmons and Metal Waveguides

16.1 Introduction

In the previous chapter, the mode analysis method for optical waveguides with a complex refractive index distribution such as semiconductor lasers was described, and calculations using Newton's method in Excel were introduced. Also, the waveguide mode in a loss-guided optical waveguide whose cladding is made of a semiconductor material with a large extinction coefficient was described.

Another typical example of a material with a large extinction coefficient is metal. Therefore, in this chapter, the modes of optical waveguides composed of metals and dielectrics will be discussed.

16.2 Surface Plasmon Polaritons

16.2.1 Waveguide modes formed at a metal/dielectric/metal interface

As mentioned in Chapter 5, the refractive index of metal is expressed using the Drude model as follows [see Eq. (5.42)]:

$$\varepsilon_{\mathrm{r}} = {n_{\mathrm{c}}}^2 = 1 - \frac{{\omega_{\mathrm{p}}}^2}{\omega^2 - i\omega/\tau}, \tag{16.1}$$

where ω_{p} is the plasma angular frequency. As can be seen from the above equation, in the region $\omega < \omega_{\mathrm{p}}$, the real part of the dielectric constant or ${n_{\mathrm{c}}}^2$ becomes negative, and the extinction coefficient κ takes a large value.

A localized mode of electromagnetic waves is formed at the interface between a dielectric material and a metal having such a negative dielectric constant.[1] This is the surface plasmon polariton.[†] It is also sometimes called a surface plasmon for short.

Let β be the effective propagation constant of the surface plasmon polariton propagating in the z direction, and let $F(x)$ be the mode function. Then $F(x)$ can be expressed as follows, similar to Eq. (15.14) in the previous chapter:

$$F(x) = \begin{cases} \exp(\gamma_1 x) & (x \leq 0) \\ \exp(-\gamma_2 x) & (x \geq 0) \end{cases}. \tag{16.2}$$

From the boundary condition for the derivative of $F(x)$,

$$\alpha_1\gamma_1 + \alpha_2\gamma_2 = 0, \tag{16.3}$$

where α_i $(i = 1, 2)$ is the same as Eq. (15.18) in the previous chapter and is given by

$$\alpha_i = \begin{cases} 1 & \text{(TE mode)} \\ 1/n_{ci}^2 & \text{(TM mode)} \end{cases}. \tag{16.4}$$

First, considering the case of the TE mode, where $\alpha_i = 1$. In this case,

$$\gamma_1 + \gamma_2 = 0. \tag{16.5}$$

Since $\mathrm{Re}(\gamma_1) > 0$ and $\mathrm{Re}(\gamma_2) > 0$, there is no solution that satisfies Eq. (16.5). On the other hand, in the TM mode,

$$\frac{\gamma_1}{n_{c1}^2} + \frac{\gamma_2}{n_{c2}^2} = 0. \tag{16.6}$$

As mentioned above, since $\mathrm{Re}(n_c^2) < 0$ for metals and $\mathrm{Re}(n_c^2) > 0$ for dielectrics, a solution exists in Eq. (16.6). The relationship between γ_i and β is given by

$$\gamma_i^2 = k_0^2 n_{ci}^2 - \beta^2. \tag{16.7}$$

From Eqs. (16.6) and (16.7),

[†]Note: Plasmons are the quantized vibrational energy of free electrons in metals, and surface plasmon polaritons are electromagnetic waves coupled to the vibrations of these free electrons. The energy of a polariton is also quantized. The quantization is not discussed here. As mentioned in Chapter 5, the reason that the real part of the permittivity of metals is negative corresponds to the fact that the macroscopic response of electric polarization to electromagnetic waves is negative.

$$\beta^2 = \frac{k_0{}^2 n_{c1}{}^2 n_{c2}{}^2}{n_{c1}{}^2 + n_{c2}{}^2} = \frac{k_0{}^2}{\alpha_1 + \alpha_2}. \tag{16.8}$$

Using the equivalent refractive index $n_{eq} = \beta/k_0$, Eq. (16.8) is expressed as follows:

$$n_{eq} = \sqrt{\frac{1}{\alpha_1 + \alpha_2}}. \tag{16.9}$$

In Eq. (16.1), when the dumping factor is set to 0 (i.e., $\tau = \infty$), $n_c{}^2$ becomes a negative real number. Therefore, in that case, β will also be a real number from Eq. (16.8). Even though metals have large extinction coefficients, there are surface plasmon polaritons that propagate without attenuation on the surface.

In actual metals, the imaginary part of the dielectric constant is not 0, so β also has an imaginary part and attenuates as it propagates. For example, in the case of gold (Au), the surface plasmon formed at the air/gold interface is calculated using the refractive index $n_c = 0.15 - 3.5i$ at the wavelength $\lambda = 0.65$ μm, as shown in Fig. 16.1. The imaginary part of $F(x)$ is shown magnified by a factor of 10 and is smaller than the real part. If the real part of n_c is 0 (that is, $\tau = \infty$), Im[$F(x)$] becomes exactly 0.

From Eq. (16.9), the equivalent refractive index for this surface plasmon polariton n_{sp} is given by

$$n_{sp} \equiv n_{eq} = 1.043 - 0.004i. \tag{16.10}$$

If the imaginary part is converted into an attenuation coefficient, it is about 77 mm^{-1}, so the distance at which 3 dB is attenuated is approximately 10 μm. At longer wavelengths, the attenuation coefficient becomes even smaller.

Because a mode that propagates over short distances can be obtained even when using metal, an optical waveguide composed of metal/dielectric/metal

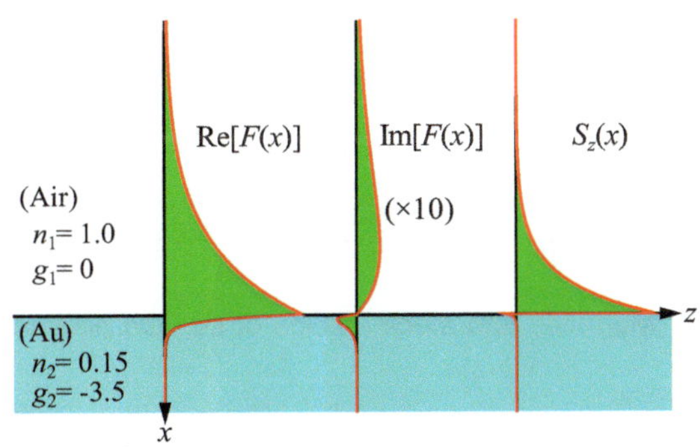

Figure 16.1 A surface plasmon polariton at the air/gold interface.

materials is sometimes called a "plasmon waveguide."† The structure shown in Fig. 16.1 produces an eigenmode, so it is also a type of optical waveguide. Optical waveguides usually require three or more layers, but in the case of metal/dielectric/metal materials, an eigenmode is formed with two layers.

16.2.2 Electric field of surface plasmon polaritons

Let's take a look at the electric and magnetic fields of a surface plasmon polariton (SPP) in dielectrics (in air) and metals. Since the SPP is in the TM mode, the magnetic field has only the y component, and its amplitude is expressed by the function of $F(x)$ in Eq. (16.2) multiplied by $\exp(-i\beta z + i\omega t)$. On the other hand, the electric field is given by Eq. (14.6) in Chapter 14. Figure 16.2 shows the z dependence of the real parts of the magnetic and electric fields. In the Excel animation, the propagation in the z direction can be seen.

As mentioned in Chapter 14, in the TM mode, the electric field has a longitudinal component (z component), that is, a longitudinal wave component, in addition to the transverse component (x component). Figure 16.3 shows the electric field components in air and metal. Each electric field oscillates along an ellipse in the figure. In air, it is an ellipse that is long in the x direction, so it vibrates almost in the x direction. On the other hand, in metals, the z component is large and is almost a longitudinal wave.

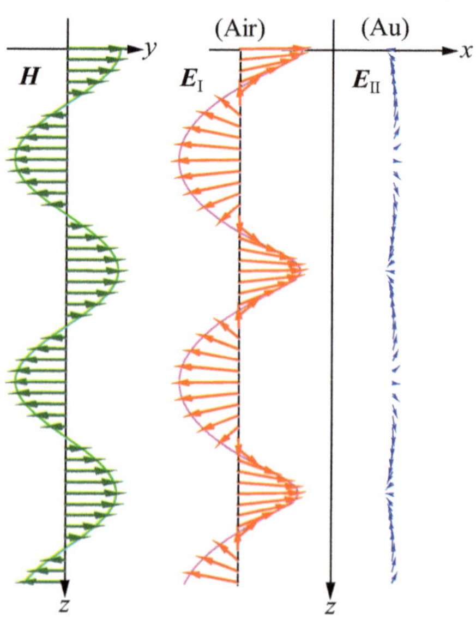

Figure 16.2 Magnetic and electric fields of surface plasmon polaritons.

†Note: Here, the chapter title uses "metal waveguides" to mean that they are equivalent to dielectric optical waveguides and semiconductor optical waveguides.

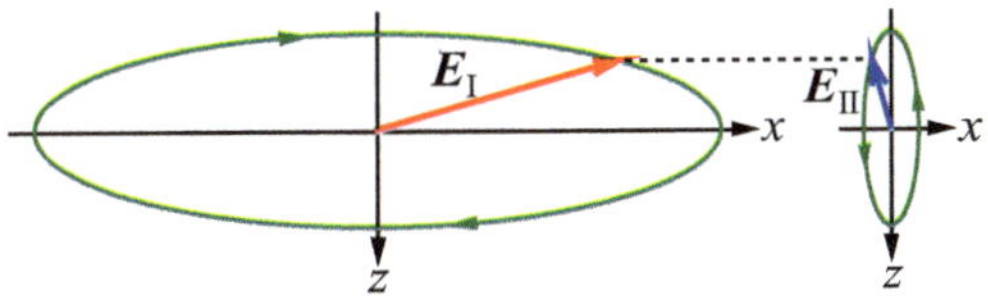

Figure 16.3 Electric field components in (left) air and (right) metals.

This figure shows the electric field component of electromagnetic waves, and electric polarization similarly oscillates in metals in response to this. In other words, compression waves (plasmons) of electrons are generated almost along the propagation direction.

Note that the fact that the electric field has a traveling direction component is not limited to surface plasmon polaritons but also applies to TM modes in dielectric optical waveguides, as described in Chapter 14.

Once each component of the electric field is known, the electric flux lines can be drawn using the methods described in Chapters 2 and 3. That is, the point (x, z) on the electric flux lines satisfies the following relational expression:

$$\frac{dz}{dx} = \frac{\mathrm{Re}(E_z)}{\mathrm{Re}(E_x)}. \tag{16.11}$$

For simplicity, consider the case where there is no damping factor (when the real part of the refractive index of the metal is 0). Then Eq. (16.11) becomes

$$\frac{dz}{dx} = \begin{cases} -\frac{\gamma_1}{\beta}\tan(\beta z - \omega t) & \text{(in air)} \\ \frac{\gamma_2}{\beta}\tan(\beta z - \omega t) & \text{(in metal)} \end{cases}. \tag{16.12}$$

From this, the formula representing the electric flux lines is represented by

$$x = -\frac{1}{\gamma_1}\{\ln|\sin(\beta z - \omega t)| + C\} \quad \text{(in air)}, \tag{16.13a}$$

$$x = \frac{1}{\gamma_2}\{\ln|\sin(\beta z - \omega t)| + C\} \quad \text{(in metal)}, \tag{16.13b}$$

where C is an integral constant.

The electric flux lines expressed by Eq. (16.13) are shown in Fig. 16.4. The direction of the electric field can be compared with that shown in Fig. 16.2.

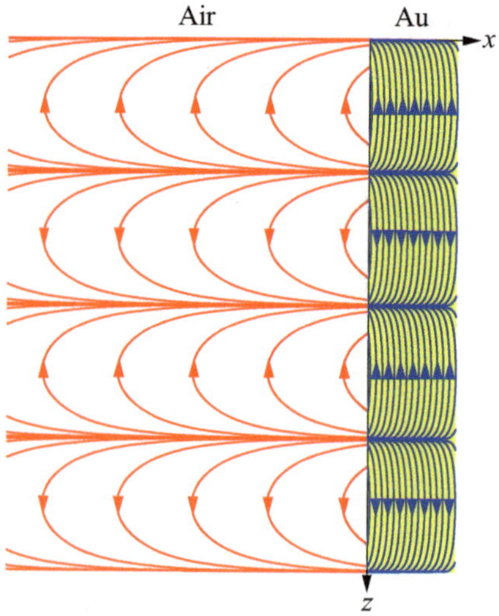

Figure 16.4 Electric flux lines of a surface plasmon polariton.

16.3 Metal/Dielectric/Metal Optical Waveguides

16.3.1 Waveguide modes in metal/dielectric/metal optical waveguides

The structure shown in Fig. 16.1 shows that eigenvalues of the TM mode exist. Therefore, let's consider what kind of waveguide modes will appear when a three-layer optical waveguide is used. A metal/dielectric/metal optical waveguide as shown in Fig. 16.5 is assumed here. The central guided layer may be made of glass or a semiconductor, but in the following discussion, an air layer is assumed for simplicity.

As in the case of Fig. 16.1, gold (Au) will be considered as the metal and the same parameters will be assumed. First, specifically the case for

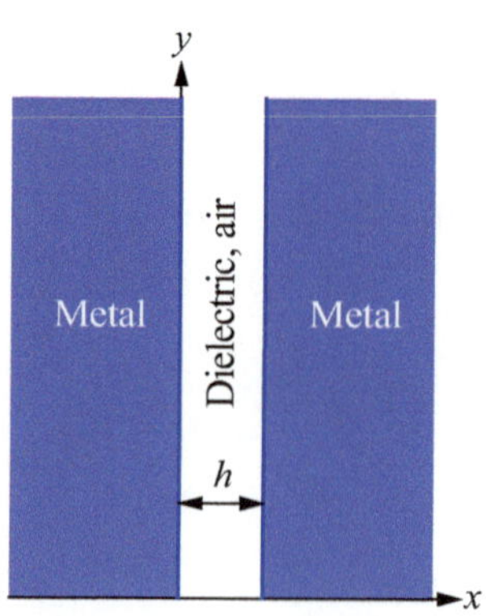

Figure 16.5 Metal/dielectric/metal optical waveguide.

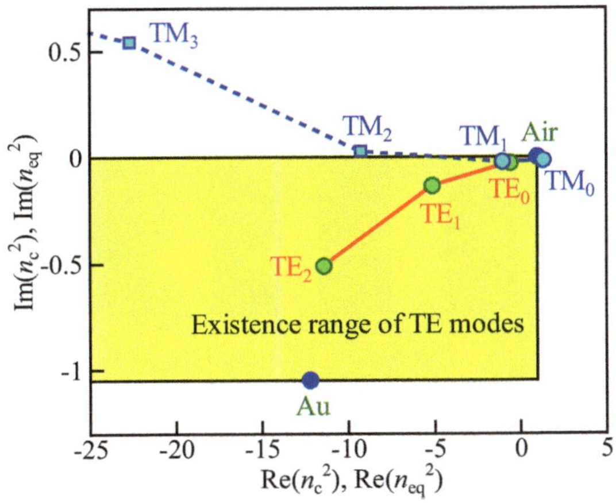

Figure 16.6 Eigenmodes in an Au/air/Au optical waveguide.

$h = 0.2$ μm is calculated, then the solutions for three TE modes and three or more TM modes are obtained, as shown in Fig. 16.6. The meaning of "three or more" will be explained in Section 16.3.2.

First, regarding the TE mode, the explanation in Chapter 15 applies exactly. In other words, all solutions fall within the range shown as the "existence range of TE modes" in Fig. 16.6. The mode function $F(x)$ $(= E_y)$ of these modes is shown in Fig. 16.7. Here, only $\mathrm{Re}[F(x)]$ is shown.

Now, the problem is the TM mode. The TM mode in a metal waveguide has a unique distribution shape, as shown in Fig. 16.8. As shown in Fig. 16.6, modes other than TM_1 are not within the solution range for the TE mode. Moreover, $\mathrm{Im}(n_{eq}^2)$ in the TM_2 mode is a positive value. If this is expressed by the effective propagation constant β, then $\mathrm{Re}(\beta) \cdot \mathrm{Im}(\beta) > 0$, resulting in a mode that is amplified in the propagation direction. This solution is physically incorrect and should be discarded. In the above, three or more TM modes were mentioned, but since $\mathrm{Im}(n_{eq}^2) > 0$ for all modes after TM_2, the only meaningful solutions in this case are TM_0 and TM_1.

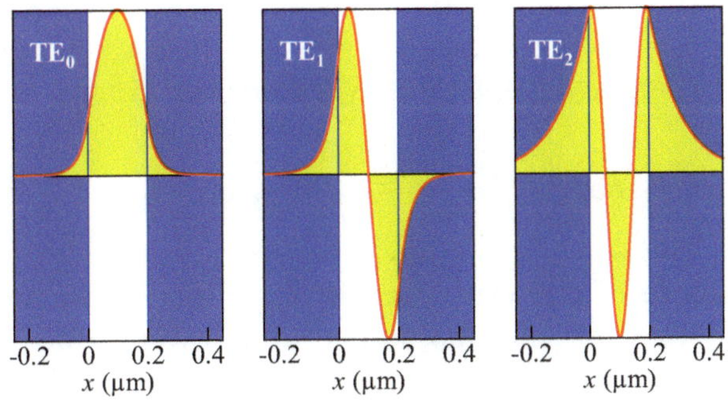

Figure 16.7 TE modes in an Au/air/Au optical waveguide.

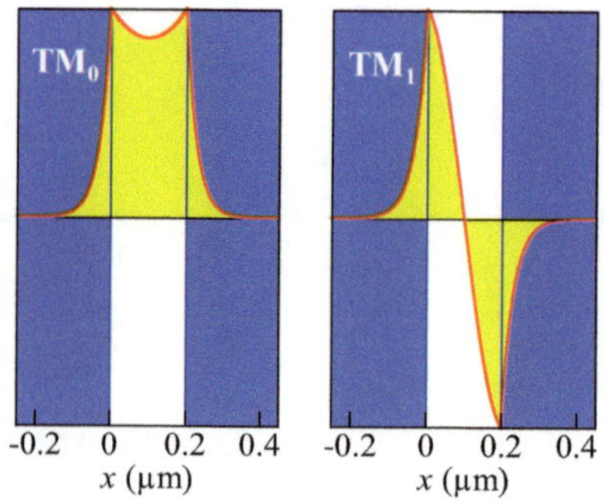

Figure 16.8 TM modes in an Au/air/Au optical waveguide.

16.3.2 Existence range of TM mode solutions

In the previous chapter, the existence range of the TE mode solutions was derived based on the wave equation. In a planar optical waveguide consisting of a layered structure of uniform media, the TM mode also satisfies the same wave equation, so Eq. (15.11) in the previous chapter is the same in the TM mode. To reproduce the formula,

$$\int_{-\infty}^{\infty} F^*(x)\frac{d^2}{dx^2}F(x)dx + \int_{-\infty}^{\infty}\{k_0{}^2 n_c{}^2(x) - \beta^2\}|F(x)|^2 dx = 0. \tag{16.14}$$

The first term on the left side becomes the following equation by integrating by parts:

$$\int_{-\infty}^{\infty} F^*(x)\frac{d^2}{dx^2}F(x)dx = \left[F^*\frac{dF}{dx}\right]_{-\infty}^{\infty} - \int_{-\infty}^{\infty}\left|\frac{dF}{dx}\right|^2 dx. \tag{16.15}$$

The first term on the right side of Eq. (16.15) becomes 0 in the TE mode but does not become 0 in the TM mode because *dF(x)/dx* is discontinuous at the boundary. Specifically, it is as follows:

$$\left[F^*\frac{dF}{dx}\right]_{-\infty}^{\infty} = \sum_i F^*(x_i)\frac{dF}{dx}(x_i)\left(1 - \frac{n_{ci+1}{}^2}{n_{ci}{}^2}\right). \tag{16.16}$$

In the case of a three-layer optical waveguide, Eq. (16.16) is represented by

$$\left[F^*\frac{dF}{dx}\right]_{-\infty}^{\infty} = \gamma_1|\cos\phi_1|^2\left(1 - \frac{n_{c2}{}^2}{n_{c1}{}^2}\right) + \gamma_3|\cos\phi_2|^2\left(1 - \frac{n_{c2}{}^2}{n_{c3}{}^2}\right). \tag{16.17}$$

Due to the existence of the term given by Eq. (16.17), the ranges of both the upper limit of $\mathrm{Re}(n_{eq}{}^2)$ and the upper and lower limits of $\mathrm{Im}(n_{eq}{}^2)$ are expanded in the case of the TM mode. Looking at Fig. 16.6, it is clear that the solution for TM_0 exceeds the upper limit of $\mathrm{Re}(n_{eq}{}^2)$ in the TE mode.

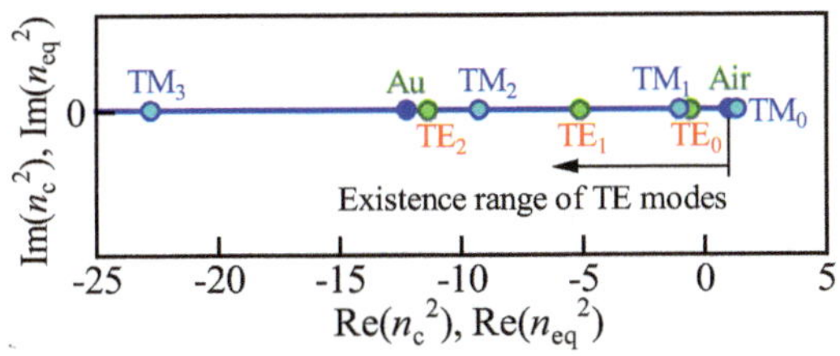

Figure 16.9 Eigenmodes of an Au/air/Au optical waveguide in the case where $\mathrm{Im}(n_c^2) = 0$.

The fact that the TM mode solution is also outside the existence range of the TE mode solution is also true when all n_c^2 are real numbers. As an example, consider the case of $\mathrm{Im}(n_c^2) = 0$ in Fig. 16.6. This corresponds to the case where the damping factor described in Section 16.2 is 0 ($\tau = \infty$). Assuming that $\mathrm{Re}(n_c^2)$ is the same as in Fig. 16.6, the solution will be as shown in Fig. 16.9. The values of $\mathrm{Re}(n_{eq}^2)$ for these solutions are almost equal to the values in Fig. 16.6. Also, since $\mathrm{Im}(n_{eq}^2) = 0$ in all solutions, TM_2, TM_3, etc., which were discarded in Fig. 16.6, are also solutions.

16.3.3 Dispersion curves for the TE mode

A dispersion curve corresponding to the b–V curve shown in Fig. 14.4 or Fig. 14.6 in Chapter 14 can also be drawn for metal/dielectric/metal optical waveguides. Here, since the refractive index is a complex number, the relationship between the square of the effective propagation constant β and the core thickness h will be plotted without normalized parameters b and V. The basic formula is Eq. (14.21), which is shown again below:

$$\kappa h = \tan^{-1}\left(\frac{\alpha_1}{\alpha_2}\frac{\gamma_1}{\kappa}\right) + \tan^{-1}\left(\frac{\alpha_3}{\alpha_2}\frac{\gamma_3}{\kappa}\right) + m\pi. \tag{16.18}$$

In this subsection and the two subsections that follow, the case of a symmetrical optical waveguide as shown in Fig. 16.5 will be considered. Let $\gamma_1 = \gamma_3$ and $\alpha_1 = \alpha_3$.

First, the dispersion curve of the TE mode in the Au/air/Au optical waveguide is shown in Fig. 16.10.[2] Since the equivalent refractive index n_{eq} and its square are complex numbers, $\mathrm{Re}(n_{eq}^2)$ and $\mathrm{Im}(n_{eq}^2)$ are shown here, respectively. Note that all solutions are within the existence range of eigenvalues of the TE mode shown in Fig. 16.6. The shape of this dispersion curve is very similar to the b–V curve in Chapter 14. However, $\mathrm{Re}(n_{eq}^2)$ is a negative value in most regions.

As shown by the circle in Fig. 16.10, there is a cutoff of core thickness (thickness of the air layer) h in the TE_1 mode and above. This cutoff is determined by the lower limit of $\mathrm{Im}(n_{eq}^2)$ in the existence range of solutions [see Fig. 16.10(b)]. When h is sufficiently large, n_{eq}^2 asymptotically approaches n_{Air}^2 ($= 1$) of air. In other words, it approaches a plane wave

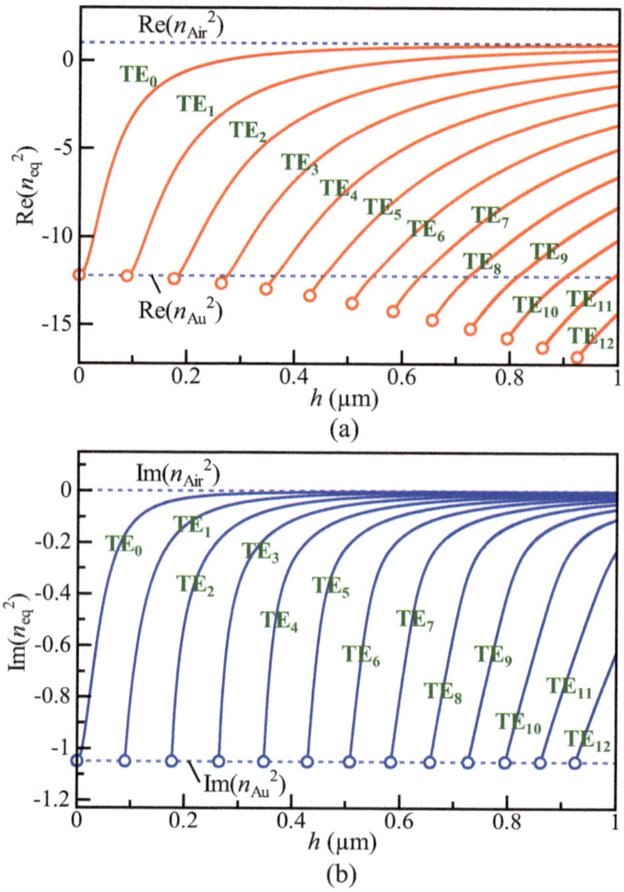

Figure 16.10 Dispersion curves of the TE mode in an Au/air/Au optical waveguide, representing the relationship between n_{eq}^2 and h: (a) Re(n_{eq}^2) and (b) Im(n_{eq}^2). Plotted with permission by modifying a figure from G. Hatakoshi and H. Furuyama[2] © IOP Publishing, Ltd.

propagating in air. This may seem quite obvious, but it will be explained in Section 16.3.4 that this is not the case in the TM mode.

Figure 16.10 can be converted to the diagram representing the relationship between n_{eq} and h, which is shown in Fig. 16.11.[2] As can be seen from the figure, the relationship between n_{eq} and h is a unique dispersion curve that is different from the b–V curve of a real refractive index optical waveguide. In particular, the real part decreases once with respect to h and then suddenly rises and increases at a certain thickness.

16.3.4 Dispersion curves for the TM mode

As shown in Fig. 16.8, the TM mode of a metal waveguide has a unique distribution shape. This reflects the mode of the surface plasmon polariton shown in Fig. 16.1.

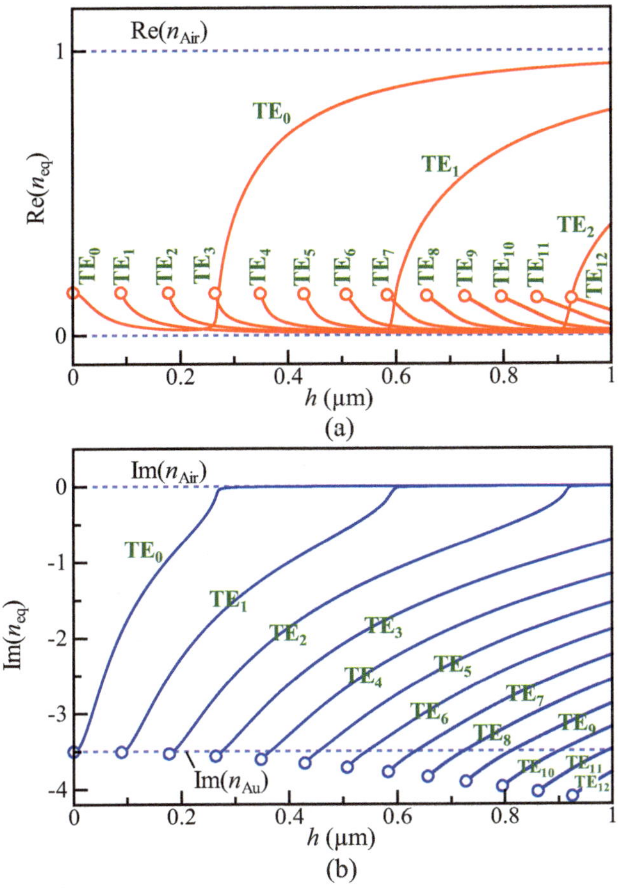

Figure 16.11 Dispersion curves of the TE mode in an Au/air/Au optical waveguide, representing the relationship between n_{eq} and h: (a) Re(n_{eq}) and (b) Im(n_{eq}). Plotted with permission by modifying a figure from G. Hatakoshi and H. Furuyama[2] © IOP Publishing, Ltd.

The dispersion curve of the TM mode in the Au/air/Au waveguide is shown in Fig. 16.12.[2] The real part looks similar to the b–V curve of a real index optical waveguide, as in Fig. 16.10, but the major difference is that only the TM_0 mode looks different from the other modes.

Even in the TM mode, there is a cutoff for the air layer thickness h in the modes above TM_1. As explained in Fig. 16.6, this cutoff is the result of discarding solutions where $\mathrm{Im}(n_{eq}{}^2) > 0$ [the part shown by the broken line in Fig. 16.12(b)]. The cutoff condition is given by[2]

$$\mathrm{Im}(n_{eq}{}^2) = 0. \tag{16.19}$$

The thickness h that satisfies Eq. (16.19) can be easily obtained from Eq. (16.18). Assuming the refractive index of air to be 1 and $n_c{}^2$ for metal to be

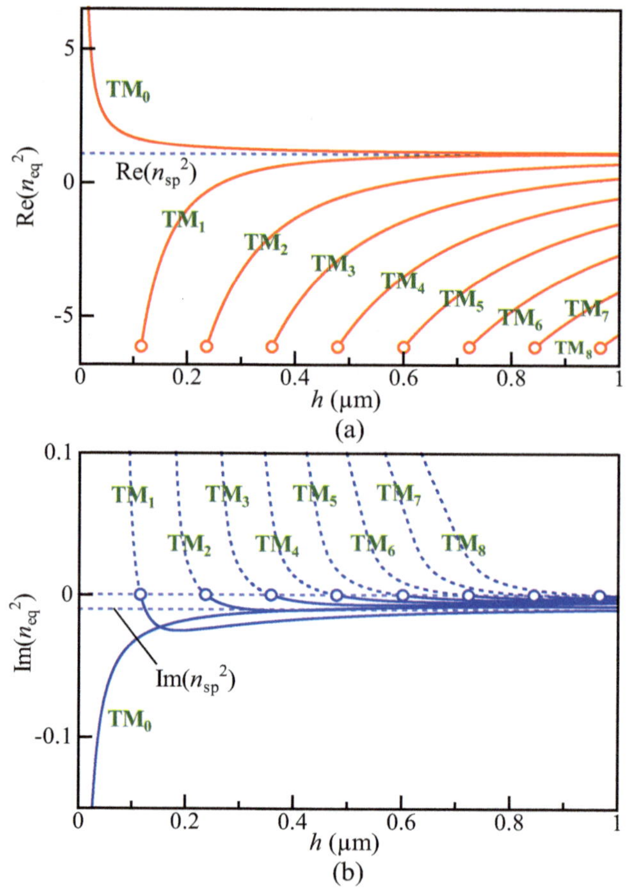

Figure 16.12 Dispersion curve of TM mode in an Au/air/Au optical waveguide, representing the relationship between n_{eq}^2 and h: (a) Re(n_{eq}^2) and (b) Im(n_{eq}^2). Plotted with permission by modifying a figure from G. Hatakoshi and H. Furuyama[2] © IOP Publishing, Ltd.

$$n_c^2 = a + bi, \tag{16.20}$$

the cutoff thickness h_{cutoff} is given by

$$h_{\text{cutoff}} = \frac{m\pi - 2\tan^{-1}\left(\dfrac{1}{\sqrt{a^2 + b^2 - 2a}}\right)}{k_0\sqrt{1 - \dfrac{a^2 + b^2}{2a}}}. \tag{16.21}$$

Note that a is negative in Eqs. (16.20) and (16.21).

Looking at Eq. (16.19), does this equation mean that the imaginary part of the equivalent refractive index n_{eq} becomes 0 at the cutoff? If this is the case, there will be a mode that propagates forever without loss. Actually, this is not the case; as shown in Fig. 16.12(a), $\mathrm{Re}(n_{\mathrm{eq}}^2)$ at this cutoff is negative. Therefore, as shown in Fig. 16.13, $\mathrm{Re}(n_{\mathrm{eq}}) = 0$ at the cutoff, and $\mathrm{Im}(n_{\mathrm{eq}})$ takes a large negative value. In other words, the mode at the cutoff is not a propagation mode but an evanescent mode that decays exponentially in the z direction.

Figure 16.13 explained above is a dispersion curve converted to the relationship between n_{eq} and h, similar to Fig. 16.11.[2] The situation of the only TM_0 mode being different from the other modes is the same as in Fig. 16.12.

Now, in the TE mode, the solution approaches a plane wave propagating in air as $h \to \infty$. On the other hand, in the TM mode, as can be seen from Figs. 16.12 and 16.13, the refractive index of air is not the point of asymptosis

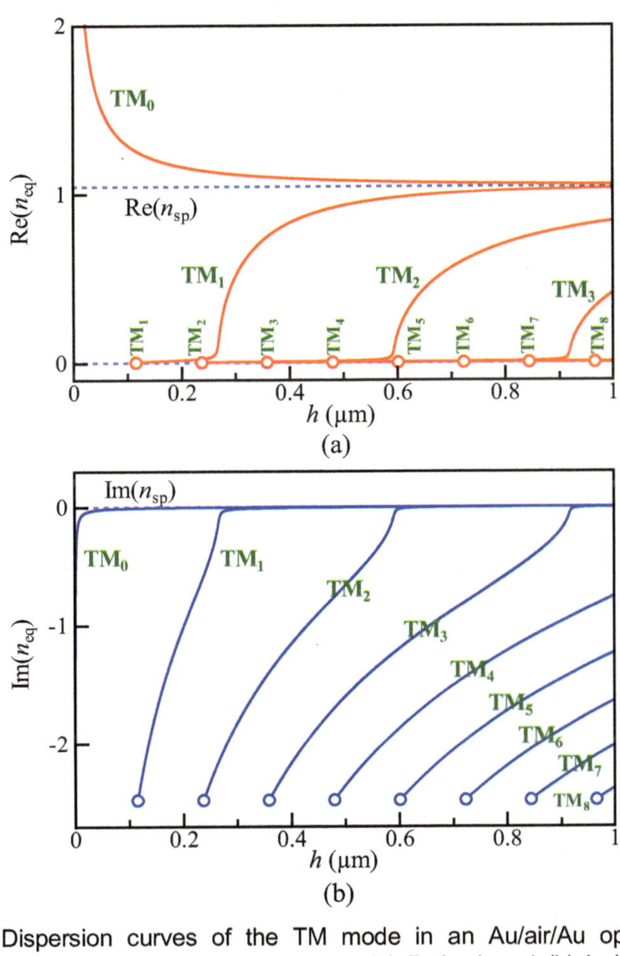

Figure 16.13 Dispersion curves of the TM mode in an Au/air/Au optical waveguide, representing the relationship between n_{eq} and h: (a) $\mathrm{Re}(n_{\mathrm{eq}})$ and (b) $\mathrm{Im}(n_{\mathrm{eq}})$. Plotted with permission by modifying a figure from G. Hatakoshi and H. Furuyama[2] © IOP Publishing, Ltd.

when $h \rightarrow \infty$. In fact, in the TM mode, it asymptotes the surface plasmon polariton shown in Fig. 16.1. That is, as shown in Fig. 16.13, the asymptotic point of the equivalent refractive index in the TM mode is as follows:

$$\lim_{h \to \infty} n_{\mathrm{eq}} = n_{\mathrm{sp}}, \tag{16.22}$$

where n_{sp} is the equivalent refractive index of the surface plasmon polariton, and its value in the case of air/gold is given by Eq. (15.10) in the previous chapter. In this case, $\mathrm{Re}(n_{\mathrm{sp}}) > 1$ and $\mathrm{Im}(n_{\mathrm{sp}}) < 0$.

The surface plasmon polariton shown in Fig. 16.1 assumes that the thickness of the air layer and the metal layer are both infinite. Equation (16.22) means that in the TM mode in a metal/air/metal optical waveguide, the same situation occurs when the air layer thickness approaches infinity.

16.3.5 TE mode and TM mode in metal micro-apertures

In Figs. 16.12 and 16.13, it was mentioned that the TM_0 mode is the exceptional mode. Although it is clear from Figs. 16.12(a) and 16.13(a) that the real part changes differently compared to the TE mode case shown in Fig. 16.10(a), what should be noted is the imaginary part. As can be seen from Fig. 16.13(b), in the TM_0 mode, the absolute value of $\mathrm{Im}(n_{\mathrm{eq}})$ does not become very large even in the region where the air layer thickness h is small. In other words, this shows that the absorption loss is small. For example, if $h = 50$ nm in an Au/air/Au optical waveguide, $\mathrm{Im}(n_{\mathrm{eq}}) = -0.023$ at $\lambda = 650$ nm, which is 0.447 $\mu\mathrm{m}^{-1}$ in terms of absorption loss. This means that, for example, if there is a 50-nm-wide slit in a 1-μm-thick gold plate, 64% of the incident light will be transmitted through the aperture area if it is p-polarized. The absorption loss of the TE_0 mode is 51 $\mu\mathrm{m}^{-1}$ for the same structure, which is two orders of magnitude larger than that of the TM_0 mode. Therefore, the transmittance for s-polarized light is almost 0. This is also the reason metal gratings act as polarizing plates.

The reason the losses in the TE mode and TM mode differ greatly in the case of a small aperture in metal[2] is that the modes penetrate the metal of the cladding differently. This situation is shown in Fig. 16.14. This figure is a plot of S_z values that correspond to the magnitude of the Poynting vector. When the aperture width $h = 50$ nm, a large amount of optical energy penetrates the metal in the TE_0 mode, while less penetrates in the TM_0 mode. This is because in the TM mode, the electric field component in the direction perpendicular to the interface divided by n_c^2 is continuous, so the electric field component becomes smaller in metals where $|n_c^2|$ is large.

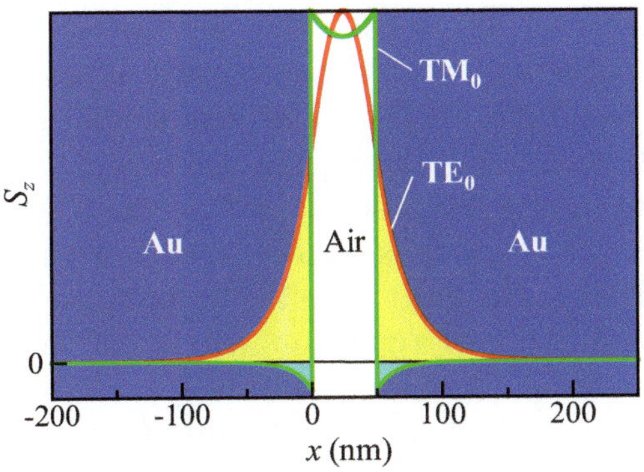

Figure 16.14 Comparison of the TE mode and TM mode in a metal micro-aperture.

16.4 Conclusion

Surface plasmon polaritons and waveguide modes in metal/dielectric/metal optical waveguides have been described. In the presence of metals, the TM mode exhibits extremely unique properties. It was shown that the existence range of eigenvalues in the TM mode is different from that in the TE mode. In particular, in metal/air/metal optical waveguides, the TM_0 mode is outside the range where the TE mode solution exists, and its core thickness dependence is different from that in the TM_1 and higher modes. In the TM_0 mode, the loss is relatively small even when the air layer thickness is small, and in the case of small metal apertures, the loss in the TM_0 mode is two orders of magnitude smaller than that in the TE_0 mode.

Side Topic: Free space and interface

In the optical waveguides introduced in three chapters including this one, the waveguide mode is determined from the boundary conditions at the interface. Although there are only two interfaces in a three-layer optical waveguide, solving this waveguide is considerably more difficult than finding a solution in free space. Still, compared to so-called "surface science," which studies solid surfaces, interfaces in optics may be relatively simple mathematically. Wolfgang Ernst Pauli, famous for his exclusion principle, said, "God made the bulk; surfaces were invented by the devil." The word "invent" not only means "to create" but also seems to mean "to forge."

In the case of a three-layer optical waveguide, the cladding thickness on both sides is assumed to be infinite. This is because if it is sufficiently thick compared to the wavelength, it can be approximated sufficiently for practical purposes. Similarly, it is common sense to think that if the interface is sufficiently far away, the solution in free space can be sufficiently approximated. However, as mentioned in the main text, in the TM mode in a metal/air/metal optical waveguide, even if the air layer thickness is made sufficiently large, the solution does not reach the free space solution but asymptotes the surface plasmon solution. In this case, the equivalent refractive index is greater than 1, as shown in Eqs. (16.22) and (16.10). This is sufficiently significant because it corresponds to a decrease in phase velocity of approximately 4%. Even if we think we are working in free space, if there is a metal wall far away, we can no longer approximate it with a free space solution. Considering this, "surface plasmons" may have been invented by the devil.

References

1. K. Kuroda, *Physical Optics—Propagation of light waves in a medium*, Asakura Publishing (2011) [in Japanese].
2. G. Hatakoshi and H. Furuyama, "Polarization dependence analysis of optical loss in small-aperture metal waveguide," *Jpn. J. Appl. Phys.* **40**(3S), 1548–1551 (2001).

Appendix 16A Excel Files

Appendix I-A contains the following Excel files. Each tab corresponds to a figure/s in the main text.

16_Plasmon.xlsm

Instruction	
SPP1	Fig. 16.1
SPP2	SPP propagation (informative)
SPP3	Fig. 16.2

16_Waveguide4.xlsm

Instruction	
Au-TE	Fig. 16.7, Fig. 16.14
Au-TM	Fig. 16.8, Fig. 16.14

Chapter 17
Optical Fiber Part 1

17.1 Introduction

Chapters 14 to 16 introduced mode analysis of a planar optical waveguide and calculation examples using Newton's method in Excel. Another typical example of an optical waveguide is an optical fiber.[1–4] Although the planar optical waveguide was analyzed using a Cartesian coordinate system, a cylindrical coordinate system will be used for the optical fiber. This chapter describes the mode analysis method for step-index optical fibers.

17.2 Modes in Optical Fibers

17.2.1 Structure of optical fiber and coordinate system

Figure 17.1 shows an example of the structure of an optical fiber and a cylindrical coordinate system. The optical fiber has a stepped refractive index structure consisting of a core and a cladding, with the core having a refractive index of n_1 and the cladding having a refractive index of n_2. As in the case of the planar optical waveguide in Chapter 14, the relative refractive index difference Δ is defined below:

$$\Delta \equiv \frac{n_1^2 - n_2^2}{2n_1^2} \cong \frac{n_1 - n_2}{n_1}. \tag{17.1}$$

This Δ is related to the numerical aperture NA of the optical fiber. NA corresponds to the divergence angle of the output light for the maximum order mode in an optical fiber and is an index when laser light is focused by a lens and coupled into an optical fiber. The divergence angle of the output light for the maximum-order mode can be approximated by the angle in air for the light propagating at the critical angle of total reflection at the core/cladding interface, and the NA is given by the following equation:

$$NA = \sqrt{n_1^2 - n_2^2} = n_1\sqrt{2\Delta}. \tag{17.2}$$

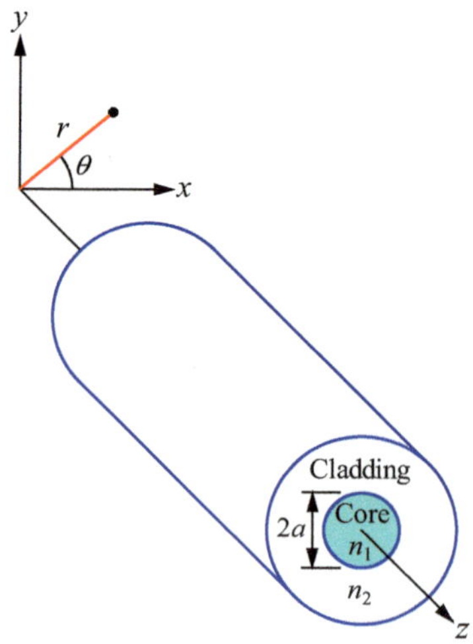

Figure 17.1 Optical fiber structure example and coordinate system.

In a single-mode optical fiber, *NA* $\approx$ 0.15 or less, and the relative refractive index difference Δ is very small, about 0.005 or less. Even multimode optical fibers have an *NA* of about 0.2, and high-*NA* optical fibers have an *NA* of about 0.35, but even in that case, when converted to Δ, it is about 0.03.

17.2.2 Wave equation in cylindrical coordinates

The modes of the optical fiber can be analyzed using the cylindrical coordinate system shown in Fig. 17.1.[1–3] The wave equation obtained from Maxwell's equations is

$$\nabla^2 \boldsymbol{E} - \varepsilon\mu \frac{\partial^2 \boldsymbol{E}}{\partial t^2} = 0. \tag{17.3}$$

Let the time dependence of the electric field $\boldsymbol{E}$ be $\exp(i\omega t)$, in which case $\boldsymbol{E}$ satisfies the following Helmholtz equation:

$$\nabla^2 \boldsymbol{E} + k^2 \boldsymbol{E} = 0, \tag{17.4}$$

where

$$k^2 \equiv \omega^2 \varepsilon\mu = \omega^2 \varepsilon_0 \mu_0 n^2 = k_0^2 n^2. \tag{17.5}$$

As mentioned above, if the refractive indices of the core and cladding are n_1 and n_2, respectively, k is given by

$$k = \begin{cases} k_0 n_1 & (r < a) \\ k_0 n_2 & (r > a) \end{cases}. \tag{17.6}$$

The wave equation in cylindrical coordinates was also treated for cylindrical waves in Chapter 2. Equation (17.4) is expressed in cylindrical coordinates as follows:

$$\frac{1}{r}\frac{\partial}{\partial r}\left(r\frac{\partial}{\partial r}E_z\right)+\frac{1}{r^2}\frac{\partial^2}{\partial\theta^2}E_z+\frac{\partial^2}{\partial z^2}E_z+k^2E_z=0, \tag{17.7}$$

where E_z is the z component of the electric field. The z component H_z of the magnetic field also satisfies a similar equation. The same goes for the other components, but let's start with the z component in the propagation direction.

Separate E_z into variables using cylindrical coordinates:

$$E_z=R(r)\cos(l\theta)e^{i(\omega t-\beta z)}. \tag{17.8}$$

Then, the following Bessel differential equation is obtained:

$$\frac{d^2R}{dr^2}+\frac{1}{r}\frac{dR}{dr^2}+\left\{(k_i^2-\beta^2)-\frac{l^2}{r^2}\right\}R=0, \tag{17.9}$$

where β is the effective propagation constant in the z direction. As in the case of planar optical waveguides,

$$k_0n_2<\beta<k_0n_1. \tag{17.10}$$

Then, Eq. (17.9) becomes

$$\frac{d^2R}{dr^2}+\frac{1}{r}\frac{dR}{dr^2}+\left\{\kappa^2-\frac{l^2}{r^2}\right\}R=0\quad(r\le a), \tag{17.11a}$$

$$\frac{d^2R}{dr^2}+\frac{1}{r}\frac{dR}{dr^2}-\left\{\gamma^2+\frac{l^2}{r^2}\right\}R=0\quad(r\ge a), \tag{17.11b}$$

where

$$\kappa=\sqrt{k_0^2n_1^2-\beta^2}, \tag{17.12a}$$

$$\gamma=\sqrt{\beta^2-k_0^2n_2^2}. \tag{17.12b}$$

The solution to Eq. (17.11) is expressed as a linear combination of two independent cylindrical functions as follows:

$$R(r)=\begin{cases}a_1J_l(\kappa r)+b_1Y_l(\kappa r) & (r\le a)\\ a_2I_l(\kappa r)+b_2K_l(\gamma r) & (r\ge a)\end{cases}, \tag{17.13}$$

where $J_l(x)$ and $Y_l(x)$ are the lth-order Bessel functions of the first kind and the second kind, respectively, and $I_l(x)$ and $Y_l(x)$ are the lth-order modified

Bessel functions of the first kind and the second, respectively. A solution to Eq. (17.13) that does not diverge at $r = 0$ and $r = \infty$ is given by

$$R(r) = \begin{cases} J_l(\kappa r) & (r \le a) \\ K_l(\gamma r) & (r \ge a) \end{cases}. \tag{17.14}$$

From the above, the solutions for E_z and H_z are expressed by the following equations:

$$E_z = \begin{cases} A_l J_l(\kappa r)\cos(l\theta)e^{i(\omega t - \beta z)} & (r \le a) \\ B_l' K_l(\gamma r)\cos(l\theta)e^{i(\omega t - \beta z)} & (r \ge a) \end{cases}, \tag{17.15a}$$

$$H_z = \begin{cases} B_l J_l(\kappa r)\sin(l\theta)e^{i(\omega t - \beta z)} & (r \le a) \\ B_l' K_l(\gamma r)\sin(l\theta)e^{i(\omega t - \beta z)} & (r \ge a) \end{cases}. \tag{17.15b}$$

Using the above E_z and H_z, the other components are given by Maxwell's equations as follows:

$$E_r = -\frac{i}{k_0^2 n^2 - \beta^2}\left[\beta\frac{\partial E_z}{\partial r} + \frac{\omega\mu_0}{r}\frac{\partial H_z}{\partial \theta}\right], \tag{17.16a}$$

$$E_\theta = -\frac{i}{k_0^2 n^2 - \beta^2}\left[\frac{\beta}{r}\frac{\partial E_z}{\partial \theta} - \omega\mu_0\frac{\partial H_z}{\partial r}\right], \tag{17.16b}$$

$$H_r = -\frac{i}{k_0^2 n^2 - \beta^2}\left[-\frac{\omega\varepsilon_0 n^2}{r}\frac{\partial E_z}{\partial \theta} + \beta\frac{\partial H_z}{\partial r}\right], \tag{17.17a}$$

$$H_\theta = -\frac{i}{k_0^2 n^2 - \beta^2}\left[\omega\varepsilon_0 n^2\frac{\partial E_z}{\partial r} + \frac{\beta}{r}\frac{\partial H_z}{\partial \theta}\right]. \tag{17.17b}$$

17.2.3 Eigenvalue equation

All components of $\boldsymbol{E}$ and $\boldsymbol{H}$ were expressed by Eqs. (17.15) to (17.17). However, unless the value of β is determined, κ and γ cannot be determined. The parameter β is determined by the boundary condition at $r = a$.

First, the following boundary conditions are given from the continuity of E_z and H_z at $r = a$:

$$A_l' K_l = A_l J_l(\kappa a)/K_l(\gamma a), \tag{17.18a}$$

$$B_l' = B_l J_l(\kappa a)/K_l(\gamma a). \tag{17.18b}$$

From this equation and the continuity of E_θ and H_θ, the following relationship can be obtained:

$$\omega\mu_0\left\{\frac{1}{\kappa}\frac{J_l'(\kappa a)}{J_l(\kappa a)}+\frac{1}{\gamma}\frac{K_l'(\gamma a)}{K_l(\gamma a)}\right\}B_l=-\frac{l\beta}{a}\left(\frac{1}{\kappa^2}+\frac{1}{\gamma^2}\right)A_l, \tag{17.19a}$$

$$\omega\varepsilon_0\left\{\frac{n_1^2}{\kappa}\frac{J_l'(\kappa a)}{J_l(\kappa a)}+\frac{n_2^2}{\gamma}\frac{K_l'(\gamma a)}{K_l(\gamma a)}\right\}A_l=-\frac{l\beta}{a}\left(\frac{1}{\kappa^2}+\frac{1}{\gamma^2}\right)B_l, \tag{17.19b}$$

where the symbol $'$ represents the differential [i.e., $J_l'(x)=dJ_l(x)/dx, K_l'(x)=dK_l(x)/dx$].

In Eq. (17.19), the condition that a solution other than $A_l=B_l=0$ can exist is given by

$$\left\{\frac{1}{\kappa}\frac{J_l'(\kappa a)}{J_l(\kappa a)}+\frac{1}{\gamma}\frac{K_l'(\gamma a)}{K_l(\gamma a)}\right\}\left\{\frac{n_1^2}{\kappa}\frac{J_l'(\kappa a)}{J_l(\kappa a)}+\frac{n_2^2}{\gamma}\frac{K_l'(\gamma a)}{K_l(\gamma a)}\right\}=\frac{l^2\beta^2}{k_0^2a^2}\left(\frac{1}{\kappa^2}+\frac{1}{\gamma^2}\right)^2. \tag{17.20}$$

This equation is the eigenvalue equation for the effective propagation constant β.

17.2.4 Normalization

Similar to the parameters used for the planar optical waveguide in Chapter 14, the following normalized frequency (or normalized thickness) V and normalized propagation constant b can be defined:

$$V\equiv k_0a\sqrt{n_1^2-n_2^2}=k_0an_1\sqrt{2\Delta}, \tag{17.21}$$

$$b\equiv\frac{n_{\mathrm{eq}}^2-n_2^2}{n_1^2-n_2^2}=\frac{\beta^2-k_0^2n_2^2}{k_0^2(n_1^2-n_2^2)}. \tag{17.22}$$

Using this, κ and γ are represented as follows:

$$\kappa a=\sqrt{1-b}V, \tag{17.23}$$

$$\gamma a=\sqrt{b}V. \tag{17.24}$$

The effective propagation constant β is expressed below:

$$\beta=k_0n_1\sqrt{1-2\Delta(1-b)}. \tag{17.25}$$

By substituting Eqs. (17.23) to (17.25) into Eq. (17.20) and using the relationships in Eqs. (17.A4a) and (17.A5a) in the appendix, Eq. (17.20) becomes as follows:

$$V\left\{\sqrt{b}\frac{J_{l-1}(\sqrt{1-b}V)}{J_l(\sqrt{1-b}V)}-\sqrt{1-b}\frac{K_{l-1}(\sqrt{b}V)}{K_l(\sqrt{b}V)}\right\}$$
$$\times\left\{\sqrt{b}\frac{J_{l-1}(\sqrt{1-b}V)}{J_l(\sqrt{1-b}V)}-c_1\sqrt{1-b}\frac{K_{l-1}(\sqrt{b}V)}{K_l(\sqrt{b}V)}\right\}$$
$$-2l\left\{\frac{c_1+(1+b)\Delta}{\sqrt{1-b}}\frac{J_{l-1}(\sqrt{1-b}V)}{J_l(\sqrt{1-b}V)}-\frac{c_1+b\Delta}{\sqrt{b}}\frac{K_{l-1}(\sqrt{b}V)}{K_l(\sqrt{b}V)}\right\}=0, \quad (17.26)$$

where

$$c_1 = 1 - 2\Delta. \quad (17.27)$$

Equation (17.26) expresses the relationship between b and V and can be used to draw a b–V dispersion curve that is similar to the case of a planar optical waveguide.

Note that if Eqs. (17.A4b) and (17.A5b) are used instead of Eqs. (17.A4a) and (17.A5a), Eq. (17.26) becomes as follows:

$$V\left\{\sqrt{b}\frac{J_{l+1}(\sqrt{1-b}V)}{J_l(\sqrt{1-b}V)}+\sqrt{1-b}\frac{K_{l+1}(\sqrt{b}V)}{K_l(\sqrt{b}V)}\right\}$$
$$\times\left\{\sqrt{b}\frac{J_{l+1}(\sqrt{1-b}V)}{J_l(\sqrt{1-b}V)}+c_1\sqrt{1-b}\frac{K_{l+1}(\sqrt{b}V)}{K_l(\sqrt{b}V)}\right\}$$
$$-2l\left\{\frac{c_1+(1+b)\Delta}{\sqrt{1-b}}\frac{J_{l+1}(\sqrt{1-b}V)}{J_l(\sqrt{1-b}V)}+\frac{c_1+b\Delta}{\sqrt{b}}\frac{K_{l+1}(\sqrt{b}V)}{K_l(\sqrt{b}V)}\right\}=0. \quad (17.28)$$

Equations. (17.26) and (17.28) have different forms but are essentially the same equation and therefore give the same eigenvalue.

17.3 Solution of the Eigenvalue Equation for $l = 0$

Equation (17.20) and its normalized Eqs. (17.26) and (17.28) are somewhat complicated, but they can be solved using Newton's method introduced in Chapters 14 to 16. However, since these equations do not include mode orders other than l, multiple solutions corresponding to the number of modes must be found, making the calculation procedure somewhat difficult. Therefore, the general case of $l \geq 1$ will be left to the next chapter, and the cases of the TE mode and TM mode, where Eq. (17.20) is a little simpler, will first be discussed in this section.

17.3.1 TE mode

In the TE mode, $E_z = 0$ ($A_l = A_l' = 0$) by definition, so the following holds from Eq. (17.19):

$$\frac{1}{\kappa}\frac{J_l'(\kappa a)}{J_l(\kappa a)} + \frac{1}{\gamma}\frac{K_l'(\gamma a)}{K_l(\gamma a)} = 0, \tag{17.29}$$

$$\frac{\beta}{a} l J_l(\kappa a) = 0. \tag{17.30}$$

From Eq. (17.30),

$$l = 0. \tag{17.31}$$

Therefore, Eq. (17.29) becomes

$$\frac{1}{\kappa}\frac{J_1(\kappa a)}{J_0(\kappa a)} + \frac{1}{\gamma}\frac{K_1(\gamma a)}{K_0(\gamma a)} = 0. \tag{17.32}$$

This is the eigenvalue equation for the TE mode. Expressed by b and V,

$$\frac{1}{\sqrt{1-b}}\frac{J_1(\sqrt{1-b}V)}{J_0(\sqrt{1-b}V)} + \frac{1}{\sqrt{b}}\frac{K_1(\sqrt{b}V)}{K_0(\sqrt{b}V)} = 0. \tag{17.33}$$

Equation (17.33) is, of course, included in the equation where l is set to 0 in Eq. (17.26) or Eq. (17.28).[†]

Rewrite Eq. (17.33) as

$$f_{\mathrm{TE}} \equiv \sqrt{b}J_1(\sqrt{1-b}V)K_0(\sqrt{b}V) + \sqrt{1-b}J_0(\sqrt{1-b}V)K_1(\sqrt{b}V) = 0. \tag{17.34}$$

When solving this using Newton's method, the required process is to calculate df_{TE}/dV to find V for a given b, or to calculate df_{TE}/db to find b for a given V. Practically speaking, the latter is the case, but for the drawing of a b–V curve, the former process can be used.

First, from Eq. (17.34), df_{TE}/dV is represented as follows:

$$\frac{df_{\mathrm{TE}}}{dV} = -J_1(\sqrt{1-b}V)K_1(\sqrt{b}V) - \frac{1}{2}\sqrt{b}\sqrt{1-b}\Big\{J_0(\sqrt{1-b}V)K_2(\sqrt{b}V) + J_2(\sqrt{1-b}V)K_0(\sqrt{b}V)\Big\}. \tag{17.35}$$

[†]Note: When setting $l = 0$ in Eq. (17.26) or Eq. (17.28), the equation becomes the product of the eigenvalue equation for the TE mode and the eigenvalue equation for the TM mode.

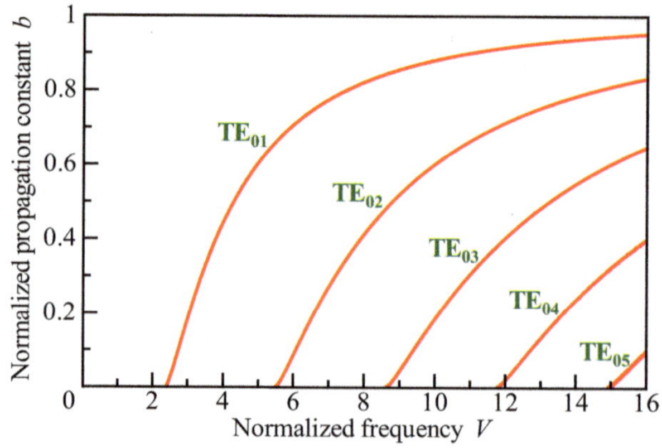

Figure 17.2 *b–V* curve for the TE mode.

The b–V dispersion curve calculated using Eq. (17.35) is shown in Fig. 17.2. Each mode is expressed as TE_{0m}. The first 0 in the mode order represents $l = 0$, and the next m means the mth mode. Note that in the TE mode and the TM mode described below, $l = 0$. Therefore, there is no such mode as TE_{1m} or TE_{2m}.

For comparison, Fig. 17.3 shows the b–V curve for the planar optical waveguide described in Chapter 14. This figure is the same as the TE mode b–V curve shown in Fig. 14.4. A comparison between Figs. 17.2 and 17.3 will be discussed in Section 17.3.4.

Instead of being defined by Eq. (17.35), df_{TE}/db, which is used for the case when V is given and b is calculated, is as follows:

$$\begin{aligned}\frac{df_{TE}}{db} &= \frac{1}{2\sqrt{b}} J_1(\sqrt{1-b}V)K_0(\sqrt{b}V) - \frac{1}{2\sqrt{b}} J_0(\sqrt{1-b}V)K_1(\sqrt{b}V) \\ &- \frac{V}{4}\left\{\left(\sqrt{\frac{b}{1-b}} + \sqrt{\frac{1-b}{b}}\right) J_0(\sqrt{1-b}V)K_0(\sqrt{b}V)\right. \\ &\left. - \sqrt{\frac{b}{1-b}} J_2(\sqrt{1-b}V)K_0(\sqrt{b}V) + \sqrt{\frac{1-b}{b}} J_0(\sqrt{1-b}V)K_2(\sqrt{b}V)\right\}.\end{aligned} \tag{17.36}$$

Although the above equation is a little more complicated than Eq. (17.35), there is no particular problem when calculating it in Excel.

17.3.2 TM mode

Since $H_z = 0$ in the TM mode, $l = 0$ in the TE mode, and the eigenvalue equation is as follows:

$$\frac{n_1^2}{\kappa}\frac{J_1(\kappa a)}{J_0(\kappa a)} + \frac{n_2^2}{\gamma}\frac{K_1(\gamma a)}{K_0(\gamma a)} = 0. \tag{17.37}$$

Expressed in terms of normalized parameters b and V, the equation corresponding to Eq. (17.34) is as follows:

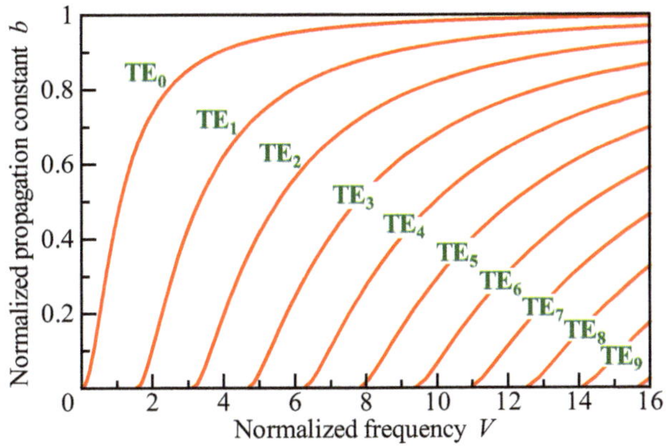

Figure 17.3 Comparison of *b*–*V* curves for the TE mode in the planar optical waveguide.

$$f_{\mathrm{TM}} \equiv \sqrt{b}J_1(\sqrt{1-b}V)K_0(\sqrt{b}V) + c_1\sqrt{1-b}J_0(\sqrt{1-b}V)K_1(\sqrt{b}V) = 0, \tag{17.38}$$

where c_1 is given by Eq. (17.27). Since the only difference from Eq. (17.34) is this coefficient c_1, equations corresponding to Eq. (17.35) and Eq. (17.36) can also be easily obtained.

As mentioned at the beginning of this chapter, in optical fibers, the relative refractive index difference Δ is a small value of about 0.005, so c_1 in Eq. (17.27) is almost unity. Then, Eq. (17.38) almost matches Eq. (17.34). When actually solving Eq. (17.38) strictly and plotting the TM mode *b*–*V* curve, it is almost the same as Fig. 17.2 and cannot be distinguished. So, let's take a look at the difference in *b* between the TE mode and TM mode: $b_{\mathrm{TE}} - b_{\mathrm{TM}}$. A plot of this will look like Fig. 17.4. In this figure, Δ is set to a relatively large value of 0.03. Even in this case, the difference in *b* is less than 0.008.

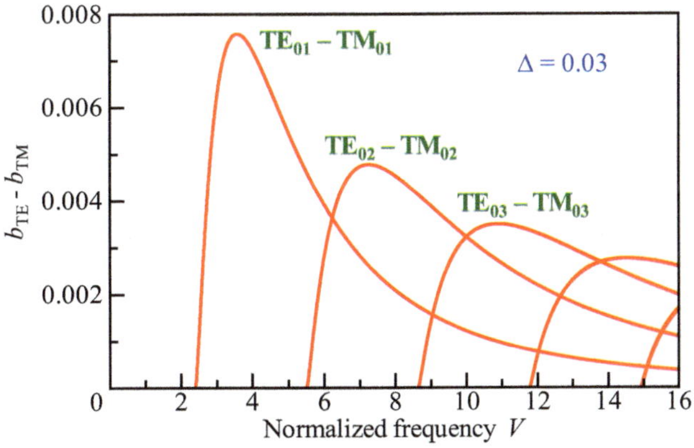

Figure 17.4 Difference in *b* between the TE mode and the TM mode.

17.3.3 Electromagnetic field distribution

The case for the TE mode will be considered below. In the case for the TM mode, almost the same explanation can be made by swapping ***E*** and ***H***.

The Eqs. (17.15) to (17.17) become as follows, in the case for the TE mode:

$$H_z = \begin{cases} J_0(\kappa r)e^{i(\omega t-\beta z)} & (r \leq a) \\ \dfrac{J_0(\kappa a)}{K_0(\gamma a)}K_0(\gamma r)e^{i(\omega t-\beta z)} & (r \geq a) \end{cases}, \tag{17.39a}$$

$$E_z = 0, \tag{17.39b}$$

$$E_r = 0, \tag{17.39c}$$

$$\frac{E_\theta}{\omega\mu_0} = \begin{cases} -\dfrac{i}{\kappa}J_1(\kappa r)e^{i(\omega t-\beta z)} & (r \leq a) \\ \dfrac{i}{\gamma}\dfrac{J_0(\kappa a)}{K_0(\gamma a)}K_1(\gamma r)e^{i(\omega t-\beta z)} & (r \geq a) \end{cases}, \tag{17.39d}$$

$$\frac{H_r}{\beta} = \begin{cases} \dfrac{i}{\kappa}J_1(\kappa r)e^{i(\omega t-\beta z)} & (r \leq a) \\ -\dfrac{i}{\gamma}\dfrac{J_0(\kappa a)}{K_0(\gamma a)}K_1(\gamma r)e^{i(\omega t-\beta z)} & (r \geq a) \end{cases}, \tag{17.39e}$$

$$H_\theta = 0. \tag{17.39f}$$

Figure 17.5 shows the calculated results of E_θ given by Eq. (17.39d). In order to exhibit higher-order modes, a multimode optical fiber, with a core

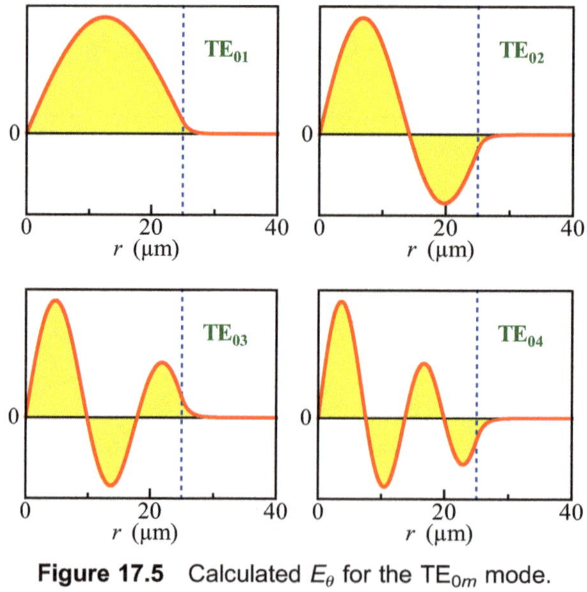

Figure 17.5 Calculated E_θ for the TE_{0m} mode.

Figure 17.6 Distribution of E_θ in the plane perpendicular to the propagation direction for the TE_{01} mode.

diameter of 50 μm and $\Delta = 0.01$, was assumed. In the case of these parameters, there are eight modes, TE_{01} to TE_{08}. The figure shows up to TE_{04}. The broken line in the figure is the core/cladding interface. For the TE_{01} mode, the distribution of E_θ in the plane perpendicular to the propagation direction is shown in Fig. 17.6.

Now that each component of the electric field and magnetic field is given by Eq. (17.39), let's calculate what the electric and magnetic flux lines will look like.

As shown in Eqs. (17.39b) to (17.39d), the electric field has only the E_θ component, so it is clear that the electric flux lines form concentric circles in the plane perpendicular to the propagation direction. On the other hand, magnetic flux lines have H_r and H_z components. Therefore, magnetic flux lines can be drawn using the methods described in Chapters 2 and 3. In other words, the point (r, z) on the magnetic flux lines satisfies the following relation:[†]

$$\frac{dr}{dz} = \frac{\text{Re}(H_r)}{\text{Re}(H_z)}. \tag{17.40}$$

[†]Note: Equation (17.40) is a ratio of the real parts, but since it is sufficient to take the ratio of terms in the same phase, it can also be a ratio of the imaginary parts. In the case of the imaginary parts, sin and cos are swapped and the formula changes, but they represent the same curves only with a phase shift.

Substituting Eqs. (17.39a) and (17.39e), Eq. (17.40) becomes as follows:

$$\frac{dr}{dz} = \begin{cases} \dfrac{\beta}{\kappa}\dfrac{J_1(\kappa r)\sin(\beta z)}{J_0(\kappa r)\cos(\beta z)} & (r \le a) \\ -\dfrac{\beta}{\gamma}\dfrac{K_1(\gamma r)\sin(\beta z)}{K_0(\gamma r)\cos(\beta z)} & (r \ge a) \end{cases}. \tag{17.41}$$

From this, for $r \le a$,

$$\frac{J_0(\kappa r)}{J_1(\kappa r)}dr = \frac{\beta}{\kappa}\frac{\sin(\beta z)}{\cos(\beta z)}dz. \tag{17.42}$$

Using the formula in the appendix, Eq. (17.A4a), the above equation becomes

$$\left\{\frac{dJ_1(\kappa r)/d(\kappa r)}{J_1(\kappa r)} + \frac{1}{\kappa r}\right\}dr = \frac{\beta}{\kappa}\frac{\sin(\beta z)}{\cos(\beta z)}dz. \tag{17.43}$$

Integrating this, the following equation is obtained:

$$\ln[|J_1(\kappa r)|] + \ln(r) = -\ln[|\cos(\beta z)|] + C, \tag{17.44}$$

where C is an integral constant. From this, βz is expressed as a function of r as follows:

$$\beta z = \pm\cos^{-1}\left(\frac{C_1}{rJ_1(\kappa r)}\right) + m\pi. \tag{17.45}$$

Similarly, for $r \ge a$,

$$\beta z = \pm\cos^{-1}\left(\frac{C_2}{rK_1(\gamma r)}\right) + m\pi. \tag{17.46}$$

Considering the connection at $r = a$,

$$C_2 = \frac{K_1(\gamma a)}{J_1(\kappa a)}C_1. \tag{17.47}$$

Using the above equations, the electric and magnetic flux lines for the TE_{01} mode can be plotted as shown in Fig. 17.7. In this figure, the direction of the electric flux lines is reversed between the front and back surfaces. The point of inversion is at the intermediate position of $z = \pi/(2\beta)$. At this position, the electric field amplitude becomes 0.

Figure 17.8 shows the case for the TE_{02} mode. The method of displaying electric and magnetic flux lines was based on Snitzer's paper.[5] In the TM mode, the diagrams are almost the same, as seen by just swapping the electric and magnetic flux lines.

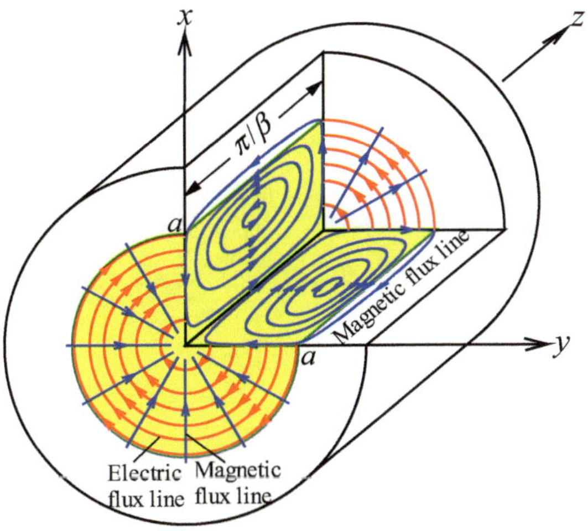

Figure 17.7 Electric and magnetic flux lines for the TE_{01} mode.

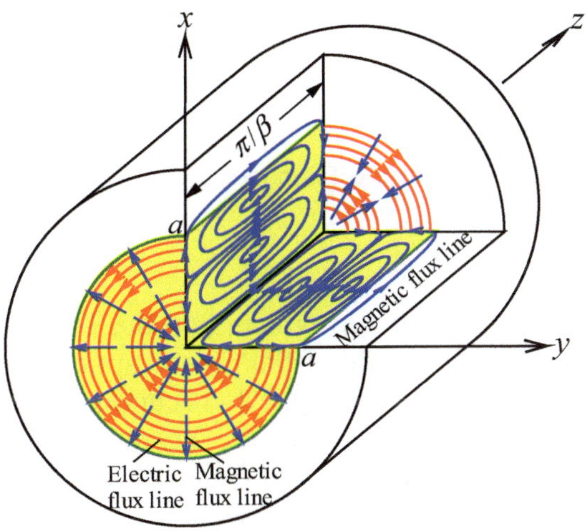

Figure 17.8 Electric and magnetic flux lines for the TE_{02} mode.

17.3.4 Comparison with planar optical waveguide

For comparison with planar optical waveguides, the b–V curves of both are shown in Figs. 17.2 and 17.3. Let's take a look at how the electromagnetic flux lines shown in the previous section behave in a planar waveguide.

Figure 17.9 shows the magnetic flux lines for TE_0 to TE_3 in a planar optical waveguide. The optical waveguide parameters are the same as those used in Fig. 14.7 of Chapter 14. Comparing this figure with Figs. 17.7 and 17.8, it can be seen that the TE_{01} mode of the optical fiber corresponds to the

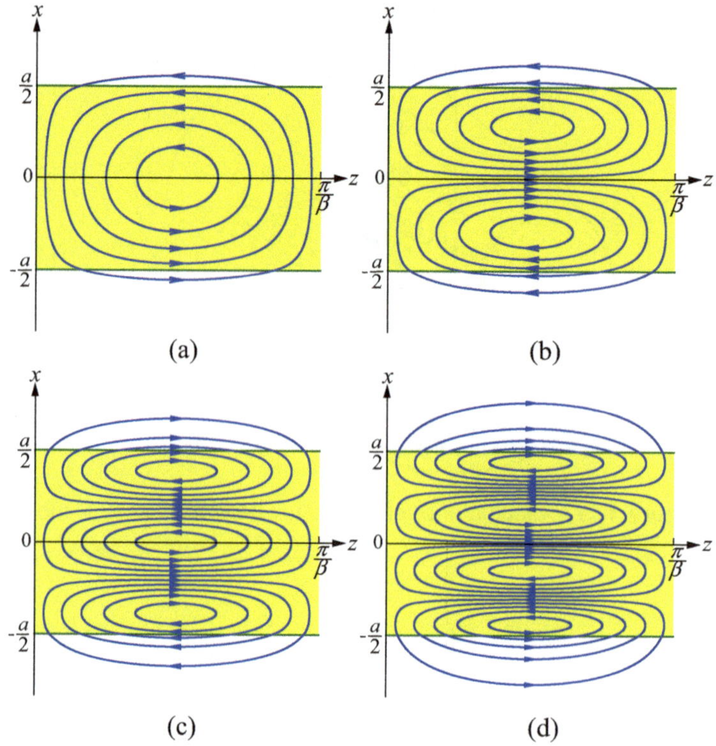

Figure 17.9 Magnetic flux lines for the TE modes in a planar optical waveguide: (a) TE_0, (b) TE_1, (c) TE_2, and (d) TE_3.

TE_1 mode of the planar optical waveguide, and the TE_{02} mode of the optical fiber corresponds to the TE_3 mode of the planar optical waveguide. This can be predicted from the comparison of the *b*–*V* curves in Figs. 17.2 and 17.3.

From the above, it can be seen that the TE mode (and TM mode) in an optical fiber corresponds to the odd-order mode in a planar optical waveguide. This means that there is no mode equivalent to the lowest fundamental mode. In fact, the lowest-order mode in optical fiber is not the TE_{01} mode or TM_{01} mode, but the HE_{11} mode, which will be discussed in the next chapter.

17.4 Conclusion

Among the solutions of the eigenvalue equations in optical fibers, the TE and TM modes where the order l is 0 were described in regard to the analysis method and the calculation method using Excel as well as the comparison with a planar optical waveguide. In the next chapter, the analysis method for the general case including $l \geq 1$ will be discussed.

Side Topic: Are the TE and TM modes of optical fibers even functions or odd functions?

As mentioned in the main text, the TE and TM modes in optical fibers correspond to the odd-order modes in a planar optical waveguide. Therefore, the TE or TM mode function in optical fibers is an odd function. No, you might think that's not true; the distribution in Fig. 17.6 is an even function. It is true that E_θ shown in Fig. 17.6 is an even function. However, what is called an odd function here is the electric field $\boldsymbol{E}$.

Figure 17.S1(a) shows the electric field in the TE mode expressed as vectors including electric field amplitude. The distributions of E_x and E_y are shown in Fig. 17.S1(b) and Fig. 17.S1(c), respectively. As can be seen from these three figures, the following formulas hold:

$$\boldsymbol{E}(-x,\ -y) = -\boldsymbol{E}(x, y) \tag{17.S1}$$

$$E_x(-x,\ -y) = -E_x(x, y) \tag{17.S2}$$

$$E_y(-x,\ -y) = -E_y(x, y). \tag{17.S3}$$

A point-symmetric function whose symmetry point is the central axis of the optical fiber is an odd function. If it is an even function, the function rotated by π will cancel the original function and becomes 0. Therefore, there is no TE/TM mode in optical fiber that corresponds to the even-order mode of a planar optical waveguide.

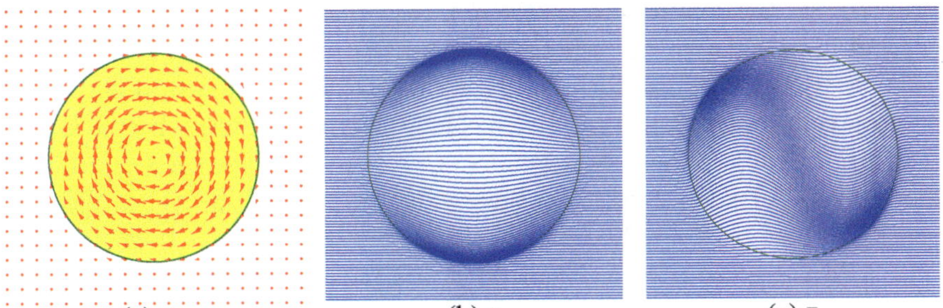

Figure 17.S1 Distributions of $\boldsymbol{E}$, E_x, and E_y.

Acknowledgments

The author would like to thank Professor Emeritus Kenichi Iga, Institute of Science Tokyo, former president of Tokyo Institute of Technology, and Professor Emeritus Yasuo Kokubun, Yokohama National University, former president of Institute of Technologists, for their guidance and discussions regarding the modal analysis theory of optical fibers.

References

1. Y. Suematsu and K. Iga, *Introduction to Optical Fiber Communications, 5th Ed.*, Ohmsha, Tokyo (2017) [in Japanese].
2. K. Iga and Y. Kokubun, *Optical Fiber*, Ohmsha, Tokyo (1993); Electronic reprint version, ITSC, Shizuoka (2014) [in Japanese].
3. K. Iga and G. Hatakoshi, Chapter 30, *Treasure Microbox of Optoelectronics II*, Advanced Communication Media, Tokyo (2018) [in Japanese].
4. G. Hatakoshi, "Optical fiber," Chapter III-4, *Optical Engineering Handbook*, T. Ose, H. Saito, S. Tanaka, J, Tsujiuchi, and T. Namioka eds., Asakura Publishing, Tokyo (1986) [in Japanese].
5. E. Snitzer, "Cylindrical dielectric waveguide modes," *J. Opt. Soc. Am.* **51**(5), 491–498 (1961).

Appendix 17A Bessel Function

The Bessel function, named after the German mathematician Friedrich Wilhelm Bessel, is a special solution of the differential equation given by the following equation and is also called a cylindrical function:

$$\frac{d^2w}{dx^2}+\frac{1}{x}\frac{dw}{dx}+\left(1-\frac{\nu^2}{x^2}\right)w=0. \tag{17.A1}$$

The solutions to Eq. (17.A1) include the Bessel function of the first kind $J_\nu(x)$, the Bessel function of the second kind $Y_\nu(x)$ [sometimes called the Neumann function $N_\nu(x)$], the Hankel function $H_\nu^{(1)}(x)$, $H_\nu^{(2)}(x)$, the modified Bessel function of the first kind, the first $I_\nu(x)$, and the modified Bessel function of the second kind $K_\nu(x)$. The index ν represents the order. The Hankel function was also introduced in Chapter 2. The modified Bessel function is the solution of the equation [corresponding to Eq. (17.11b) in the text] when x is replaced by ix in Eq. (17.A1). The functions $J_\nu(x)$, $Y_\nu(x)$, $I_\nu(x)$, and $K_\nu(x)$ are all built-in functions in Excel, and can be easily calculated using a spreadsheet. Figure 17.AP1 shows $J_\nu(x)$, $Y_\nu(x)$, $I_\nu(x)$, and $K_\nu(x)$ for $\nu=0$ to 7.

The following recurrence formula holds true for $J_\nu(x)$ and $K_\nu(x)$:

$$\frac{2\nu}{x}J_\nu(x)=J_{\nu-1}(x)+J_{\nu+1}(x), \tag{17.A2a}$$

$$2J_\nu'(x)=J_{\nu-1}(x)-J_{\nu+1}(x), \tag{17.A2b}$$

$$\frac{2\nu}{x}K_\nu(x)=-K_{\nu-1}(x)+K_{\nu+1}(x), \tag{17.A3a}$$

$$2K_\nu'(x)=-K_{\nu-1}(x)-K_{\nu+1}(x). \tag{17.A3b}$$

From Eqs. (17.A2a) and (17.A2b), the following formulas can be obtained:

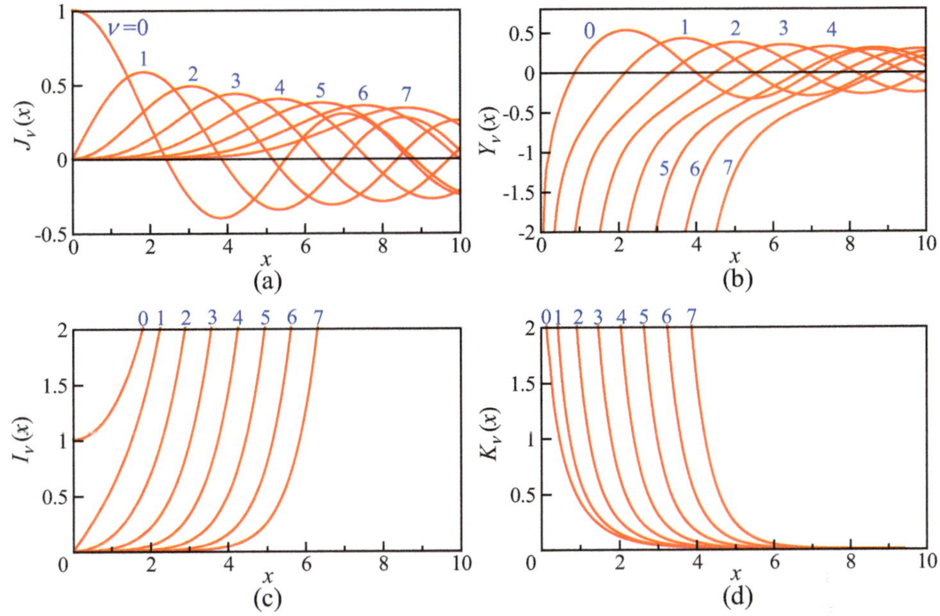

Figure 17.AP1 (a) $J_\nu(x)$, (b) $Y_\nu(x)$, (c) $I_\nu(x)$, and (d) $K_\nu(x)$.

$$\frac{J_\nu'(x)}{J_\nu(x)} = \frac{J_{\nu-1}(x)}{J_\nu(x)} - \frac{\nu}{x}, \tag{17.A4a}$$

$$\frac{J_\nu'(x)}{J_\nu(x)} = -\frac{J_{\nu+1}(x)}{J_\nu(x)} + \frac{\nu}{x}. \tag{17.A4b}$$

Similarly, from Eqs. (17.A3a) and (17.A3b),

$$\frac{K_\nu'(x)}{K_\nu(x)} = -\frac{K_{\nu-1}(x)}{K_\nu(x)} - \frac{\nu}{x}, \tag{17.A5a}$$

$$\frac{K_\nu'(x)}{K_\nu(x)} = -\frac{K_{\nu+1}(x)}{K_\nu(x)} + \frac{\nu}{x}. \tag{17.A5b}$$

Appendix 17B Excel Files

Appendix I-A contains the following Excel files. Each tab corresponds to a figure/s in the main text. As mentioned in the main text, in Newton's method using Eqs. (17.34) to (17.36), the order m cannot be determined, so it is necessary to change the initial value and find the eigenvalue of each order. In 17_Fiber1.xlsx, the initial value is set using an approximate formula that calculates the eigenvalue by giving m. This will be explained in the next chapter.

17_Fiber1.xlsx	
Instruction	
TE0m	Fig. 17.5, Fig. 17.6
TM0m	(For information)
TE01_Field	Fig. 17.7
TE02_Field	Fig. 17.8
17_Waveguide5.xlsx	
Instruction	
WG1	
TE0_Field	Fig. 17.9(a)
TE1_Field	Fig. 17.9(b)
TE2_Field	Fig. 17.9(c)
TE3_Field	Fig. 17.9(d)

Chapter 18
Optical Fiber Part 2

18.1 Introduction

In the previous chapter, among the solutions of the eigenvalue equations in optical fibers,[1–4] the analysis method for the TE mode and TM mode where the order l is 0 was described. The calculation method using Excel and the comparison with the planar optical waveguide were introduced. This chapter describes the classification of modes with $l \geq 1$ and their analysis method.

18.2 Classification of Optical Fiber Modes

The eigenvalue equation for the effective propagation constant β of a mode in a step-index optical fiber was introduced in the previous chapter. This will be discussed again in Section 18.3. This eigenvalue equation can be numerically solved using Newton's method as introduced so far. Before finding the formula used in Newton's method, let's consider the classification of solutions.

In Eq. (17.15) in the previous chapter, the coefficients of E_z and H_z are set as A_l and B_l, respectively. E. Snitzer describes the mode of an optical fiber using the following parameter P, which is proportional to the ratio B_l/A_l:[4]

$$P \equiv -\frac{\omega\mu_0}{\beta}\frac{B_l}{A_l}. \tag{18.1}$$

Here again, this P will be used to represent each component of the electric field $\boldsymbol{E}$ and magnetic field $\boldsymbol{H}$. In Section 18.2, $A_l = 1$, and the term $e^{i(\omega t - \beta z)}$ is omitted. Also, in order to simplify the notation of the formula, Bessel function of the first kind J_l, modified Bessel function of the second kind K_l, etc., will be represented by simplified descriptions as shown below:

$$J_l \equiv J_l(\kappa r), \tag{18.2a}$$

$$K_l \equiv K_l(\gamma r), \tag{18.2b}$$

$$c_l \equiv \cos(l\theta), \tag{18.2c}$$

$$s_l \equiv \sin(l\theta). \tag{18.2d}$$

Then, Eqs. (17.15) to (17.17) in the previous chapter become as follows:†

$$E_z = \begin{cases} J_l c_l & (r \le a) \\ a_l K_l c_l & (r \ge a) \end{cases}, \tag{18.3a}$$

$$E_r = \begin{cases} -\dfrac{i\beta}{2\kappa}\{(1-P)J_{l-1} - (1+P)J_{l+1}\}c_l & (r \le a) \\ -\dfrac{i\beta}{2\gamma}a_l\{(1-P)K_{l-1} + (1+P)K_{l+1}\}c_l & (r \ge a) \end{cases}, \tag{18.3b}$$

$$E_\theta = \begin{cases} \dfrac{i\beta}{2\kappa}\{(1-P)J_{l-1} + (1+P)J_{l+1}\}s_l & (r \le a) \\ \dfrac{i\beta}{2\gamma}a_l\{(1-P)K_{l-1} - (1+P)K_{l+1}\}s_l & (r \ge a) \end{cases}, \tag{18.3c}$$

$$H_z = -\frac{\beta}{\omega\mu_0}P \times \begin{cases} J_l s_l & (r \le a) \\ a_l K_l s_l & (r \ge a) \end{cases}, \tag{18.4a}$$

$$H_r = \begin{cases} -iQ_1\{(1-P_1)J_{l-1} + (1+P_1)J_{l+1}\}s_l & (r \le a) \\ -iQ_2 a_l\{(1-P_2)K_{l-1} - (1+P_2)K_{l+1}\}s_l & (r \ge a) \end{cases}, \tag{18.4b}$$

$$H_\theta = \begin{cases} -iQ_1\{(1-P_1)J_{l-1} - (1+P_1)J_{l+1}\}c_l & (r \le a) \\ -iQ_2 a_l\{(1-P_2)K_{l-1} + (1+P_2)K_{l+1}\}c_l & (r \ge a) \end{cases}, \tag{18.4c}$$

where

$$a_l = \frac{J_l(\kappa a)}{K_l(\gamma a)}, \tag{18.5}$$

$$Q_1 = \frac{k_0^2 n_1^2}{2\kappa\omega\mu_0}, \tag{18.6}$$

$$Q_2 = \frac{k_0^2 n_2^2}{2\gamma\omega\mu_0}, \tag{18.7}$$

$$P_1 = \frac{\beta^2}{k_0^2 n_1^2}P, \tag{18.8}$$

†Note: Some signs are different from those in Snitzer's paper,[4] where the time dependence is defined as $e^{-i\omega t}$. In this book, the time dependence is assumed to be $e^{+i\omega t}$, so the signs of each component of the electric field and magnetic field are also different.

$$P_2 = \frac{\beta^2}{k_0^2 n_2^2} P. \tag{18.9}$$

Now, let's take a look at the value of P defined by Eq. (18.1). Using Eq. (17.19a) from the previous chapter, the following equation is obtained:

$$P = \frac{l\left\{\dfrac{1}{(\kappa a)^2} + \dfrac{1}{(\gamma a)^2}\right\}}{\left\{\dfrac{1}{\kappa a}\dfrac{J_l'(\kappa a)}{J_l(\kappa a)} + \dfrac{1}{\gamma a}\dfrac{K_l'(\gamma a)}{K_l(\gamma a)}\right\}}. \tag{18.10}$$

Similarly, using Eq. (17.19b) from the previous chapter,

$$P = \frac{k_0^2 n_1^2\left\{\dfrac{1}{\kappa a}\dfrac{J_l'(\kappa a)}{J_l(\kappa a)} + \dfrac{1-2\Delta}{\gamma a}\dfrac{K_l'(\gamma a)}{K_l(\gamma a)}\right\}}{l\beta^2\left\{\dfrac{1}{(\kappa a)^2} + \dfrac{1}{(\gamma a)^2}\right\}}. \tag{18.11}$$

From Eqs. (18.10) and (18.11),

$$P^2 = \frac{k_0^2 n_1^2\left\{\dfrac{1}{\kappa a}\dfrac{J_l'(\kappa a)}{J_l(\kappa a)} + \dfrac{1-2\Delta}{\gamma a}\dfrac{K_l'(\gamma a)}{K_l(\gamma a)}\right\}}{\beta^2\left\{\dfrac{1}{\kappa a}\dfrac{J_l'(\kappa a)}{J_l(\kappa a)} + \dfrac{1}{\gamma a}\dfrac{K_l'(\gamma a)}{K_l(\gamma a)}\right\}}. \tag{18.12}$$

As mentioned in the previous chapter, the relative refractive index difference Δ in optical fibers is a very small value. From Eq. (17.25) in the previous chapter, when $\Delta \to 0$, $\beta \to k_0 n_1$, so from this and Eq. (18.12), it can be seen that when $\Delta \to 0$, $P^2 \to 1$. Therefore, P in Eq. (18.1), P_1 in Eq. (18.8), and P_2 in Eq. (18.9) all take values close to $+1$ or -1. For this reason, the coefficients of the terms $J_{l-1}(\kappa r)$ and $J_{l+1}(\kappa r)$, and $K_{l-1}(\gamma r)$ and $K_{l+1}(\gamma r)$ that appear in Eqs. (18.3b), (18.3c), (18.4b), and (18.4c) are dominated by one of them, and the other becomes almost 0. The mode of the optical fiber can be classified depending on the sign of this parameter P. As described in Section 18.3, the mode with $P \approx -1$ is called the HE mode, and the mode with $P \approx +1$ is called the EH mode.

18.3 Eigenvalue Equation

The eigenvalue Eq. (17.20) introduced in the previous chapter is reproduced below:

$$\left\{\frac{1}{\kappa}\frac{J_l'(\kappa a)}{J_l(\kappa a)}+\frac{1}{\gamma}\frac{K_l'(\gamma a)}{K_l(\gamma a)}\right\}\left\{\frac{n_1^2}{\kappa}\frac{J_l'(\kappa a)}{J_l(\kappa a)}+\frac{n_2^2}{\gamma}\frac{K_l'(\gamma a)}{K_l(\gamma a)}\right\}=\frac{l^2\beta^2}{k_0^2a^2}\left(\frac{1}{\kappa^2}+\frac{1}{\gamma^2}\right)^2. \tag{18.13}$$

In addition, Eq. (17.26) from the previous chapter, which is a modification of the above equation and further expressed using normalized parameters, is reproduced below:

$$\begin{aligned}&V\left\{\sqrt{b}\frac{J_{l-1}(\sqrt{1-b}V)}{J_l(\sqrt{1-b}V)}-\sqrt{1-b}\frac{K_{l-1}(\sqrt{b}V)}{K_l(\sqrt{b}V)}\right\}\\&\times\left\{\sqrt{b}\frac{J_{l-1}(\sqrt{1-b}V)}{J_l(\sqrt{1-b}V)}-c_1\sqrt{1-b}\frac{K_{l-1}(\sqrt{b}V)}{K_l(\sqrt{b}V)}\right\}\\&-2l\left\{\frac{c_1+(1+b)\Delta}{\sqrt{1-b}}\frac{J_{l-1}(\sqrt{1-b}V)}{J_l(\sqrt{1-b}V)}-\frac{c_1+b\Delta}{\sqrt{b}}\frac{K_{l-1}(\sqrt{b}V)}{K_l(\sqrt{b}V)}\right\}=0,\end{aligned} \tag{18.14}$$

where

$$V\equiv k_0a\sqrt{n_1^2-n_2^2}=k_0an_1\sqrt{2\Delta}, \tag{18.15}$$

$$b\equiv\frac{n_{\mathrm{eq}}^2-n_2^2}{n_1^2-n_2^2}=\frac{\beta^2-k_0^2n_2^2}{k_0^2(n_1^2-n_2^2)}, \tag{18.16}$$

$$c_1=1-2\Delta, \tag{18.17a}$$

$$c_2=1-\Delta(1-b), \tag{18.17b}$$

$$c_3=1-\Delta(2-b). \tag{18.17c}$$

By multiplying Eq. (18.14) by $\sqrt{b}\sqrt{1-b}J_l(\sqrt{1-b}V)K_l(\sqrt{b}V)$, the function f used in Newton's method (the function that should be 0) is given as follows:

$$f=V(A-B)(A-c_1B)-2lC(c_2A-c_3B), \tag{18.18}$$

where

$$A=b\sqrt{1-b}J_{l-1}(\sqrt{1-b}V)K_l(\sqrt{b}V), \tag{18.19}$$

$$B=\sqrt{b}(1-b)J_l(\sqrt{1-b}V)K_{l-1}(\sqrt{b}V), \tag{18.20}$$

$$C=J_l(\sqrt{1-b}V)K_l(\sqrt{b}V). \tag{18.21}$$

The function f in Eq. (18.18) is a function of b and V. First, let's look at how f in Eq. (18.18) changes with respect to V when b is fixed. Note that, the larger V is, the smaller the absolute value of f becomes. Therefore, in order to make it easier to see on the graph, the value multiplied by a_1V^4 (a_1 is a constant) will be plotted. In this case, the value of V at the zero point to be found does not change.

Figure 18.1(a) shows the change in $a_1 V^4 f$ with respect to V calculated with $b = 0.1$ and $\Delta = 0.01$. The name of the mode corresponding to this zero point is shown in the figure, but as can be seen from the figure, the graph appears to be almost the tangential line to the $f = 0$ axis, and the zero point is not clearly understood as viewed here. Therefore, by enlarging the vertical axis scale, it becomes as shown in Fig. 18.1(b).

The zero point of the graph for $l = 1$ corresponds to the HE_{1m} mode and EH_{1m} mode ($m = 1, 2, 3, \ldots$), and the zero point of the graph for $l = 2$ corresponds to the HE_{2m} mode and EH_{2m} mode ($m = 1, 2, 3, \ldots$).

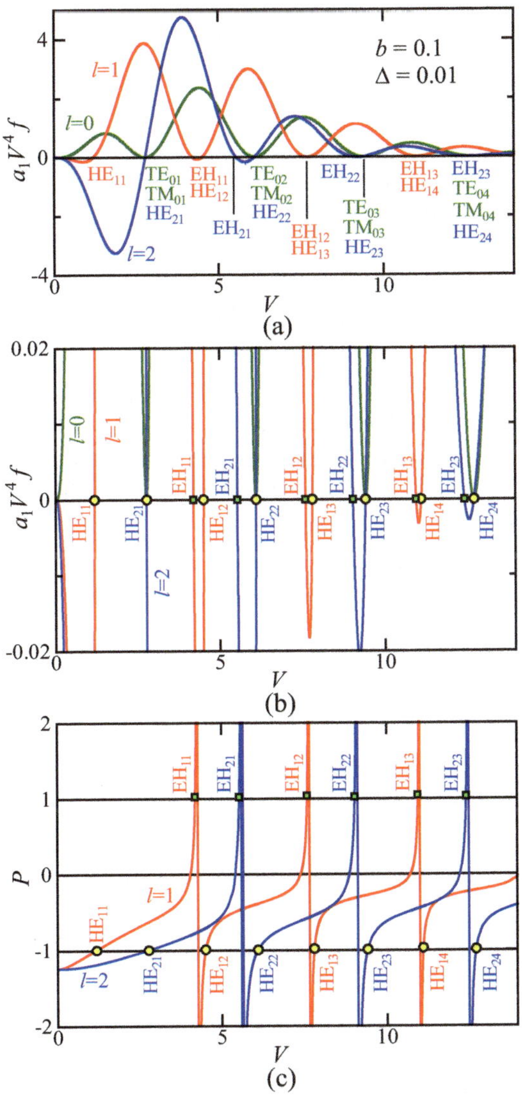

Figure 18.1 (a), (b) V dependence of f and (c) V dependence of P.

The zero point of the graph for $l=0$ corresponds to the TE_{0m} mode and TM_{0m} mode, but as can be seen from Fig. 18.1(b), even if the vertical axis is enlarged to this extent, it still appears to be the tangential line to the $f=0$ axis. As mentioned in the previous chapter, this is due to the fact that the difference in eigenvalues between the TE_{0m} mode and TM_{0m} mode is extremely small. Therefore, for the TE_{0m} mode and TM_{0m} mode, it is more reliable to use f_{TE} or f_{TM} described in the previous chapter to find the eigenvalues, rather than using f in Eq. (18.18).

Now, the names HE_{lm} mode and EH_{lm} mode were used unilaterally above, but whether it is the HE mode or EH mode depends on whether E_z and H_z have the same sign or different signs; that is, it's determined by the sign of P defined by Eq. (18.1).† The values of P for $l=1$ and $l=2$ are plotted in Fig. 18.1(c). Although P changes greatly as a function of V, it can be seen that $P\approx-1$ at the zero point corresponding to the HE mode, and $P\approx+1$ at the zero point corresponding to the EH mode. As can be seen from Fig. 18.1(a), the TE_{01} mode or TM_{01} mode is not the lowest-order mode, but the HE_{11} mode is the lowest-order mode in optical fibers.

The relationship between V and b can be determined by Newton's method using Eq. (18.18). As in the previous chapter, the required differential formula is df/dV to find V for a given b, and df/db to find b for a given V. Since it is easy to obtain df/dV or df/db from Eq. (18.18), the specific equation will be omitted here. As in the case of the TE/TM mode, the df/dV formula is relatively simple and can be used to draw the b–V dispersion curve.

The b–V curve for the HE_{lm} mode is shown in Fig. 18.2. It was calculated using Newton's method in Excel using Eq. (18.18). Only modes with $l=1$, 2, and 3 are shown here. As mentioned in the previous chapter, both $J_\nu(x)$ and $K_\nu(x)$ that appear in Eq. (18.18) are built in as functions in Excel and can be easily calculated using a spreadsheet.

Next, the b–V curve of the EH_{lm} mode is shown in Fig. 18.3. Similarly, only modes with $l=1$, 2, and 3 are shown. It should be noted that Figs. 18.2 and 18.3 both use the same formula (18.18). Equation (18.18) is an equation that can be applied to all of the following: HE_{lm} mode, EH_{lm} mode, TE_{0m} mode, and TM_{0m} mode. Therefore, the mode of the obtained solution must be determined by looking at the value of P, for example, in Eq. (18.1). Note that in Figs. 18.2 and 18.3, the relative refractive index difference Δ was set to a relatively large value of 0.03.

Additionally, in Figs. 18.2 and 18.3, only modes with $l=1$, 2, and 3 are shown, but there are also higher-order modes in this range. Of course, the TE_{0m}

†Note: The HE mode and the EH mode are hybrid modes in which neither the E_z component nor the H_z component becomes 0. The convention of which is called HE or EH and is based on the name used in microwave waveguides. In other words, since the lowest-order mode of a circular cross-section microwave waveguide was called the HE_{11} mode, the same name was adopted for optical fibers.[4]

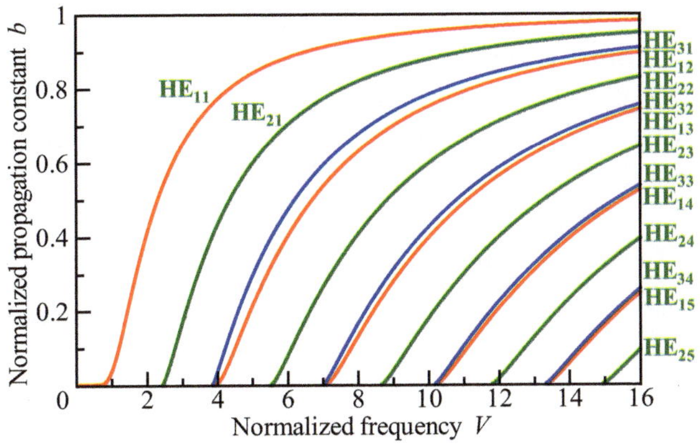

Figure 18.2 b–V curve for the HE_{lm} mode (modes with $l = 1, 2, 3$ only).

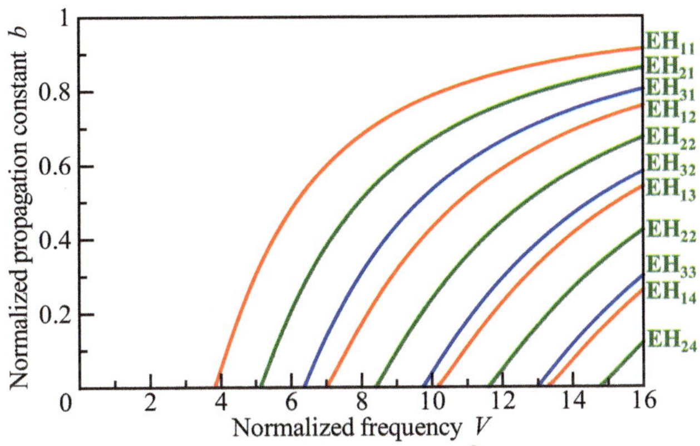

Figure 18.3 b–V curve for the EH_{lm} mode (modes with $l = 1, 2, 3$ only).

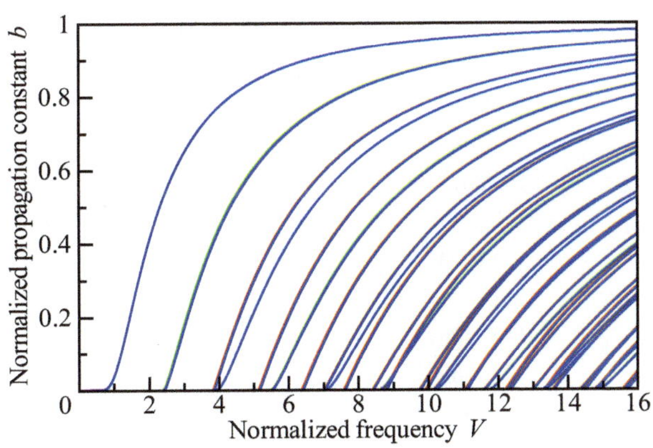

Figure 18.4 All modes of TE_{0m}, TM_{0m}, HE_{lm}, and EH_{lm} ($b = 0.03$).

mode and TM_{0m} mode described in the previous chapter also fall within the range of this diagram. If all the modes are plotted, the diagram becomes like Fig. 18.4. Although some modes overlap and cannot be distinguished at the resolution shown in Fig. 18.4, there are a total of 70 modes in this range. Table 18.1 shows these modes and the cutoff V value (described in Section 18.4)

Table 18.1 Cutoff V value for the modes in Fig. 18.4.

Mode	V_{cutoff}	Mode	V_{cutoff}
TE_{01}	2.404826	HE_{11}	0
TE_{02}	5.520078	HE_{12}	3.831706
TE_{03}	8.653728	HE_{13}	7.015587
TE_{04}	11.791534	HE_{14}	10.173468
TE_{05}	14.930918	HE_{15}	13.323692
TM_{01}	2.404826	HE_{21}	2.430936
TM_{02}	5.520078	HE_{22}	5.531605
TM_{03}	8.653728	HE_{23}	8.661094
TM_{04}	11.791534	HE_{24}	11.796943
TM_{05}	14.930918	HE_{25}	14.935191
EH_{11}	3.831706	HE_{31}	3.864312
EH_{12}	7.015587	HE_{32}	7.033665
EH_{13}	10.173468	HE_{33}	10.185977
EH_{14}	13.323692	HE_{34}	13.333256
EH_{21}	5.135622	HE_{41}	5.171982
EH_{22}	8.417244	HE_{42}	8.439778
EH_{23}	11.619841	HE_{43}	11.636238
EH_{24}	14.795952	HE_{44}	14.808854
EH_{31}	6.380162	HE_{51}	6.419077
EH_{32}	9.761023	HE_{52}	9.786865
EH_{33}	13.015201	HE_{53}	13.034684
EH_{41}	7.588342	HE_{61}	7.629151
EH_{42}	11.064709	HE_{62}	11.093141
EH_{43}	14.372537	HE_{63}	14.394552
EH_{51}	8.771484	HE_{71}	8.813772
EH_{52}	12.338604	HE_{72}	12.369138
EH_{53}	15.700174	HE_{73}	15.724320
EH_{61}	9.936110	HE_{81}	9.979596
EH_{62}	13.589290	HE_{82}	13.621577
EH_{71}	11.086370	HE_{91}	11.130853
EH_{72}	14.821269	HE_{92}	14.855047
EH_{81}	12.225092	$HE_{10,1}$	12.270421
EH_{91}	13.354300	$HE_{11,1}$	13.400360
$EH_{10,1}$	14.475501	$HE_{12,1}$	14.522198
$EH_{11,1}$	15.589848	$HE_{13,1}$	15.637110

for each mode. Although some modes have the same cutoff V value, they all have different b–V characteristics, and their electric and magnetic field distributions are also different. The cutoff conditions will be discussed in the next section.

18.4 Cutoff Conditions

A mode in an optical fiber becomes cut off when $\gamma = 0$ in the eigenvalue equation, which corresponds to the case for $b = 0$ when expressed in terms of normalized parameters.

In the case of the TE_{0m} mode and TM_{0m} mode (when $l = 0$), by setting $b = 0$ in Eqs. (17.34) and (17.38) in the previous chapter, it can be seen that the following is the cutoff condition:

$$J_0(V) = 0. \tag{18.22}$$

The value of V that satisfies Eq. (18.22) is the cutoff V value V_{cutoff}. V_{cutoff} can be determined using Newton's method. The cutoff conditions for the TE_{0m} mode and TM_{0m} mode are determined by the same Eq. (18.22), but as mentioned in the previous chapter, the b–V characteristics are not the same.

Next, consider what the cutoff conditions will be in the HE_{lm} mode and EH_{lm} mode. First, as preparation, Eq. (18.14) is rewritten as follows:

$$b(1-b)V^2(\xi_1 - \xi_2)(\xi_1 - c_1\xi_2) - 2l(c_2\xi_1 - c_3\xi_2) = 0, \tag{18.23}$$

where ξ_1 and ξ_2 are parameters used in Ref. 4 and are defined below.

$$\xi_1 = \frac{J_{l-1}(\sqrt{1-b}V)}{(\sqrt{1-b}V)J_l(\sqrt{1-b}V)}, \tag{18.24}$$

$$\xi_2 = \frac{K_{l-1}(\sqrt{b}V)}{(\sqrt{b}V)K_l(\sqrt{b}V)}. \tag{18.25}$$

First, by simply setting $b = 0$ in Eq. (18.23), the following equation is obtained:

$$c_2\xi_1 - c_3\xi_2 = 0. \tag{18.26}$$

There is no doubt that this is one of the cutoff conditions, but let's take a closer look.

As shown in Eq. (18.25), $K_l(\sqrt{b}V)$ and $K_{l-1}(\sqrt{b}V)$ appear in the denominator and numerator of ξ_2, respectively. All of these become infinity when $b \to 0$. Therefore, it is necessary to find the limit value of ξ_2 when $b \to 0$. As shown in Eqs. (18.A8), (18.A9), and (18.A11) in Appendix 18A, the following holds:

$$\lim_{b\to 0}\xi_2 = \lim_{x\to 0}[\ln(2/x) - \gamma_E] = \infty \quad (l = 1), \tag{18.27a}$$

$$\lim_{b\to 0}\xi_2 = \frac{1}{2(l-1)} \quad (l \geq 2), \tag{18.27b}$$

$$\lim_{b\to 0}[bV^2\xi_2] = 0 \quad (l \geq 1). \tag{18.27c}$$

Equation (18.23) is transformed as follows:

$$\begin{aligned} &b(1-b)V^2\xi_1^2 - \{b(1-b)V^2(1+c_1)\xi_2 + 2lc_2\}\xi_1 \\ &+ \{b(1-b)V^2c_2^2\xi_2 + 2lc_3\}\xi_2 = 0. \end{aligned} \tag{18.28}$$

According to the condition of Eq. (18.27c), Eq. (18.28) becomes the following equation when $b \to 0$:

$$b(1-b)V^2\xi_1^2 - 2lc_2\xi_1 + 2lc_3\xi_2 = 0, \tag{18.29}$$

or

$$2lc_3\xi_2/\xi_1 = 2lc_2 - b(1-b)V^2\xi_1. \tag{18.30}$$

In the case where ξ_1 is finite in Eq. (18.30), the right-hand side is only $2lc_2$ when $b \to 0$. Therefore,

$$\xi_1 = \frac{c_3}{c_2}\xi_2. \tag{18.31}$$

Naturally, this equation agrees with Eq. (18.26). Note that Eq. (18.26) holds true even if $\xi_2 \to \infty$, so Eq. (18.31) is always a cutoff condition.

By rewriting Eq. (18.31) in the case of $l \geq 2$, the following equation is obtained from Eqs. (18.24) and (18.27b):

$$\frac{J_{l-1}(V)}{VJ_l(V)} = \frac{1-2\Delta}{1-\Delta}\frac{1}{2(l-1)}. \tag{18.32}$$

Using Eq. (17.A2a) in Appendix 17A of the previous chapter, the above equation becomes as follows:

$$\frac{J_{l-2}(V) + J_l(V)}{J_l(V)} = 1 - \frac{\Delta}{1-\Delta}. \tag{18.33}$$

This becomes a simple formula as follows:

$$\frac{J_{l-2}(V)}{J_l(V)} = -\frac{\Delta}{1-\Delta}. \tag{18.34}$$

This cutoff condition can be solved numerically using Newton's method, and V_{cutoff} can be determined.

In the case where $l = 1$, $\xi_2 \to 0$ from Eq. (18.27a), so the condition of Eq. (18.31) is $\xi_1 \to \infty$. From Eq. (18.24), this cutoff condition is as follows:

$$J_1(V) = 0. \tag{18.35}$$

The minimum V that satisfies Eq. (18.35) is $V = 0$, and as described in Table 18.2, this is the cutoff V value in the HE_{11} mode.

On the other hand, Eq. (18.30) also holds when $\xi_1 \to \infty$ and the following equation is satisfied:

$$\xi_1 = \frac{2lc_2}{b(1-b)V^2} \to \infty. \tag{18.36}$$

From Eq. (18.24), the following equation is obtained in this case:

$$J_l(V) = 0. \tag{18.37}$$

In other words, Eq. (18.37) is a solution to another cutoff condition.

From the above, it was found that the cutoff condition is given by Eq. (18.31) or Eq. (18.36). Whether this is a cutoff condition for the HE mode or EH mode can be determined by examining the value of P in Eq. (18.10). Expressing Eq. (18.10) using normalized parameters, P is represented by

$$P = \frac{l}{b(1-b)V^2(\xi_1 - \xi_2) - l}. \tag{18.38}$$

Equations (17.A4a) and (17.A5a) from the previous chapter were used to derive the above equation. Under the cutoff condition of Eq. (18.31), and from Eq. (18.27c), Eq. (18.38) becomes as follows when $b \to 0$:

$$\lim_{b \to 0} P = -1. \tag{18.39}$$

Therefore, it can be seen that Eq. (18.31) is the cutoff condition for the HE mode.

In the case where Eq. (18.36) is the cutoff condition, Eq. (18.38) becomes as follows when $b \to 0$:

$$\lim_{b \to 0} P = \frac{l}{2lc_2 - l} = \frac{1}{1 - 2\Delta} = \frac{1}{c_1} \cong 1. \tag{18.40}$$

Therefore, Eq. (18.36) is the cutoff condition for the EH mode.

Table 18.2 summarizes the above cutoff conditions and P values at the cutoff.

Table 18.2 Cutoff conditions and P value at the cutoff.

l	Mode	Cutoff condition	P
$l=0$	TE_{0m}	$J_0(V)=0$	0
	TM_{0m}		∞
$l=1$	HE_{1m}	$J_1(V)=0$	-1
	EH_{1m}		$1/(1-2\Delta)$
$l\geq 2$	HE_{lm}	$\dfrac{J_{l-2}(V)}{J_l(V)}=-\dfrac{\Delta}{1-\Delta}$	-1
	EH_{lm}	$J_l(V)=0$	$1/(1-2\Delta)$

18.5 Conclusion

This chapter describes the classification of modes in optical fibers and the cutoff conditions for each mode. The TE_{0m} mode and TM_{0m} mode described in the previous chapter, and the HE_{lm} mode and EH_{lm} mode described in this chapter are all modes in a step-index optical fiber. The eigenvalues of these modes are all solutions of one eigenvalue equation. This chapter also introduced the method for solving eigenvalue equations using Newton's method in Excel and the b–V curve obtained from it. The electromagnetic field distribution of each mode is expressed using the formulas described in Section 18.2, but the specific method of expressing them will be introduced in the next chapter.

Acknowledgments

The author would like to thank Professor Emeritus Kenichi Iga, Institute of Science Tokyo, former president of Tokyo Institute of Technology, and Professor Emeritus Yasuo Kokubun, Yokohama National University, former president of Institute of Technologists, for their guidance and discussions regarding the modal analysis theory of optical fibers.

References

1. K. Iga and Y. Kokubun, *Optical Fiber*, Ohmsha, Tokyo (1993); Electronic reprint version, ITSC, Shizuoka (2014) [in Japanese].
2. Y. Suematsu and K. Iga, *Introduction to Optical Fiber Communications, 5th Ed.*, Ohmsha, Tokyo (2017) [in Japanese].
3. K. Iga and G. Hatakoshi, Chapter 30, *Treasure Microbox of Optoelectronics II*, Advanced Communication Media, Tokyo (2018) [in Japanese].
4. E. Snitzer, "Cylindrical dielectric waveguide modes," *J. Opt. Soc. Am.* **51**(5), 491–498 (1961).

5. S. Moriguchi, K. Udagawa, and S. Hitotsumatsu, *Mathematical Formula III, Special Functions, 27th Impress.*, Iwanami Shoten, Tokyo (2016) [in Japanese].

Side Topic: Optical fiber analysis in Excel

As introduced in the main text, all exact solutions for optical fiber modes can be found using Newton's method in Excel. Of course, the same results can be obtained using other programs such as Fortran, C/C++, and Mathcad, but Excel has the advantage of being easily usable by anyone with a PC. Another convenient point is that the Bessel functions and modified Bessel functions that appear in the modes of optical fibers are both incorporated as functions in Excel.

As calculation examples in Excel, the electric field, magnetic field, and $|E|^2$ distribution in the HE_{31} mode and EH_{32} mode are shown in Figs. 18.S1 and 18.S2, respectively. A similar diagram for the TE mode was also shown in the previous side topic. Examples of these calculations will be introduced in the next chapter.

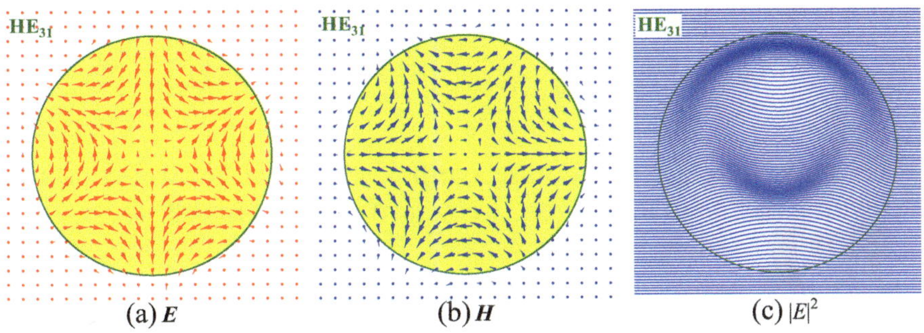

Figure 18.S1 Calculated examples of electric field, magnetic field, and $|E|^2$ distribution in the HE_{31} mode using Excel.

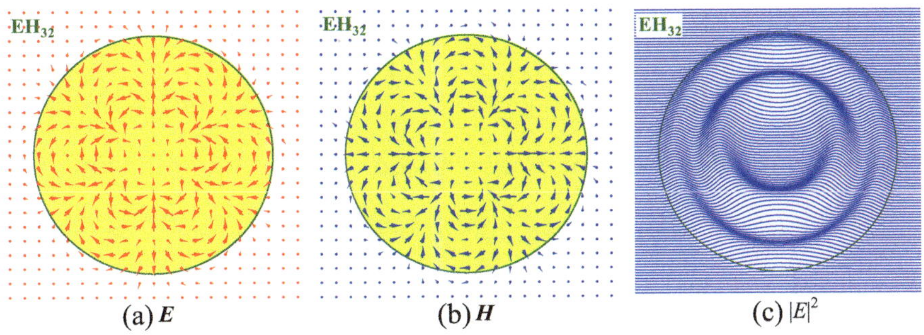

Figure 18.S2 Calculated examples of electric field, magnetic field, and $|E|^2$ distribution in the EH_{32} mode using Excel.

Appendix 18A Series Expansion of Bessel Functions

The Bessel function and the modified Bessel function can be expressed as the following Taylor expansion around $x = 0$:[5]

$$J_\nu(x) = \sum_{m=0}^{\infty} \frac{(-1)^m}{m!(m+\nu)!}\left(\frac{x}{2}\right)^{2m+\nu}, \tag{18.A1}$$

$$I_\nu(x) = \sum_{m=0}^{\infty} \frac{1}{m!(m+\nu)!}\left(\frac{x}{2}\right)^{2m+\nu}, \tag{18.A2}$$

$$K_\nu(x) = (-1)^{\nu+1} I_\nu(x)\{\gamma_\mathrm{E} + \ln(x/2)\} + \frac{(-1)^\nu}{2}\sum_{k=0}^{\infty}\frac{(x/2)^{2k+\nu}}{k!(k+\nu)!}\left\{\sum_{m=1}^{k}\frac{1}{m} + \sum_{m=1}^{k+\nu}\frac{1}{m}\right\}$$
$$+\frac{1}{2}\sum_{k=0}^{\nu-1}\frac{(-1)^k(\nu-k-1)!}{k!}\left(\frac{x}{2}\right)^{2k-\nu}, \tag{18.A3}$$

where γ_E is called Euler's constant or the Euler–Mascheroni constant and is defined below:

$$\gamma_\mathrm{E} = \lim_{n\to\infty}\left[\sum_{k=1}^{n}\frac{1}{k} - \ln(n)\right] = 0.5772156649\ldots \tag{18.A4}$$

Since Lorenzo Mascheroni used the symbol γ to represent this constant, it is also called Euler's γ. However, Leonhard Euler himself seems to have expressed this constant with the symbol C. γ is normally used, but in this chapter γ_E is used because γ is the propagation constant of the cladding region of the optical fiber. Note that the definition of Euler's constant is different from this in Ref. 4.

Now, from Eq. (18.A2),

$$\lim_{x\to 0} I_0(x) = 1, \tag{18.A5a}$$

$$\lim_{x\to 0} I_\nu(x) = 0 \quad (\nu \geq 1). \tag{18.A5b}$$

Therefore, from the above equations and Eq. (18.A3), the following holds:

$$\lim_{x\to 0} K_0(x) = -\{\gamma_\mathrm{E} + \ln(x/2)\}, \tag{18.A6}$$

$$\lim_{x\to 0}[x^\nu K_\nu(x)] = 2^{\nu-1}(\nu-1)!. \tag{18.A7}$$

From Eqs. (18.A6) and (18.A7),

$$\lim_{x\to 0}\left[\frac{K_0(x)}{xK_1(x)}\right] = -\{\gamma_{\mathrm{E}} + \ln(x/2)\} \to \infty. \tag{18.A8}$$

On the other hand, in the case of $\nu \geq 2$,

$$\lim_{x\to 0}\left[\frac{K_{\nu-1}(x)}{xK_\nu(x)}\right] = \frac{1}{2(\nu-1)}. \tag{18.A9}$$

Also, from Eqs. (17.A5a) and (17.A5b) in the previous chapter,

$$\frac{K_{\nu-1}(x)}{K_\nu(x)} = \frac{K_{\nu+1}(x)}{K_\nu(x)} - \frac{2\nu}{x}. \tag{18.A10}$$

Therefore, the following equation also holds:

$$\lim_{x\to 0}\left[x\frac{K_{\nu-1}(x)}{K_\nu(x)}\right] = \lim_{x\to 0}\left[\left(\frac{K_\nu(x)}{xK_{\nu+1}(x)}\right)^{-1} - 2\nu\right] = 0. \tag{18.A11}$$

This is clear from Eq. (18.A9) when $\nu \geq 2$, but it also holds true when $\nu = 1$ from Eq. (18.A10).

Appendix 18B Excel File

Appendix I-A contains the following Excel file. Each tab corresponds to a figure/s in the text.

18_Fiber2.xlsm

Instruction	
b–V(TE–TM)	Fig. 17.2 in the previous chapter
b–V(HE–EH)	Fig. 18.2, Fig. 18.3

Note that this Excel file uses VBA, but since VBA does not have a Bessel function, a worksheet function is used as shown below.

```
bessjj = WorksheetFunction.BesselJ(x, n)
besskk = WorksheetFunction.BesselK(x, n)
```

Chapter 19
Optical Fiber Part 3

19.1 Introduction

In the previous two chapters, the classification of modes in step-index optical fibers and the cutoff conditions for each mode were described, and the solution method for the eigenvalue equation using Newton's method in Excel and the b–V curve obtained from it were introduced. This chapter describes an example of calculating the specific electromagnetic field distribution of each mode using Excel and an approximate solution that provides the initial value when solving the eigenvalue equation.

19.2 Calculation Examples for the Optical Fiber Mode

Calculation examples for the electromagnetic field distributions[1–5] in optical fibers using Excel are shown in Figs. 19.1 to 19.3. Assuming a multimode optical fiber, the core diameter was set as 50 μm and the relative refractive index difference was set as $\Delta = 0.01$, both of which were the same as were used in Figs. 17.5 to 17.8 in Chapter 17. Figure 19.1 shows the radial distribution, where the TE mode is reproduced from Chapter 14. In Figs. 19.2 and 19.3, (a) and (b) are the distributions of the electric field $\boldsymbol{E}$ and magnetic field $\boldsymbol{H}$, respectively, and (c) is the distribution of $|E|^2$. For reference, the distribution of $|E|$ is shown in (d). In terms of intensity distribution, it should be $|E|^2$, but if $|E|^2$ is displayed in (d), the contrast becomes difficult to understand, so the gamma value was set to 0.5 (therefore it becomes an $|E|$ distribution). If (d) is present, (c) is not necessary, but (c) is included here as a figure that can be plotted in Excel.

In Figs. 19.2 and 19.3, the x-axis is oriented upward and y-axis is oriented rightward (see the TE_{01} mode diagram in Fig. 19.2). Therefore, the z-axis (propagation direction) is the direction from the front to the back of the page.

As shown in the equations in the previous chapter, E_r, E_θ, H_r, and H_θ in the HE_{lm} mode and EH_{lm} mode have $\cos(l\theta)$ or $\sin(l\theta)$ dependence in the angular direction. The mode order l represents this. On the other hand, the other mode order m corresponds to the number of nodes (zero points) in the

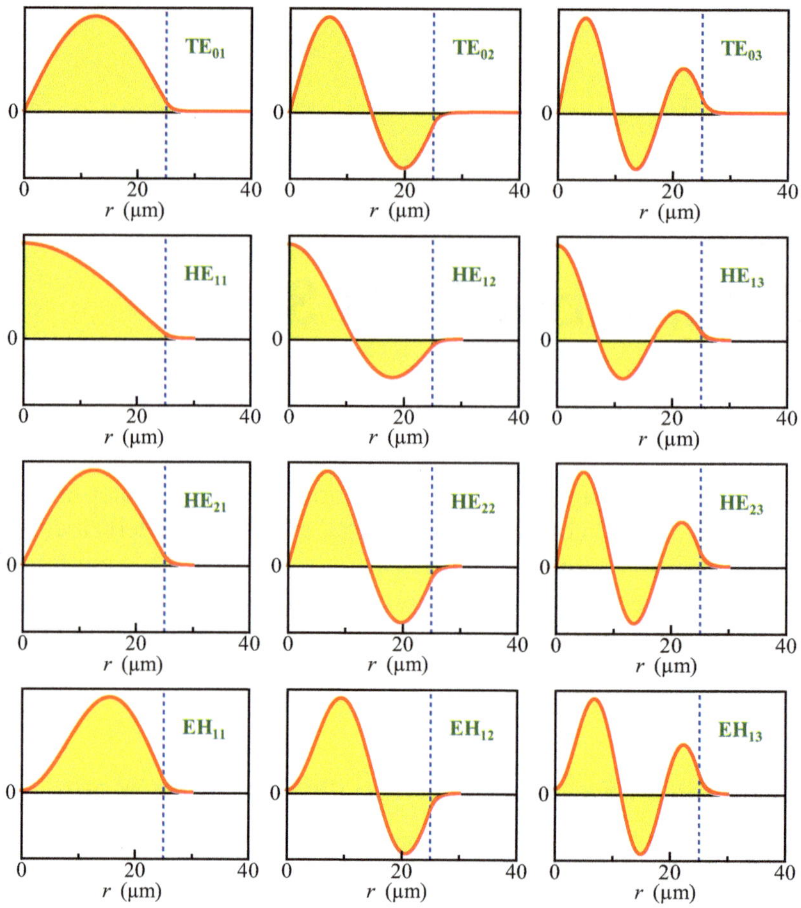

Figure 19.1 Radial amplitude distribution of modes in optical fiber. [E_θ for the TE mode, and E_r ($\theta = 0$) for the HE mode and EH mode.]

radial direction. There are modes in which the amplitude becomes 0 when $r = 0$ and modes in which it does not become 0. Excluding the zero point at $r = 0$, the number of radial nodes in each mode is $m - 1$.

19.3 Approximate Formula for the HE Mode and the EH Mode

The equations for the mode in optical fibers are shown in Eqs. (18.3) and (18.4) in the previous chapter, and they are rigorous solutions.[1–4] Although these equations are exhaustive, an approximate formula for qualitatively explaining the electromagnetic field distribution is shown in the next two subsections.

As mentioned in the previous chapter, the parameter P appears in the mode equation and has a value of $P \approx -1$ in HE mode and $P \approx +1$ in EH mode.[4] Let's consider an approximate formula when this becomes just $P = -1$ or $P = +1$.

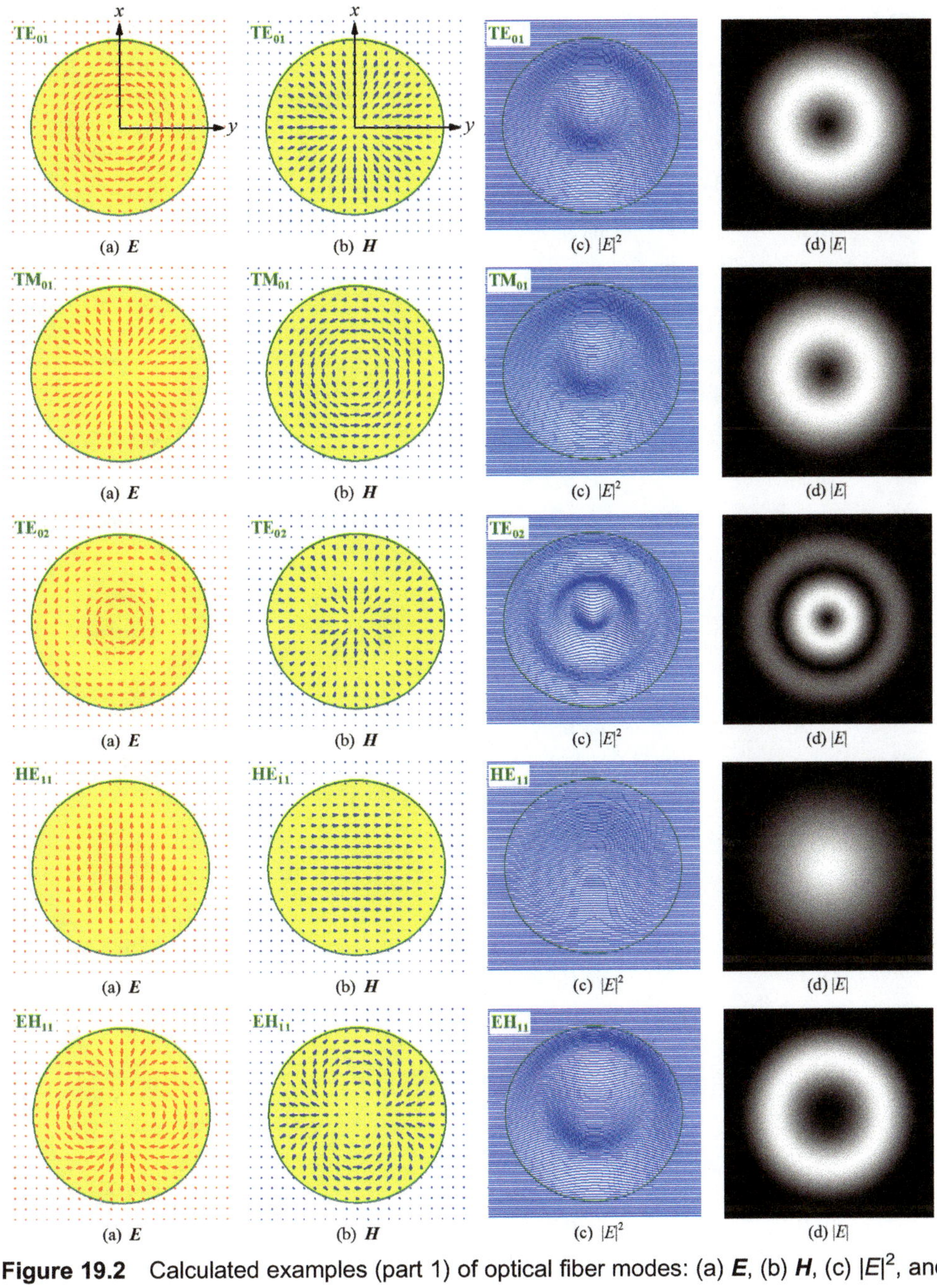

Figure 19.2 Calculated examples (part 1) of optical fiber modes: (a) ***E***, (b) ***H***, (c) $|E|^2$, and (d) $|E|$.

19.3.1 HE mode

First, in the case of the HE mode, set $P = P_1 = P_2 = -1$ in Eqs. (18.3) and (18.4) in the previous chapter, then the following equations are obtained:

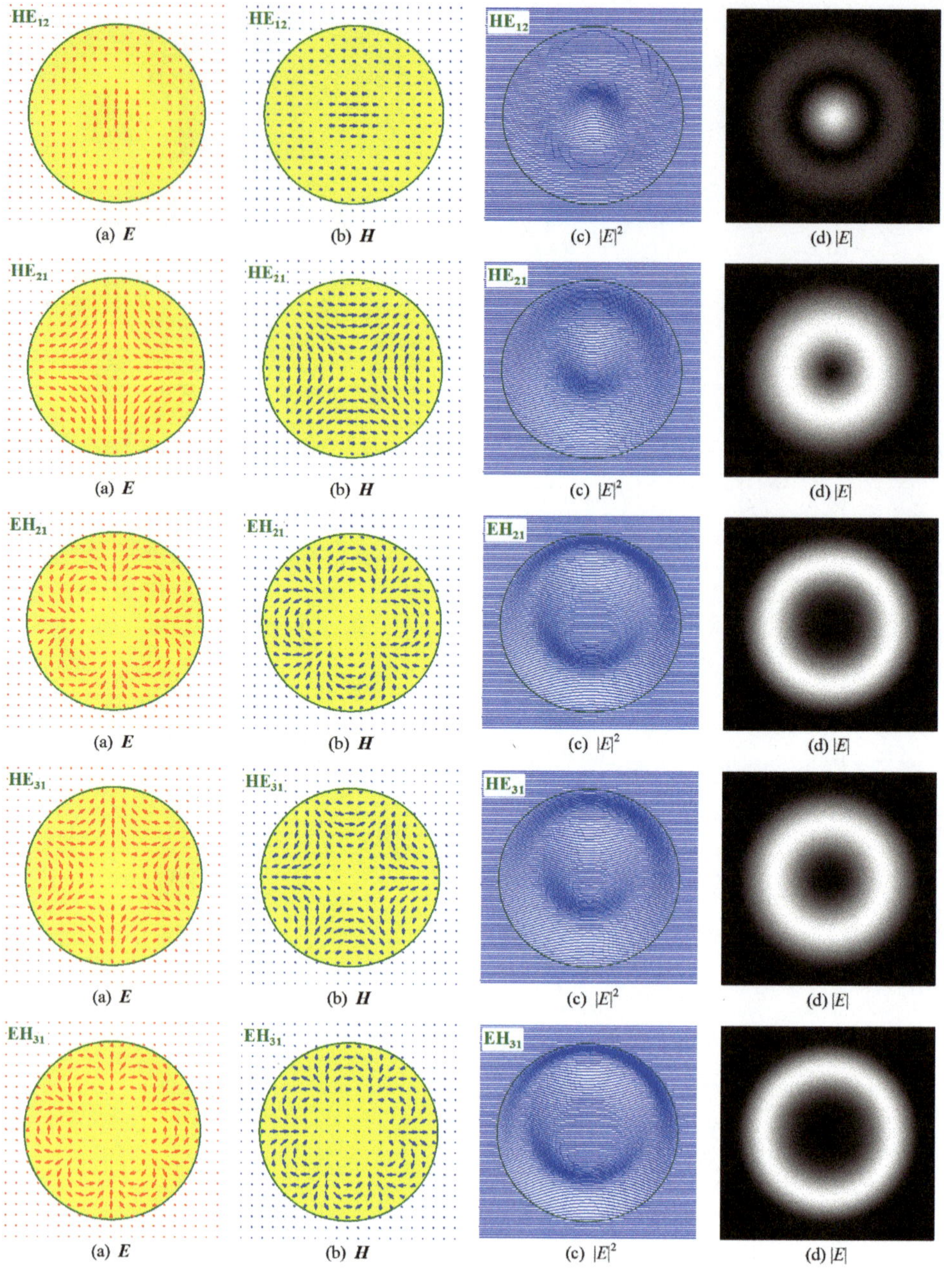

Figure 19.3 Calculated examples (part 2) of optical fiber modes: (a) ***E***, (b) ***H***, (c) $|E|^2$, and (d) $|E|$.

$$E_z = \begin{cases} J_l(\kappa r)\cos(l\theta) & (r \le a) \\ a_l K_l(\gamma r)\cos(l\theta) & (r \ge a) \end{cases}, \tag{19.1a}$$

$$E_r = \begin{cases} -\dfrac{i\beta}{\kappa} J_{l-1}(\kappa r)\cos(l\theta) & (r \le a) \\ -\dfrac{i\beta}{\gamma} a_l K_{l-1}(\gamma r)\cos(l\theta) & (r \ge a) \end{cases}, \tag{19.1b}$$

$$E_\theta = \begin{cases} \dfrac{i\beta}{\kappa} J_{l-1}(\kappa r)\sin(l\theta) & (r \le a) \\ \dfrac{i\beta}{\gamma} a_l K_{l-1}(\gamma r)\sin(l\theta) & (r \ge a) \end{cases}, \tag{19.1c}$$

$$H_z = \frac{\beta}{\omega\mu_0} \times \begin{cases} J_l(\kappa r)\sin(l\theta) & (r \le a) \\ a_l K_l(\gamma r)\sin(l\theta) & (r \ge a) \end{cases}, \tag{19.2a}$$

$$H_r = \begin{cases} -i2Q_1 J_{l-1}(\kappa r)\sin(l\theta) & (r \le a) \\ -i2Q_2 a_l K_{l-1}(\gamma r)\sin(l\theta) & (r \ge a) \end{cases}, \tag{19.2b}$$

$$H_\theta = \begin{cases} -i2Q_1 J_{l-1}(\kappa r)\cos(l\theta) & (r \le a) \\ -i2Q_2 a_l K_{l-1}(\gamma r)\cos(l\theta) & (r \ge a) \end{cases}, \tag{19.2c}$$

$$a_l = \frac{J_l(\kappa a)}{K_l(\gamma a)}, \tag{19.3}$$

$$Q_1 = \frac{k_0^2 n_1^2}{2\kappa\omega\mu_0}, \tag{19.4a}$$

$$Q_2 = \frac{k_0^2 n_2^2}{2\gamma\omega\mu_0}. \tag{19.4b}$$

Although the exponential term is omitted in the above equations, these equations are actually multiplied by $\exp(-i\beta z + i\omega t)$.

19.3.2 EH mode

Next, in the case of the EH mode, set $P = P_1 = P_2 = +1$ in Eqs. (18.3) and (18.4) in the previous chapter; then the following equations are obtained:

$$E_z = \begin{cases} J_l(\kappa r)\cos(l\theta) & (r \le a) \\ a_l K_l(\gamma r)\cos(l\theta) & (r \ge a) \end{cases}, \tag{19.5a}$$

$$E_r = \begin{cases} \dfrac{i\beta}{\kappa} J_{l+1}(\kappa r)\cos(l\theta) & (r \le a) \\ -\dfrac{i\beta}{\gamma} a_l K_{l+1}(\gamma r)\cos(l\theta) & (r \ge a) \end{cases}, \tag{19.5b}$$

$$E_\theta = \begin{cases} \dfrac{i\beta}{\kappa} J_{l+1}(\kappa r)\sin(l\theta) & (r \le a) \\ -\dfrac{i\beta}{\gamma} a_l K_{l+1}(\gamma r)\sin(l\theta) & (r \ge a) \end{cases}, \tag{19.5c}$$

$$H_z = -\frac{\beta}{\omega\mu_0} \times \begin{cases} J_l(\kappa r)\sin(l\theta) & (r \le a) \\ a_l K_l(\gamma r)\sin(l\theta) & (r \ge a) \end{cases}, \tag{19.6a}$$

$$H_r = \begin{cases} -i2Q_1 J_{l+1}(\kappa r)\sin(l\theta) & (r \le a) \\ i2Q_2 a_l K_{l+1}(\gamma r)\sin(l\theta) & (r \ge a) \end{cases}, \tag{19.6b}$$

$$H_\theta = \begin{cases} i2Q_1 J_{l+1}(\kappa r)\cos(l\theta) & (r \le a) \\ -i2Q_2 a_l K_{l+1}(\gamma r)\cos(l\theta) & (r \ge a) \end{cases}. \tag{19.6c}$$

As shown in the next section, by using these approximate formulas, the equations for the electric and magnetic flux lines can be easily expressed.

19.4 Electric and Magnetic Flux Lines

In both the HE mode and EH mode, both $\boldsymbol{E}$ and $\boldsymbol{H}$ have a z component (a component in the propagation direction). Therefore, electric or magnetic flux lines cannot be displayed only in the xy plane as in the case of the TE mode or TM mode but must be represented in the three-dimensional xyz space. However, for projection onto the xy plane ($r\theta$ plane) or rz plane, the equations for the electric and magnetic flux lines introduced so far can be used.

The projection of electric flux lines onto the $r\theta$ plane and the rz plane is expressed by the following equations, respectively:

$$\frac{dr}{rd\theta} = \frac{\mathrm{Re}(E_r)}{\mathrm{Re}(E_\theta)}, \tag{19.7}$$

$$\frac{dr}{dz} = \frac{\mathrm{Re}(E_r)}{\mathrm{Re}(E_z)}. \tag{19.8}$$

Similarly, the projection of magnetic flux lines onto the $r\theta$ plane and the rz plane is expressed by the following equations, respectively:

$$\frac{dr}{rd\theta} = \frac{\mathrm{Re}(H_r)}{\mathrm{Re}(H_\theta)}, \tag{19.9}$$

$$\frac{dr}{dz} = \frac{\mathrm{Re}(H_r)}{\mathrm{Re}(H_z)}. \tag{19.10}$$

As can be seen from Eqs. (19.1), (19.2), (19.5), and (19.6), Eqs. (19.7) to (19.10) can be easily obtained. Here, only the results are shown in Table 19.1. The formula on the rz plane is the same for electric flux lines and magnetic flux lines. From this table, for example, the electric flux line in the HE_{lm} mode can be obtained by finding a curve connecting the points (r, θ, z) that satisfies both Eqs. (19.17) and (19.19).

The table shows specific formulas for $l = 1$ to 3 in the HE_{lm} mode and EH_{lm} mode. As can be seen from Eqs. (19.17a) and (19.18a), the projection of the electric and magnetic flux lines in the HE_{1m} mode onto the xy plane becomes a straight line. Also, from Eqs. (19.17b) and (19.18b), this projection becomes a hyperbola in the HE_{2m} mode.[4] Similarly, from Eqs. (19.21a) and (19.22a), the projection of the electric and magnetic flux lines in the EH_{1m} mode onto the xy plane becomes a circle.[4]

The electric and magnetic flux lines in the TE mode are shown in Figs. 17.7 and 17.8 in Chapter 17. Similarly, the electric and magnetic flux lines in the HE_{11} mode are shown in Fig. 19.4. The circle shown in the figure is the core region. As mentioned above, in the HE mode or EH mode, it is not possible to display electric or magnetic flux lines only in the xy plane. However, in the HE_{11} mode, there are planes where the electric and magnetic flux lines are extremely close to straight lines in the xy plane. Figure 19.4 shows such a case. In reality, the curve in Fig. 19.4 is a curve expressed by Eq. (19.19) in Table 19.1, so strictly speaking, the straight line on the front surface in Fig. 19.4 is a projection onto the xy plane.

Next, the electric and magnetic flux lines in the HE_{21} mode are shown in Fig. 19.5. If electric and magnetic flux lines were displayed in the same space, the plot would be complicated and difficult to understand, so the electric flux lines were shown on the right half and the magnetic flux lines on the left. In reality, both are symmetrical about the xz and yz planes. The semicircle in the foreground shows the projection onto the xy plane. As shown by Eqs. (19.17b) and (19.18b) in Table 19.1, this projection is hyperbolic.

Figure 19.6 shows the electric and magnetic flux lines for the EH_{11} mode. Similar to Fig. 19.5, the right half shows the electric flux lines, and the left half shows the magnetic flux lines. These are also curves in 3D space, so the plot may be difficult to understand. The projection onto the xy plane and the projection onto the rz plane are shown in Fig. 19.7. All projections of the curves plotted in Fig. 19.6 are shown as solid lines in Fig. 19.7. As shown in Eqs. (19.21a) and (19.22a) in Table 19.1, the projection onto the xy plane is a circle. The dashed curve in Fig. 19.7(b) is not included in the curve group in Fig. 19.6 but is an example of the case when the value of constant C_1 is different from that in Fig. 19.6. As can be seen from this dashed curve, the

Table 19.1 Projection of electric and magnetic flux lines onto the $r\theta$ and rz planes. For HE modes and EH modes, equations are approximate expressions using Eqs. (19.1), (19.2), (19.5), and (19.6).

	$r\theta$ plane		rz plane
Mode	Electric flux line	Magnetic flux line	Electric flux line, Magnetic flux line
TE_{0m}	$r = C_r$ (19.11)	$\theta = C_\theta$ (19.12)	$\beta z = \begin{cases} \pm\cos^{-1}\left(\frac{C_1}{rJ_1(\kappa r)}\right) + m\pi & (r \le a) \\ \pm\cos^{-1}\left(\frac{C_2}{rK_1(\gamma r)}\right) + m\pi & (r \ge a) \end{cases}$ (19.15)
TM_{0m}	$\theta = C_\theta$ (19.13)	$r = C_r$ (19.14)	$C_2 = \frac{K_1(\gamma a)}{J_1(\kappa a)} C_1$ (19.16)
HE_{lm}	$r = \frac{C}{[\sin(l\theta)]^{1/l}}$ (19.17)	$r = \frac{C}{[\cos(l\theta)]^{1/l}}$ (19.18)	$\beta z = \begin{cases} \pm\cos^{-1}\left(\frac{C_1 r^{l-1}}{J_{l-1}(\kappa r)}\right) + m\pi & (r \le a) \\ \pm\cos^{-1}\left(\frac{C_2 r^{l-1}}{K_{l-1}(\gamma r)}\right) + m\pi & (r \ge a) \end{cases}$ (19.19)
HE_{1m}	$y = C$ (19.17a)	$x = C$ (19.18a)	
HE_{2m}	$xy = C^2/2$ (19.17b)	$x^2 - y^2 = C^2$ (19.18b)	
HE_{3m}	$3x^2y - y^3 = C^3$ (19.17c)	$x^3 - 3xy^2 = C^3$ (19.18c)	$C_2 = \frac{K_{l-1}(\gamma a)}{J_{l-1}(\kappa a)} C_1$ (19.20)
EH_{lm}	$r = C[\sin(l\theta)]^{1/l}$ (19.21)	$r = C[\cos(l\theta)]^{1/l}$ (19.22)	$\beta z = \begin{cases} \pm\cos^{-1}\left(\frac{C_1}{r^{l+1} J_{l+1}(\kappa r)}\right) + m\pi & (r \le a) \\ \pm\cos^{-1}\left(\frac{C_2}{r^{l+1} K_{l+1}(\gamma r)}\right) + m\pi & (r \ge a) \end{cases}$ (19.23)
EH_{1m}	$x^2 + y^2 = Cy$ (19.21a)	$x^2 + y^2 = Cx$ (19.22a)	
EH_{2m}	$r^4 = 2C^2xy$ (19.21b)	$r^4 = C^2(x^2 - y^2)$ (19.22b)	
EH_{3m}	$r^6 = C^3(3x^2y - y^3)$ (19.21c)	$r^6 = C^3(x^3 - 3xy^2)$ (19.22c)	$C_2 = \frac{K_{l+1}(\gamma a)}{J_{l+1}(\kappa a)} C_1$ (19.24)

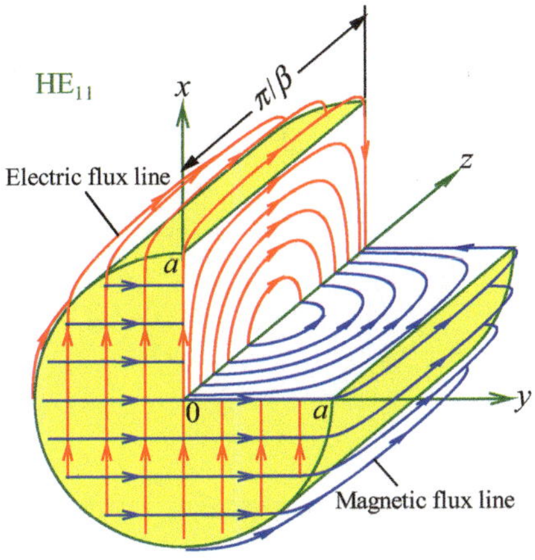

Figure 19.4 Electric and magnetic flux lines for the HE_{11} mode.

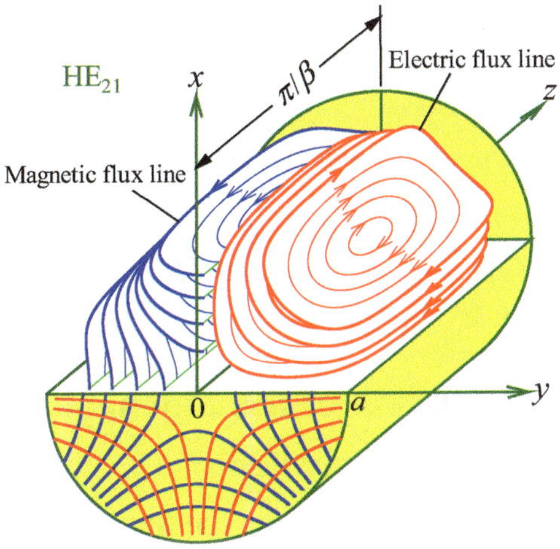

Figure 19.5 Electric and magnetic flux lines for the HE_{21} mode.

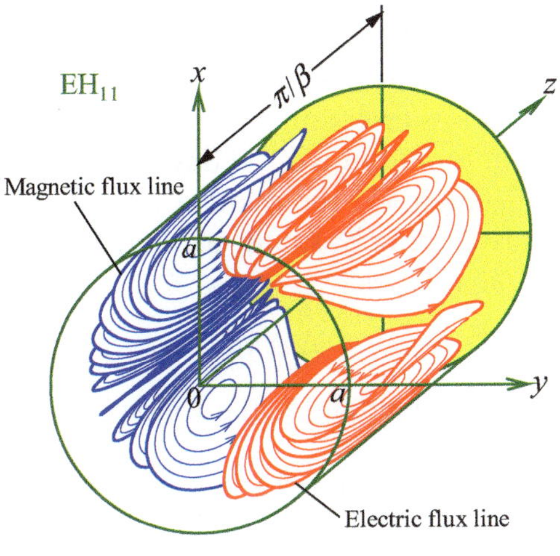

Figure 19.6 Electric and magnetic flux lines for the EH_{11} mode.

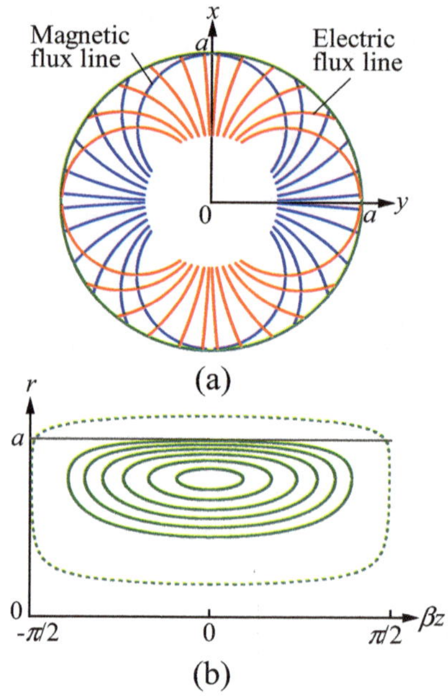

Figure 19.7 Projection of the electric and magnetic flux lines shown in Fig. 19.6 onto the *xy* and *rz* planes.

electric and magnetic flux lines are continuous between the core region ($r \leq a$) and the cladding region ($r \geq a$).

19.5 Approximate Equation of the Eigenvalue Equation

19.5.1 Weakly guiding approximation

The eigenvalue equations for determining the optical fiber modes are shown in Eq. (18.13) or Eq. (18.14) in the previous chapter and can be solved numerically using Newton's method. There is no problem with this, but if it is compared with the eigenvalue equation for a planar optical waveguide [Eq. (14.21) in Chapter 14], the order m is not included in the equation. The eigenvalue equation for optical fibers only gives another order l but cannot give m. Therefore, in this subsection, an approximate formula to find the eigenvalue equation containing m will be described.

As mentioned in Chapter 17, the relative refractive index difference Δ in normal optical fibers is extremely small, around 0.005 or less. Therefore, it may be approximated as follows:

$$\Delta \cong 0. \tag{19.25}$$

Then, from Eq. (17.25) in Chapter 17, the following approximation holds:

$$\beta = k_0 n_1 \sqrt{1 - 2\Delta(1 - b)} \cong k_0 n_1. \tag{19.26}$$

Using this approximation, the eigenvalue equation is expressed as follows:

$$\left\{\frac{1}{\kappa}\frac{J_l'(\kappa a)}{J_l(\kappa a)}+\frac{1}{\gamma}\frac{K_l'(\gamma a)}{K_l(\gamma a)}\right\}^2=\frac{l^2}{a^2}\left(\frac{1}{\kappa^2}+\frac{1}{\gamma^2}\right)^2. \tag{19.27}$$

This approximation is called the weakly guiding approximation.[6]

From Eq. (19.27),

$$\frac{1}{\kappa a}\frac{J_l'(\kappa a)}{J_l(\kappa a)}+\frac{1}{\gamma a}\frac{K_l'(\gamma a)}{K_l(\gamma a)}=\pm l\left\{\frac{1}{(\kappa a)^2}+\frac{1}{(\gamma a)^2}\right\}. \tag{19.28}$$

The parameter P that determines whether the mode is the HE mode or EH mode was given by Eq. (18.10) in the previous chapter, which is repeated as follows:

$$P=\frac{l\left\{\frac{1}{(\kappa a)^2}+\frac{1}{(\gamma a)^2}\right\}}{\left\{\frac{1}{\kappa a}\frac{J_l'(\kappa a)}{J_l(\kappa a)}+\frac{1}{\gamma a}\frac{K_l'(\gamma a)}{K_l(\gamma a)}\right\}}. \tag{19.29}$$

The mode with $P\approx -1$ is defined as the HE mode, and the mode with $P\approx +1$ is defined as the EH mode. Therefore, when the double sign on the right side of Eq. (19.28) is negative, it is the HE mode, and when positive, it is the EH mode.

Rewriting Eq. (19.28) using Eqs. (17.A4) and (17.A5) in Appendix 17A of Chapter 17 and further using the normalization parameters, the eigenvalue equation can be expressed as follows:

(For HE mode)

$$\frac{1}{\sqrt{1-b}}\frac{J_{l-1}(\sqrt{1-b}V)}{J_l(\sqrt{1-b}V)}-\frac{1}{\sqrt{b}}\frac{K_{l-1}(\sqrt{b}V)}{K_l(\sqrt{b}V)}=0, \tag{19.30}$$

(For EH mode)

$$\frac{1}{\sqrt{1-b}}\frac{J_{l+1}(\sqrt{1-b}V)}{J_l(\sqrt{1-b}V)}+\frac{1}{\sqrt{b}}\frac{K_{l+1}(\sqrt{b}V)}{K_l(\sqrt{b}V)}=0. \tag{19.31}$$

Now, by applying the relationship of Eqs. (17.A2a) and (17.A3a) in Appendix 17A of Chapter 17 to the numerator of Eq. (19.30), Eq. (19.30) is transformed as follows:

$$J_{l-2}(\sqrt{1-b}V)K_l(\sqrt{b}V)+J_l(\sqrt{1-b}V)K_{l-2}(\sqrt{b}V)=0. \tag{19.32}$$

Similarly, by using the relationships of Eqs. (17.A2a) and (17.A3a) in Chapter 17 for the numerator of Eq. (19.31),

$$J_{l+2}(\sqrt{1-b}V)K_l(\sqrt{b}V) + J_l(\sqrt{1-b}V)K_{l+2}(\sqrt{b}V) = 0. \tag{19.33}$$

Comparing these two equations, it can be seen that Eq. (19.32) in which ν is replaced with $l-1$ and Eq. (19.33) in which ν is replaced with $l+1$ are exactly the same as the following equation:

$$J_{\nu+1}(\sqrt{1-b}V)K_{\nu-1}(\sqrt{b}V) + J_{\nu-1}(\sqrt{1-b}V)K_{\nu+1}(\sqrt{b}V) = 0, \tag{19.34}$$

$$\nu = \begin{cases} l-1 & \text{(HE mode)} \\ l+1 & \text{(EH mode)} \end{cases}. \tag{19.35}$$

Then, returning to Eq. (19.30) and rewriting it with ν in Eq. (19.35), the following equation is obtained:

$$\frac{J_{\nu+1}(\sqrt{1-b}V)}{J_\nu(\sqrt{1-b}V)} = \sqrt{\frac{b}{1-b}}\frac{K_{\nu+1}(\sqrt{b}V)}{K_\nu(\sqrt{b}V)}. \tag{19.36}$$

Using Eqs. (19.A3) and (19.A4) in Appendix 19A, Eq. (19.36) becomes

$$\tan\left(\sqrt{1-b}V - \frac{\nu\pi}{2} - \frac{\pi}{4}\right) = \sqrt{\frac{b}{1-b}}. \tag{19.37}$$

Therefore,

$$\sqrt{1-b}V = \tan^{-1}\sqrt{\frac{b}{1-b}} + \frac{\nu\pi}{2} + \frac{\pi}{4} + (m-1)\pi. \tag{19.38}$$

The principal mode number[1–3] is defined below:

$$N = \nu + 2(m-1). \tag{19.39}$$

Then Eq. (19.38) becomes

$$\sqrt{1-b}V = \tan^{-1}\sqrt{\frac{b}{1-b}} + \frac{N\pi}{2} + \frac{\pi}{4}. \tag{19.40}$$

In this way, the b-V relation is expressed in a form similar to the eigenvalue equation [Eq. (14.21) in Chapter 14] in a planar optical waveguide.

Equation (19.40) is an approximation under the weakly guiding approximation, but unlike Eq. (18.14) in the previous chapter, the order m is explicitly included in the equation. Therefore, Eq. (19.40) can be used as the initial value when solving the rigorous eigenvalue equation. Figures 19.1 to 19.7 introduced in this chapter were all calculated using Excel, and Eq. (19.40) is used as the initial value when solving the rigorous eigenvalue equation using Newton's method. Since the solution b is found by giving V, Newton's method can also be used to find b by giving V in Eq. (19.40).

19.5.2 LP mode

In the weakly guiding approximation, if the ν in Eq. (19.35) is used, the eigenvalue Eq. (19.34) [or Eq. (19.36)] becomes the same as Eq. (19.35). Therefore, the effective propagation constants β of the $HE_{\nu+1,m}$ mode and $EH_{\nu-1,m}$ mode (the TE_{0m} mode or TM_{0m} mode when $\nu = 1$) are approximately equal. When these two modes are superimposed to form linearly polarized light, the mode is called a LP (linearly polarized) mode.[6]

As examples, Fig. 19.8 shows the case of $\nu = 2$ and $m = 1$, that is, the superposition of the EH_{11} mode and HE_{31} mode. The parameters used are the same as in Figs. 19.1 to 19.3. As can be seen from the figure, both the electric field $\boldsymbol{E}$ and the magnetic field $\boldsymbol{H}$ show linear polarization.

It should be noted here that the mode shown in Fig. 19.8 is a superposition of two modes with different eigenvalues (i.e., different effective propagation constants) and is not the "eigenmode" of the optical fiber, which is represented by a single eigenvalue.[7,8] With the parameters used in the calculation, the effective propagation constant difference between the two modes is estimated as $\Delta\beta = 3.42\times10^{-7}$ μm^{-1}. Therefore, when propagating in the z direction by $\pi/\Delta\beta = 9.2$ m, the relative phases of the two modes shift by π. Specifically, when the electromagnetic field distribution in the upper figure of Fig. 19.8 propagates 9.2 m, it becomes the electromagnetic field

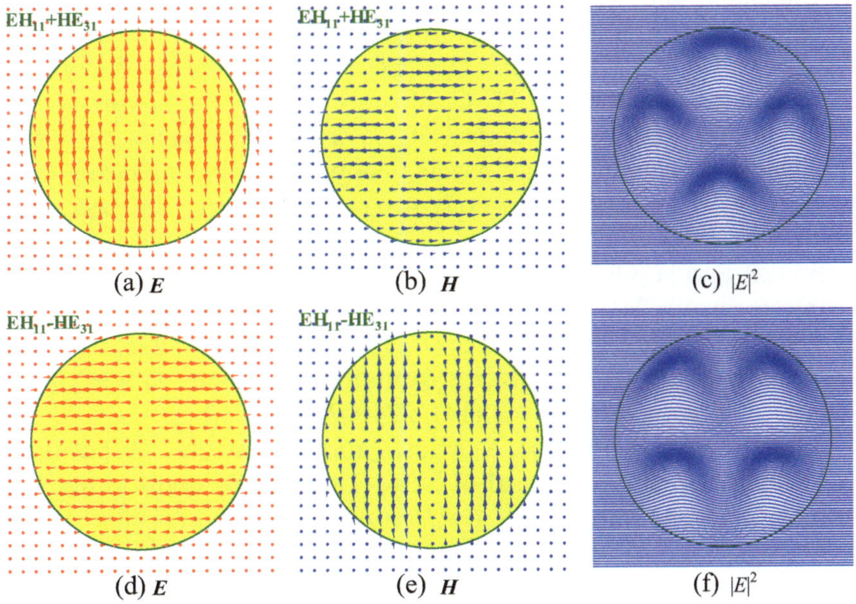

Figure 19.8 Superposition of the EH_{11} mode and HE_{31} mode.

distribution in the lower figure of Fig. 19.8. Furthermore, since the group velocities of the two modes are also different, a difference in group delay will occur in the case of pulse transmission.[7,8]

19.6 Conclusion

An example of calculating the electromagnetic field distribution of modes in an optical fiber using Excel was introduced. Approximate formulas for electric and magnetic flux lines and display examples using them were also shown. The rigorous solution to the electromagnetic field distribution can be found using Newton's method in Excel, but the mode order m cannot be given using the eigenvalue equation, and the mode order found by calculation depends on the initial value. Weakly guiding approximations are useful for providing initial values for exact solutions.

Acknowledgments

The author would like to thank Professor Emeritus Kenichi Iga, Institute of Science Tokyo, former president of Tokyo Institute of Technology, and Professor Emeritus Yasuo Kokubun, Yokohama National University, former president of Institute of Technologists, for their guidance and discussions regarding the modal analysis theory of optical fibers.

References

1. Y. Suematsu and K. Iga, *Introduction to Optical Fiber Communications, 5th Ed.*, Ohmsha, Tokyo (2017) [in Japanese].
2. K. Iga and Y. Kokubun, *Optical Fiber*, Ohmsha, Tokyo (1993); Electronic reprint version, ITSC, Shizuoka (2014) [in Japanese].
3. K. Iga and G. Hatakoshi, Chapter 30, *Treasure Microbox of Optoelectronics II*, Advanced Communication Media, Tokyo (2018) [in Japanese].
4. E. Snitzer, "Cylindrical dielectric waveguide modes," *J. Opt. Soc. Am.* **51**(5), 491–498 (1961).
5. E. Snitzer and H. Osrerberg, "Observed dielectric waveguide modes in the visible spectrum," *J. Opt. Soc. Am.* **51**(5), 499–505 (1961).
6. D. Gloge, "Weakly guiding fibers," *Appl. Opt.* **10**(10), 2252–2258 (1971).
7. H. Kogelnik and P. J. Winzer, "Modal birefringence in weakly guiding fibers," *J. Lightwave Tech.* **30**(14), 2240–2245 (2012).
8. Y. Kokubun, T. Watanabe, S. Miura, and R. Kawata, "What is a mode in few mode fibers?: Proposal of MIMO-free mode division multiplexing using true eigenmodes," *IEICE Electronics Express* **13**(18), 1–12 (2016).
9. S. Moriguchi, K. Udagawa, and S. Hitotsumatsu, *Mathematical Formula III, Special Functions, 27th Impress.*, Iwanami Shoten, Tokyo (2016) [in Japanese].

Side Topic: Optical fiber mode order and number of nodes

As mentioned in the main text, for the mode order in an optical fiber, l corresponds to the periodicity in the angular direction, and m corresponds to the number of nodes (zero points) in the radial direction. It was stated in the main text that, excluding the zero point at $r = 0$, the number of nodes in the radial direction for each mode is $m - 1$. So, what happens if the zero point at $r = 0$ is included?

As shown in Eq. (18.3) in the previous chapter, each component of the electric field $\boldsymbol{E}$ in the optical fiber is as follows in the region $r \leq a$:

$$E_z \propto J_l(\kappa r), E_r, E_\theta \propto (1 - P)J_{l-1}(\kappa r) - (1 + P)J_{l+1}(\kappa r). \tag{19.S1}$$

First, considering the HE mode, the HE_{1m} mode includes the term $J_0(\kappa r)$ from Eq. (19.S1), so the amplitude does not become 0 at $r = 0$. This is clear from Fig. 19.1. On the other hand, $J_0(\kappa r)$ does not appear when $l \geq 2$. Therefore, the number of radial zero points in the HE mode is $m - 1$ in the HE_{1m} mode and m in the HE_{lm} mode ($l \geq 2$).

So, what happens in the EH mode? Looking at Figs. 19.1 to 19.3, it appears that all EH modes have zero amplitude at $r = 0$. This seems natural since the Bessel function takes a value of $J_l(0) = 0$ for $l \geq 1$. However, this is actually a mistake. In the EH mode, $P = +1$, so from Eq. (19.S1), the E components are approximated as E_r, $E_\theta \propto -2J_{l+1}(\kappa r)$, and only the Bessel function with $l \geq 1$ appears for all E_z, E_r, and E_θ. However, this is an approximation, and P is not exactly 1. Therefore, the term $(1 - P)J_{l-1}(\kappa r)$ is always included. Thus, in the EH_{1m} mode, the amplitude does not become 0 when $r = 0$.

To summarize the above, the number of zero points in the radial direction is $m - 1$ in the HE_{1m} mode and EH_{1m} mode, and m in other modes (including TE_{0m} and TM_{0m}).

Appendix 19A Approximate Representation of the Bessel Function

The Taylor expansion formulas for the Bessel function and modified Bessel function around $x = 0$ are shown in Appendix 18A of the previous chapter. On the other hand, when x is sufficiently large, $J_\nu(x)$ and $K_\nu(x)$ are expanded asymptotically as follows:[9]

$$J_\nu(x) \cong \sqrt{\frac{2}{\pi x}} \cos\left(x - \frac{\nu\pi}{2} - \frac{\pi}{4}\right), \tag{19.A1}$$

$$K_\nu(x) \cong \sqrt{\frac{\pi}{2x}} e^{-x}\left(1 + \frac{4\nu^2 - 1}{8x}\right). \tag{19.A2}$$

From this, the following approximate formula can be obtained when x is large:

$$\frac{J_{\nu+1}(x)}{J_\nu(x)} \cong \tan\left(x - \frac{\nu\pi}{2} - \frac{\pi}{4}\right), \tag{19.A3}$$

$$\frac{K_{\nu+1}(x)}{K_\nu(x)} \cong 1. \tag{19.A4}$$

Appendix 19B Excel Files

Appendix I-A contains the following Excel files. Each tab corresponds to a figure/s in the text.

19_Fiber3.xlsx.

Instruction	
TE0m,TM0m	Fig. 19.1, Fig. 19.2, Fig. 19.3 (a) to (c)
HElm,EHlm	Fig. 19.1, Fig. 19.2, Fig. 19.3 (a) to (c)

19_Fiber4.xlsx

HE11_Field	Fig. 19.4

About the Online Supplemental Appendices

This book contains Excel files (Appendix I-A) and animations of the figures in the text (Appendix I-B) in each chapter. These files can be downloaded from the following URL:

https://optronics-ebook.com/download/appendix/9784902312744/SPIE_Appendix1.zip

The Excel files in Appendix I-A allow you to perform animations of the figures shown in the text, multilayer film characteristic analysis, and mode calculations of optical waveguides and optical fibers. Operating instructions are given on the first tab of each Excel file, so refer to them along with the explanations in the text. To run macros in Excel, you need to enable macros in the settings.

The figures in Appendix I-B can be viewed from index.html in the Appendix I-B folder. Clicking on index.html will display the following screen.

Click on each chapter number on the left side of the screen above to see animations for that chapter.

The animated GIF files used here can be viewed by, for example, pasting them into PowerPoint (ppt) and switching to presentation mode. Those who have purchased this book may reproduce or quote the appendix in ppt format. In such cases, please be sure to clearly indicate the source.

Index

Dr. Hatakoshi completed a doctoral course at the University of Tokyo, Graduate School of Engineering, and he received a Dr. Eng. degree from the University of Tokyo in 1980. The same year, he joined the Research and Development Center, Toshiba Corporation, where he engaged in R&D of optical devices, semiconductor lasers, and LEDs.

He was a Fellow at Toshiba Research Consulting Corporation from 2003 to 2014; a Fellow at Research and Development Strategy Center, Japan Science and Technology Agency from 2006 to 2009; and a part-time lecturer at Waseda University from 2015 to 2020. From 2011 to 2023, he served as the Japanese Domestic Committee Chair of ISO/TC 172/SC 9.

He received the Optical Paper Award from the Japan Society of Applied Physics in 1982; the 39th Okochi Memorial Prize in 1993; the Optical Design Special Award in 2001; and the Prize for Science and Technology, the Commendation for Science and Technology by MEXT Japan in 2005.

He is a Fellow and a Life Member of the Japan Society of Applied Physics; Organizing Chair of Microoptics Group; Life Member of IEICE Japan; SPIE Life Member; ISO/TC 172/SC 9 Member; and IEC/SC 47E/WG 9 Member.